THE PARADOX OF PLENTY

HUNGER IN A BOUNTIFUL WORLD

EDITED BY DOUGLAS M. BOUCHER

FOOD FIRST BOOKS
OAKLAND, CALIFORNIA

Printed in Canada

Text and cover design by Colored Horse Studios

Grateful appreciation is extended to the following for permission to reprint copyrighted materials:

Bullfinch Press for "Ode to Bread" from *Odes to Common Things* by Pablo Neruda. Copyright ©1994 Pablo Neruda and Pablo Neruda Fundacion (odes in Spanish); Copyright ©1994 Ferris Cook (compilation and illustrations); Copyright ©1994 Ken Krabbenhoft (English translation). By permission of Little, Brown and Company.

Grove Press for "Beyond Guilt and Fear" and "The Green Revolution is the Answer" from *World Hunger: Twelve Myths* by Frances Moore Lappé, Joseph Collins, and Peter Rosset with Luis Esparza. Copyright ©1998 Institute for Food and Development Policy.

Frances Moore Lappé for "Like Driving a Cadillac" from *Diet for a Small Planet.* Copyright ©1971, 1975, 1982 Frances Moore Lappé.

University of Nebraska Press for "The Faustian Bargain: Technology and the Price Issue" from *Family Farming: A New Economic Vision* by Marty Strange. Copyright ©1988 Institute for Food and Development Policy.

Grove Press for "Latin America: Going to Extremes" from *A Fate Worse Than Debt: The World Financial Crisis and the Poor* by Susan George. Copyright ©1988, 1990 Susan George.

Grove Press for "Getting Off the Pesticide Treadmill" from *Nicaragua: What Difference Could a Revolution Make?* by Joseph Collins with Frances Moore Lappé, Nick Allen, and Paul Rice. Copyright ©1982, 1985, 1986 Institute for Food and Development Policy.

Library of Congress Catalog-in-Publication Data

The paradox of plenty: hunger in a bountiful world/edited by
Douglas M. Boucher.
p. cm.
Includes bibliographical references.
ISBN 0-935028-71-4
1. Peasantry—Developing countries—Economic conditions. 2. Poverty—Developing Countries. 3. Food supply—Developing countries. 4. Income Distribution—Developing Countries. 5. Agriculture—Economic aspects—Developing countries. 6. Economic assistance. 7. International trade—Moral and ethical aspects.
I. Boucher, Douglas M.
HD1542.P37 1999
363.8'09172'4—dc21 99-18244
CIP

Cover Photo by David and Gregory Frankel

1 2 3 4 5 — 03 02 01 00 99

THE PARADOX OF PLENTY

Hunger in a Bountiful World

Table of Contents

Part Three: Global Policies and Hungry People

Part Four: The Free Market Path

Part Five: Alternatives

Acknowledgments

This book came about, as many do, because I wanted to read it and give it to others to read. In preparing to teach Hood College's interdisciplinary course on "Hunger, Population, and the Environment" (INST 307) in the spring of 1997, I was looking for writings on the world food system that went beyond the conventional analyses of economists and ecologists. What I sought was a way to let students see how economic and political issues, such as free trade, have a critical impact on whether people in poor countries live or die. Also, I didn't want my students to have to buy a whole lot of different books to be able to get this global picture.

Searching for such a volume, I called up my friend Peter Rosset at the Food First Institute and asked if they had published anything recently that gave a good overall view of "the Food First vision." His answer was no, but would you like to do it? The Institute had been trying to assemble a reader along the lines I described, but needed someone to edit it. After some thinking, I agreed to take on the job, and this collection is the result.

My thanks are due above all to the authors whose writings are included. Their hard work in field, library, and office, and their dedication to spreading the word about the reality of world hunger and its causes, are what really made this book happen.

Peter Rosset, Marilyn Borchardt, and Kathleen McClung at Food First got the project started and gave me strong support and understanding throughout the process. Their comments and suggestions have been unfailingly helpful, even when I've ignored them. I'm especially grateful for their willingness to be patient when deadlines slipped by and nothing showed up in their mailboxes.

Emilio Font was a great help with the copying, printing, cutting, and pasting necessary to assemble the manuscript.

The students in INST 307 at Hood College in the spring of 1997 were a great inspiration for this reader, although neither they nor I knew it at the time. My thanks to them, and to my colleagues at Hood for creating the supportive atmosphere in which I could work on this volume. My family—Charlotte, Johnny and Nellie—made it possible for me to write and think on this and other subjects, in relative peace and quiet, for many years. They have also endured my many absences (physical and mental) with tolerance, for which I'm deeply grateful.

Frances Moore Lappé and Joseph Collins deserve the gratitude of all of us for first developing 'the Food First vision,' and for founding the Institute for Food and Development Policy which has carried on their work for almost a quarter century. Finally, a special note of thinks to my friends and fellow members of the New World Agriculture and Ecology Group (NWAEG), with whom I've collaborated, debated, worked, and struggled on these issues for more than two decades. This book is dedicated to them.

Introduction

Food, Politics, and Us

In the past quarter-century, the debate about world hunger has been fundamentally transformed. During the 1960s and 70s, it was an argument about food versus population. The world was seen as unable to produce enough to feed all its people. Agricultural production was increasing, but populations were growing even faster —exponentially, at faster rates than ever before in human history. It was a race which we were losing and the inevitable result was famine, starvation, and misery.

In such best-selling books as *The Population Bomb*[1] and *Famine 1975!*,[2] experts explained how Thomas Malthus' predictions were finally coming true. Although technological progress had increased the world's output of food, it had also allowed more and more people to survive and have children. Modern medicine, ironically, was aggravating the age-old struggle to provide for all. Because of the immense power of exponential growth, even countries with small populations would soon find themselves overpopulated. And for those Third World countries such as India or China, which already had hundreds of millions of people and their population doubling in just twenty or thirty years—it was probably already too late.

The argument was basically between optimists and pessimists. The more hopeful observers felt that with the application of modern agricultural technologies such as fertilizers, pesticides, irrigation, and improved seeds, we might be able to stave off mass starvation for at least a few decades more. The more pessimistic felt that the planet had already exceeded its carrying capacity and that a population crash was inevitable.

Our only real choice, ecologists said, was whether the population would be controlled by famine or by birth control. Human populations were seen as being biologically the same as animal populations; they had a specific carrying capacity, and as they approached it, the members of the population would compete for scarcer resources. Each person would have less of what they needed to live, so their nutritional status, health, and ability to reproduce would decline.

Death rates would increase and birth rates would decrease, until they finally came into balance. The process might be a cruel one, but it was inexorable; to continue growing beyond carrying capacity would only make the final crash that much more horrible.

Both optimists and pessimists agreed on the basic metaphor: a race between population and food. They agreed that Malthus had been fundamentally correct and that people were starving because the world wasn't producing enough food. The problem was seen as residing in poor, overpopulated Third World countries, where agriculture was mostly carried out by backward, uneducated peasants with little knowledge of either scientific methods of farming or effective ways to control reproduction.

To the extent that the rich nations were involved, it was as altruistic helpers. We, the wealthy countries, could share all our scientific outlook, our knowledge of farming, our birth control techniques, and our realization of the importance of using them, to assist the Third World in avoiding large-scale famine. Politically, the foreign aid programs of the United States Agency for International Development (USAID) were sustained by their importance to the Cold War. Liberals might see USAID's work as humanitarian assistance to our 'less developed' global neighbors, while conservatives could support it as an effective weapon against the Soviet Union in the worldwide struggle for hearts and minds.

The debate was about which strategy would be ultimately be more helpful—short-term food aid until poor countries could control their populations, or a 'tough-love' refusal to share resources which would only have the effect of allowing more people to survive and have babies. There was no argument that foreign aid would help people survive in the short run; the question was whether, in the long run, that was the most humane thing to do.[3]

Given the urgency of the situation, being humane could be a luxury that we just couldn't afford. It was admitted that some coercion would probably be necessary to get Third World countries to reduce their populations. Perhaps withholding food aid from countries that wouldn't undertake population control programs would be sufficient. Or perhaps more repressive measures, such as forced sterilizations, might regrettably be necessary.

Peasant farming and traditional village structures would have to be eliminated if agricultural production were to increase. Small farms worked by uneducated peasants couldn't take advantage of the economies of scale offered by modern agricultural technology; whether by Stalinist collectivization or massive World Bank development schemes, they would have to be consolidated into larger, more efficient productive units. The legacy of the rural past would have to be swept aside.

The fierce debates of the late 1960s and early 1970s concealed an underlying consensus. There was agreement on the fundamental nature of the problem: more mouths to feed and not enough food to feed them. One side would emphasize the need for a 'Green Revolution' which would rapidly increase crop yields; the other would point to the awesome growth potential of populations growing by compound interest and insist on reducing fertility by drastic means. But both agreed that it was a question of too many people and too little food in the Third World; that those overpopulated nations needed a rapid infusion of technology (agricultural or contraceptive); and that the problem was ultimately theirs, not ours.

The New Vision

As we enter the new millennium, it is no longer possible to see world hunger in these terms. Like bell bottoms and disco, the food-versus-population metaphor is a phenomenon whose time has come and gone. It is not that one side or the other—modernizing agricultural or controlling the fertility of the poor—won out in the debate. We have realized that the very terms of the argument do not make sense.

First of all, we now recognize that much of the planet's crops do not go to feed people at all—they are fed to livestock. In the process, enormous amounts of protein, energy, water, and land are wasted. If our food system were truly directed towards feeding humanity, it would only need to redirect a small amount of the earth's grain from animal production to the people who are hungry, to satisfy their needs more than adequately.

The same is the case with many of the other resources used in agriculture. Pesticides are used to protect fruits from blemishes and cotton plants against boll weevils which have evolved resistance. Rain forests are cut down in order to make cattle pasture to export beef for fast-food hamburgers and pet food. Water is carried by complex irrigation systems to produce costly winter vegetables. We know how to produce abundant food for the hungry if we wanted to; but we do not.

We now see that it is not simply a matter of whether food is available in the market; people must have the money to buy it. In a world economy in which food is a commodity, poverty will lead to starvation no matter how productive agriculture becomes. The basic problem for hungry people is not a scarcity of food, but a scarcity of income.

This realization turns the argument about farm size and modern technology on its head. Driving peasants off their farms and consolidating the farms in the hands of large landowners will only make the situation worse. The now landless peasants will have neither the ability to raise their own food nor the income to buy it, while the landowners, with access to the tractors and combines which allow them to produce crops with fewer workers, will have no incentive to produce basic foodstuffs if there is no market for them. Even though large farms could in theory produce large amounts of food for the hungry, in reality much of their land lies idle.

Hunger is now seen not simply as a matter of how many million tons the world produces, but rather as a question of food security. Do the poor have enough income to buy what they need? Do they have land on which to grow crops? The access to water with which to irrigate them? The grain storage facilities to protect the crops from rodents and insects after harvest? The transport and marketing system to carry the food to those who are hungry? In other words, does the system guarantee that farmers will have what they need to produce food and consumers will have what they need to buy it?

On the population side, the assumptions of the 1960s have also been superseded. Ecologists have come to understand that human beings differ from other animal species in a fundamental way; as their health, nutrition, survival prospects, and access to resources increase, their birth rates go *down*, not up. Human population growth tends to slow, not accelerate, as conditions improve for the majority of the population.

The result is the well-known phenomenon of the demographic transition[4]—the fall in birth rates as societies develop and the economic status of their populations improves. Rather than continued exponential growth (which would continue if the growth rate stayed constant over time), countries around the world have shown declines in growth rates over the past few decades, and in most industrialized nations birth rates are now down to replacement level or below.

We not only recognize that population growth rates can change with social and economic development; there is also an emerging consensus on the factors which are most important in reducing fertility rates. The status of women—their education, health care, control over economic resources, access to reproductive technology, and relative power in the family and the society—is clearly a critical variable.[5] There has been a shift in the thinking of environmentalists, who formerly promoted population control even if it might lead to forced sterilization and other kinds of suppression of women's reproductive choices. They now see that the best way to reduce fertility rates is by supporting feminism. The new ideological convergence was clearly evident in the International Conference on Population and Development, which met in Cairo in September of 1994. Just as the recognition of the demographic transition had led to the earlier slogan that "development is the best contraceptive," so now there is growing acknowledgment that women's rights can have an important impact on birth rates as well.

The emphasis on the rights of women is part of a new view that recognizes the critical role of democracy in dealing with world food problems. Contrary to earlier viewpoints that saw inequality, centralized control, and repression as perhaps unfortunate but necessary to increase food production and control population growth, we now see that both food security and fertility limitation are promoted by a broad recognition of human rights. Land reform, income security, universal health care, reproductive choice, and equal access to education are among the components of a democratic society that are critical to achieving food security for all.

Finally, the new vision of the 1990s recognizes that world hunger is truly a problem of the world food system, not just a question for poor countries alone. The operation of that system produces transfers of food, resources, wealth, and income from the Third World to the rich countries, and creates market demands for export crops and meats which orient agriculture in poor nations towards satisfying the wants of far-off consumers, rather than the vital needs of the hungry at home. Grain surpluses produced in wealthy nations are dumped in poor nations, undercutting the production of peasant farmers and driving them out of business and off their land. It is ironic that transfers of food from rich to poor nations actually serve to undermine their agricultural capacity and weaken the food security of their people in the long run.

Although much of the negative impact on world hunger comes from impersonal market forces, these are reinforced by explicit policy decisions. USAID has for decades used food aid as a way of opening up markets in the Third World for American agribusiness. The international lending institutions (World Bank, International Monetary Fund, Interamerican Development Bank) regularly impose conditions on economic assistance which include cutting support for local farmers, eliminating barriers to food exports and imports, and promoting export crops with which to earn the dollars necessary to pay back loans to foreign banks. In the view of the ideology of 'free trade,' protection of the people of poor nations—whether in the form of labor rights, environmental standards, national health care, or guarantees of the right to eat—is seen as an indirect trade subsidy and therefore illegitimate. With the debt crisis of the 1980s and the market and currency collapses of the late 1990s, international lenders had a new power to impose this ideology on the Third World under the rubric of 'structural adjustment programs.' It's now clear that hunger and starvation are not caused by market forces alone; the decisions of bankers, economists, and politicians play a major role in the workings of the world food system.

The Role of Food First

A considerable part of the credit for the transformation of the consensus on hunger, from the food versus population viewpoint of the sixties to a more sophisticated analysis of the world food system in the nineties, is due to the work of Food First. Founded by Frances Moore Lappé and Joseph Collins, whose books on world hunger in the 1970s were influential best-sellers, Food First/Institute for Food and Development Policy has had a major impact on the thinking of citizens and experts alike. Rejected as radical and naïve at first, Lappé and Collins saw their ideas become grudgingly accepted and in some cases even assimilated into the mainstream.

This success was due both the force of the Institute's analyses and the impact of real world events. More and more developments undermined the food versus population description of the problem—the failure of coercive population control programs; the inability of the Green Revolution to eliminate hunger even as it increased yields; the growth of the feminist and environmental movements worldwide; and the simple fact that while world hunger persisted, the predicted planetary famines and population crashes did not take place. It became clearer and clearer that Malthus had been wrong.

But Food First's success was also due to its fundamentally new approach. From the very beginning, in *Diet for a Small Planet* (published in 1971), Frances Moore Lappé showed that world hunger was not an issue that had to be left to economists, politicians, and professors. She demonstrated that the experts' consensus had ignored key elements of the problem, such as the abundance of food produced worldwide and the enormous waste of resources caused by livestock production. She was willing to challenge the conventional wisdom—from the population bomb and the Green Revolution to the assumption that eating meat was necessary for good health. She wrote in a clear, straightforward style; her books were data filled but jargon-free. Not the least of her innovations was to use two different ways to rebut the carnivore consensus: with evidence from nutritional science, and with a collection of good-tasting, amino-acid-balanced vegetarian recipes.

In just a few years, Lappé's analysis had gone beyond the question of vegetarianism to the global political issues, and Food First was born. Her next book *Food First: Beyond the Myth of Scarcity* (published in 1977), written with Joseph Collins and Cary Fowler, used a question-and-answer format to confront the prevailing myths about hunger head-on. Transcending the focus of *Diet for a Small Planet,* it criticized agribusiness, foreign aid, and the Green Revolution, and told not only what was wrong with the world food system, but what readers could do to change it. *Food First* made it clear that food was a political issue.

In the years since, Food First has expanded its original vision to look at many aspects of the world food system. It has analyzed the pesticide trade, tropical deforestation, and the population issue in detail, and undertaken hard-hitting critiques of USAID, the World Bank, and the IMF. It has connected Third World issues to what's happening to agriculture in the U.S., with studies of the decline of the family farm, the situation of farm workers, and the impact of exports on the American farm sector. Through a series of in-depth studies of particular countries and regions, looking at how food policies impact the daily lives of people in rural villages, it has shown how the consequences of policy decisions extend to the farthest corners of the world food system. Beyond research, writing, and speaking, the Institute has also been a focus for organizing links among those wanting to make a difference on issues of hunger. Its publications also include a very useful book, now in its eighth edition, showing *Alternatives to the Peace Corps* for volunteer service.[6]

In this book, I have collected together some of the best writings from Food First over the past quarter of a century. These are critical analyses which have changed how we view world hunger, but I have not chosen them just to document the history of the Institute or even of the ebb and flow of past debates about the world food system. My selections have been guided by the issues of today. Inevitably, then, there is a concentration of pieces published in recent years, especially in the decade of the 1990s.

I have emphasized readings which confront the new realities of our time. The Cold War is over, and free-market, free trade policies are supported by Republicans and Democrats alike. Population growth rates have declined, so much that demographers predict that the world's population will stabilize at about 9.4 billion at the middle of the 21st century. Yet the urgent problems of past decades continue: hundreds of millions are malnourished, family farms keep disappearing every year, destruction of tropical forests continues apace, and the specter of famine is never far from our door. These selections try to explain why, despite the planet's bounty and wealth, we are still unable to ensure people their basic human right to survive.

I have reprinted the authors' own words without modifying them, even when the text contains minor anachronisms or even views that they later may have changed. On the other hand, when authors modified their original work in later editions (the revised 1978 edition of *Food First,* or the 20th anniversary edition of *Diet for a Small Planet),* I have used the more recent version.

The book is organized in five sections. The first, "The World Food System," describes why hunger persists in a world of plenty, and analyzes how the food system interacts with issues of population, racism, and destruction of the environment. Part two, "The American Connection," links hunger overseas to agriculture and consumption in the U.S., showing the importance of meat-based diets and the roles of family farms, farm workers, and the pesticide industry. The third section, "Global Policies and Hungry People," examines the international political issues which are most important to hunger at the end of the twentieth century: export-based economies, free trade, foreign aid, the debt crisis, and the 'structural adjustment' policies imposed on poor countries by the World Bank and the International Monetary Fund.

The two final sections look in detail at countries, regions, and communities where different food policies are being implemented, and contrast the effects of today's prevailing approach with new ways to build food security. Part four, "The Free-Market Path," examines the impact of current policies on several different countries, including some which have often presented as great success stories. The fifth section, "Alternatives," shows the accomplishments of those who are working to build positive examples of how the world food system could be changed so as to finally eliminate hunger from the globe. The book concludes with an essay on the issues of the next millennium by the Food First Institute's director, Peter Rosset, and finally with the words of a Honduran peasant leader, Elvia Alvarado. The story of her struggle for justice and a better life, and her efforts to build alliances with people in the United States and around the globe, exemplifies the lesson of this book: how we can work for a world in which everyone has the right to survive, and why we must.

1 Ehrlich, Paul. *The Population Bomb.* (New York: Ballantine Books, 1968)

2 Paddock, William and Paul Paddock. *Famine 1975!* (Boston: Little, Brown and Company, 1967)

3 Hardin, Garrett. "Living on a Lifeboat." *BioScience* 24: 561–568, 1974.

4 Murdoch, William. "World Hunger and Population." pp. 3–20 in Ronald C. Carroll, John H. Vandermeer and Peter Rosset, eds. *Agroecology.* (New York: McGraw-Hill Publishing, 1990)

5 Mazur, Laurie Ann, ed. *Beyond the Numbers: A Reader on Population, Consumption and the Environment.* (Washington, DC: Island Press, 1994)

6 Giese, Filomena, with Marilyn Borchardt and Martha Fernandez. *Alternatives to the Peace Corps,* 8th edition. (Oakland, CA: Food First Books, 1999)

Then
life itself
will have the shape of bread,
deep and simple,
immeasurable and pure.
Every living thing
will have its share
of soil and life,
all the bread we eat each morning,
everyone's daily bread,
will be hallowed
and sacred,
because it will have been won
by the longest and costliest
of human struggles.

Pablo Neruda
"Ode to Bread"
1954

Part One

The World Food System

How the World Food System Works, and for Whom

Why are people hungry? Why do hundreds of millions, all around the globe, suffer and die from a lack of food to eat? Despite massive increases in agricultural production, why do men, women, and especially children lack even the bare necessities for survival?

The conventional answer is simple: Too many people, not enough food. This has been the explanation for world hunger for decades, accepted by experts and lay people alike. Just as Malthus predicted two centuries ago, the growth of population has outrun the growth of agriculture, so that despite all our technological progress, there is still not enough for all to eat. Until this balance is set right, according to this answer, there will inevitably be some of our fellow humans who must starve.

From this simple diagnosis, there follows an equally simple prescription. If the problem is too many people and not enough food, the solution is: fewer people, more food. On the one hand, we urgently need to control the world's population, which is still described as growing exponentially (ignoring overwhelming evidence of the demographic transition). On the other hand, we need to expand agricultural production through Green Revolution technology, expanded use of inputs such as chemical fertilizers and pesticides, and the application of new developments in biotechnology. And if these 'solutions' will have some unfortunate side effects—forced sterilizations, poisoning of farm workers, or the conversion of rain forests to plantations—well, that's too bad, but the situation is too urgent to be soft-hearted. We have to act.

Except that 'we' turns out to mean mostly 'them.' If the cause of hunger is the imbalance between growing populations and inadequate agriculture, then its clearly in the worlds poor nations—the

Third World, the South, the 'underdeveloped countries'—that the answer must be found. They are the ones with the uncontrolled populations and inadequate food production. The role of the wealthy countries is to export our technology, knowledge, birth control techniques, and surplus grain—but ultimately, it's their problem, not ours.

This is the analysis that Food First has insightfully and effectively challenged for more than two decades. With their best-selling books *World Hunger: Twelve Myths* and *Food First*, Lappé and Collins showed that the simple, conventional answer is wrong. The world produces abundant food, more than enough to feed all its people. Even the countries from which come the agonizing scenes of famine and starvation we periodically see on our TV screens—the supposedly overpopulated 'basket cases'—have adequate supplies. Fundamentally the problem is not an excess of people or a lack of food, it is a lack of power and democracy. It is an unjust world food system, not an imbalance between populations and production, that causes thousands of children to die of starvation every day.

Seen from this new perspective, the Green Revolution is not a solution to world hunger—indeed, it's part of the problem. By benefiting large landowners at the expense of the much more numerous small farmers and landless workers, it accentuates the differences in power between rich and poor which cause starvation. Contrary to the assumptions on which agricultural policy is based, small farmers are not ignorant, backward, or primitive. They are actually more productive and efficient than the rich landowners—if they have the power to control their land, resources, and harvests.

Does this analysis then mean that population is irrelevant? No, say Lappé and Schurman in *Taking Population Seriously*, but it is as much an consequence of the problem as a cause. Viewed from a power-structure perspective, high fertility is intricately linked to a lack of democracy in institutions ranging from the intimate confines of the family up to the global network of world trade. Relations of power—of husbands over wives, landowners over farm workers, cities over the countryside, dictatorial governments over citizens, and rich nations over poor—are the basic causes of population growth.

Vandermeer and Perfecto's *Breakfast of Biodiversity* broadens the Food First analysis to include the environmental dimensions of hunger, population, and the lack of democracy. They too show how the conventional Malthusian perspective leads us to blame peasants and the landless for rain forest destruction, ignoring the underlying web of causality which links together export agriculture, overwhelming debt, and the undemocratic distribution of land and power. Focusing on the Atlantic coast of Costa Rica, they show how even in a nation that is a showcase for conservation, deforestation will continue unchecked unless the fundamental power relations that cause it are challenged.

The readings in this section ask us to rethink our view of world hunger—to see it not as simply an imbalance between population and food in other countries, but rather as a consequence of the way the world food system works. In this system some people have power, and for them the system works well. They are not the ones who see their crops sold to pay a never-ending debt, lose their farms, or hear their children cry from hunger in the night. But for those who have no power, the specter of famine is always close by. To banish it, we need to go beyond promoting pills and pesticides. We need to build a democratic world food system that truly works for all.

From *World Hunger: Twelve Myths* by Frances Moore Lappé, Joseph Collins, and Peter Rosset, with Luis Esparza

Beyond Guilt and Fear

For over twenty-five years we have sought to understand why there is hunger in a world of plenty. For us, learning had to begin with unlearning. Cutting through the simplistic and scary clichés about hunger, we arrived at some surprising findings:

- No country in the world is a hopeless case. Even countries many people think of as impossibly overcrowded have the resources necessary for people to free themselves from hunger.
- Increasing a nation's food production may not help the hungry. Food production per person can increase while more people go hungry.
- Our government's foreign aid often hurts rather than helps the hungry. But in a multitude of other ways we can help.
- The poor are neither a burden on us nor a threat to our interests. Unlikely as it may seem, the interests of the vast majority of Americans have much in common with those of the world's hungry.

These surprising findings and many more have freed us from a response to hunger motivated by guilt and fear. But first we must ask the seemingly grade-school question, just what is hunger? Many people assume they know—they've felt it, they've read about it, they've been touched by images of hungry people on television. But the greatest obstacle to grasping the causes and solutions to world hunger is that few of us stop to ponder this elemental question.

What is Hunger?

Television images haunt us. Stunted, bony bodies. Long lines waiting for a meager bowl of gruel. This is famine hunger in its acute form, the kind no one could miss.

But hunger comes in another form. It is the day-in-day-out hunger that almost 800 million people suffer.[1] While chronic hunger doesn't make the evening news, it takes more lives than famine.

Every day this largely invisible hunger, and its related preventable diseases, kill as many as thirty-four thousand children under the age of five.[2] That's 12 million children per year—more than the total number of people who died each year during World War II. This death toll is equivalent to the number killed instantly by a Hiroshima bomb every three days.

Statistics like this are staggering. They shock and alarm. Several years ago, however, we began to doubt the usefulness of such numbers. Numbers can numb. They can distance us from what is actually very close to us. So we asked ourselves, what really is hunger?

Is it the gnawing pain in the stomach when we miss a meal? The physical depletion of those suffering chronic undernutrition? The listless stare of a dying child in the television hunger appeal? Yes, but it is more. And we became convinced that as long as we conceive of hunger only in physical measures, we will never truly understand it, certainly not its roots.

What, we asked ourselves, would it mean to think of hunger in terms of universal human emotions, feelings that all of us have experienced at some time in our lives? We'll mention only four such emotions, to give you an idea of what we mean.

A friend of ours, Dr. Charles Clements, is a former Air Force pilot and Vietnam veteran who spent a year treating peasants in El Salvador. He wrote of a family he tried to help whose son and daughter had died of fever and diarrhea. "Both had been lost," he writes, "in the years when Camila and her husband had chosen to pay their mortgage, a sum equal to half the value of their crop, rather than keep the money to feed their children. Each year, the choice was always the same. If they paid, their children's lives were endangered. If they didn't, their land could be repossessed."[3]

Being hungry means anguish. The anguish of impossible choices. But it is more.

In Nicaragua some years ago, we met Amanda Espinoza, a poor rural woman, who until then had never had enough to feed her family. She told us that she had endured six stillbirths and watched five of her children die before the age of one.

To Amanda, being hungry means watching people you love die. It is grief.

Throughout the world, the poor are made to blame themselves for their poverty. Walking into a home in the Philippine countryside, the first words we heard were an apology for the poverty of the dwelling. Being hungry also means living in humiliation.

Anguish, grief, and humiliation are a part of what hunger means. But increasingly throughout the world, hunger has a fourth dimension.

In Guatemala we met two poor highland peasants who, with the help of World Neighbors, an Oklahoma City-based voluntary aid group, were teaching their neighbors how to reduce erosion on the steep slopes onto which they had been pushed by wealthy landowners monopolizing the flat valley land. Two years later, we learned that one had been forced into hiding, the other had been killed. In the eyes of the wealthy their crime was teaching their neighbors better farming techniques. Guatemala's oligarchy feels threatened by any change that makes the poor less dependent on low-paying jobs on their plantations.

Often, then, a fourth dimension of hunger is fear.

Anguish, grief, humiliation, and fear. What if we refused to count the hungry and instead tried to understand hunger in terms of such universal emotions?

We discovered that how we understand hunger determines what we think are its solutions. If we think of hunger only as numbers—numbers of people with too few calories—the solution also appears to us in numbers—numbers of tons of food aid, or numbers of dollars in economic assistance. But once we begin to understand hunger as real people coping with the most painful of human emotions, we can perceive its roots. We need only ask, When have we experienced any of these emotions ourselves? Hasn't it been when we have felt out of control of our lives—powerless to protect ourselves and those we love?

Hunger has become for us the ultimate symbol of powerlessness.

The Causes of Powerlessness

Understanding that hunger tells us that a person has been robbed of the most basic power—the power to protect ourselves and those we love—is a first step. Peeling back the layers of misunderstanding, we must then ask, If powerlessness lies at the very root of hunger, what are hunger's causes?

Certainly, it is not scarcity. The world is awash with food. Neither are natural disasters to blame. Put most simply, the root cause of hunger isn't a scarcity of food or land; it's a *scarcity of democracy.*

Wait a minute! What does democracy have to do with hunger? In our view—everything. Democracy carries within it the principle of accountability. Democratic structures are those in which people have a say in decisions that most affect their well-being. Leadership can be kept accountable to the needs of the majority. Antidemocratic structures are those in which power is so tightly concentrated that the majority of people are left with no say at all. Leaders are accountable only to the powerful minority.

In the United States, we think of democracy as a strictly political concept, so it may seem contrived to apply it to the economic questions of land, food, jobs, and income. Political democracy helps us as citizens to protect certain rights—to reside where we will, to vote, to have our civil liberties upheld, and so on. Unlike many societies, here such universal political citizenship is taken for granted.

But along with many other societies, we lack a concept of economic citizenship. To parallel our universal political rights, we have not yet established universal economic rights, such as the right to life-sustaining resources or the right to participate in economic decision making.

What we hope to show is that as long as this fundamental concept of democracy—accountability to those most affected by decisions—is absent from economic life, people will continue to be made powerless. From the family, to the neighborhood or the village, through the national level, to the level of international commerce and finance, we will witness the continued concentration of decision making over all aspects of economic life, including what it takes to grow and distribute that on which all life depends: food. Poverty and hunger will go on destroying the lives of millions each year and scarring the lives of hundreds of millions more.

Let us look briefly at how antidemocratic decision making robs people of power over their lives on each of the levels mentioned above.

First, within the family, who controls food resources? Women are responsible for growing at least half the world's food. The resources women have to grow staple foods largely determine their family's nutritional well-being. But many women are losing authority over land use—the result of the privatization of land ownership and a focus on export crops that began under colonialism. Credit for growing cash crops goes overwhelmingly to men, and food crops have stagnated. This dynamic within the family helps explain growing hunger.[4]

Second, at the village level, who controls the land—and how many families have none at all? In most countries, a consistent pattern emerges: Fewer and fewer people control more and more farm and pasture land. With fewer families controlling an ever greater share of the land, more and more people have none at all. A 1993 study reported alarming percentages of rural families who are landless or have insufficient land to support themselves. In Peru the number of landless or land-poor was seventy-five percent, in Ecuador seventy-five percent, sixty-six percent in Colombia, thirty-two percent in Kenya, and ninety-five percent in Egypt, among many others.[5]

Third, at the national level, how are public resources allocated? Wherever people have been made hungry, power is in the hands of those unaccountable to their people. These antidemocratic governments answer only to elites, lavishing them with credit, subsidies, and other assistance. With increasing brutality, such governments fight any reform that would make control over food-producing resources more equitable. The Landless Workers Movement (*Movimento dos Trabalhadores Rurais Sem Terra*, or MST) in Brazil is struggling to turn over land left idle on huge estates to landless families. In 1995 and 1996 at least eighty-six landless workers, family members, and MST activists were assassinated, most by the military police acting at the behest of wealthy landowners.[6]

There is yet a fourth level on which democracy is scarce—the international arena of commerce and finance. A handful of corporations dominate world trade in those commodities that are the lifeblood of third world economies. Efforts by third world governments to bargain for higher commodity prices have repeatedly failed in the face of the preeminent power of the giant trading corporations and the government trade policies of the industrial countries. Industrial countries import $60 billion a year worth of food from the Third World,[7] but traders, processors, and marketers reap most of the profit. For every dollar a U.S. consumer spends to buy cantaloupes grown in El Salvador, less than a penny goes to the farmer, while traders, shippers, and retailers receive eighty-eight cents.[8]

Heavily indebted to international aid agencies and private banks, third world nations are also at the mercy of policies decided upon in the capitals of the industrial nations, policies leading only to further impoverishment.[9]

In attempting to capsulize the antidemocratic roots of hunger, we have traveled from the level of the family to that of international commerce and finance. Let us complete the circle by returning to the family.

As economic decisions are made by those unaccountable to the majority, insecurity deepens for millions of people. Economic pressures tear family bonds asunder as men are forced to leave home in search of work, and joblessness leads to family violence and dissolution. More and more women shoulder family responsibilities alone; worldwide, perhaps as many as one-third of all households are now headed by women. On top of the weight of poverty, they confront barriers of discrimination against women. The breakdown of the traditional family structure does not bring liberation for them; it simply means greater hardship. Most of the hungry in the world are women and the children they care for. Most of those who die from hunger every year are children.

In our effort to grasp the roots of hunger, we have identified the problem: the ever greater scarcity not of food or land but of democracy, democracy understood to include the life-and-death matter of economics. But we must dig deeper. Why have we allowed this process to happen at the cost of millions of needless deaths each year?

How We Think about Hunger

Especially in troubled times, people seek ways to make sense of the world. We grasp for organizing principles to help us interpret the endlessly confusing rush of world events. It's a natural human process—perhaps as natural as eating itself. But living effectively depends on how well our organizing principles reflect reality.

Unfortunately, the principles around which many of us have come to organize our thinking about world hunger block our grasp of real solutions. We call them myths, to suggest that the views embodied may not be totally false. Many have some validity. It is as organizing principles that they fail. Not only do they prevent us from seeing how we can help the hungry, they obfuscate our own legitimate interests as well. Some fail us because they describe but don't explain, some are so partial that they lead us down blind alleys, and some simply aren't true.

What we want to do is to probe the underlying assumptions people have about world hunger's causes and cures. For we've come to believe that *the way people think about hunger is the greatest obstacle to ending it.*

After reading the following, we hope you will find that you no longer have to block out bad news about hunger but can face it squarely because a more realistic framework of understanding—to be repeatedly tested against your own experience—enables you to make real choices, choices that can contribute to ending this spreading but needless human suffering.

We may shake your most dearly held beliefs or it may confirm your deepest intuitions and experiences. Most of all, we hope that we convince you that until humanity has solved the most basic human problem—how to ensure that every one of us has food for life—we cannot consider ourselves fully human.

Myth One: There's Simply Not Enough Food

MYTH: With food-producing resources in so much of the world stretched to the limit, there's simply not enough food to go around. Unfortunately, some people will just have to go hungry.

OUR RESPONSE: The world today produces enough grain alone to provide every human being on the planet with thirty-five hundred calories a day.[10] That's enough to make most people fat! And this estimate does not even count many other commonly eaten foods—vegetables, beans, nuts, root crops, fruits, grass-fed meats, and fish. In fact, if all foods are considered together, enough is available to provide at least 4.3 pounds of food per person a day. That includes two and a half pounds of grain, beans, and nuts; about a pound of fruits and vegetables; and nearly another pound of meat, milk, and eggs.[11]

Abundance, not scarcity, best describes the supply of food in the world today. Increases in food production during the past thirty-five years have outstripped the world's unprecedented population growth by about sixteen percent.[12] Indeed, mountains of unsold grain on world markets have pushed prices strongly downward over the past three and a half decades.[13] Grain prices rose briefly during the early 1990s, as bad weather coincided with policies geared toward reducing overproduction, but remained well below the highs observed in the early sixties and mid-seventies.[14]

All well and good for the global picture, you might be thinking, but doesn't such a broad stroke tell us little? Aren't most of the world's hungry living in countries with food shortages—countries in Latin America, in Asia, and especially in Africa?

Hunger in the face of ample food is all the more shocking in the third world. According to the Food and Agriculture Organization (FAO) of the United Nations, gains in food production since 1950 have kept ahead of population growth in every region except Africa.[15] The American Association for the Advancement of Science (AAAS) found in a 1997 study that seventy-eight percent of all malnourished children under five in the developing world live in countries with food *surpluses*.[16]

Even most 'hungry countries' have enough food for all their people right now. This finding is based on official statistics even though experts warn us that newly modernizing societies invariably underestimate farm production—just as a century ago at least a third of the U.S. wheat crop went uncounted.[17] Moreover, many nations can't realize their full food production potential because of the gross inefficiencies caused by inequitable ownership of resources.

Finally, many of the countries in which hunger is rampant export much more in agricultural goods than they import. Northern countries are the main food importers, their purchases representing 71.2 percent of the total value of food items imported in the world in 1992.[18] Imports by the thirty lowest-income countries, on the other hand, accounted for only 5.2 percent of all international commerce in food and farm commodities.[19]

Looking more closely at some of the world's hunger-ravaged countries and regions confirms that scarcity is clearly not the cause of hunger.

India. India ranks near the top among third world agricultural exporters. While at least 200 million Indians go hungry,[20] in 1995

India exported $625 million worth of wheat and flour and $1.3 billion worth of rice (five million metric tons), the two staples of the Indian diet.[21]

Bangladesh. Beginning with its famine of the early 1970s, Bangladesh came to symbolize the frightening consequences of people overrunning food resources. Yet Bangladesh's official yearly rice output alone—which some experts say is seriously underreported[22]—could provide each person with about a pound of grain per day, or two thousand calories.[23] Adding to that small amounts of vegetables, fruits, and legumes could prevent hunger for everyone. Yet the poorest third of the people in Bangladesh eat at most only fifteen hundred calories a day, dangerously below what is needed for a healthy life.[24]

With more than 120 million people living in an area the size of Wisconsin, Bangladesh may be judged overcrowded by any number of standards, but its population density is not a viable excuse for its widespread hunger. Bangladesh is blessed with exceptional agricultural endowments, yet its 1995 rice yields fell significantly below the all-Asia average.[25] The extraordinary potential of Bangladesh's rich alluvial soils and plentiful water has hardly been unleashed. If the country's irrigation potential were realized, experts predict its rice yields could double or even triple.[26] Since the total calorie supply in Bangladesh falls only six percent short of needs,[27] nutritional adequacy seems an achievable goal.

Brazil. While Brazil exported more than $13 billion worth of food in 1994 (second among developing countries), 70 million Brazilians cannot afford enough to eat.[28]

Africa. It comes as a surprise for many of us to learn that the countries of sub-Saharan Africa, home to some 213 million chronically malnourished people (about twenty-five percent of the total in developing countries),[29] continue to export food. Throughout the 1980s exports from sub-Saharan Africa grew more rapidly than imports,[30] and in 1994, eleven countries of the region remained net exporters of food.[31]

The Sahelian countries of West Africa, known for recurrent famines, have been net exporters of food even during the most severe droughts. During one of the worst droughts on record, in the late 1960s and early 1970s, the value of the region's agricultural exports—$1.25 billion—remained three times greater than the value of grain imported,[32] and such figures did not even take into account significant unreported exports.[33] Once again, during the 1982–85 drought food was exported from these countries.[34]

Nevertheless, by 1990, food production per person had apparently been declining for almost two decades,[35] despite the productive capacity suggested by Africa's agricultural exports, and in 1995 over one-third of the continent's grain consumption depended on imports.[36] We use the word 'apparently' because official statistics notoriously underreport, or ignore all together, food grown for home consumption, especially by poor women, as well as food informally exchanged within family and friendship networks, making a truly accurate assessment impossible.[37] In fact, the author of the AAAS report referred to earlier argues that hunger is actually *less* severe in sub-Saharan Africa than in South Asia.[38]

Repeated reports about Africa's failing agriculture and growing dependence on imports have led many to assume that simply too many people are vying for limited resources. Africa's food crisis is real, as evidenced by moderately high rates of childhood malnutrition, but how accurate is this assumption as to why the crisis exists?

Africa has enormous, still unexploited, potential to grow food, with potential grain yields twenty-five to thirty-five percent higher than maximum potential yields in Europe and North America.[39] Beyond yield potential, ample arable land awaits use. In Chad, for example, only ten percent of the farmland rated as having no serious production constraints is actually farmed. In countries notorious for famines—Ethiopia, Sudan, Somalia, and Mali—the area of unused good-quality farmland is many times greater than the area actually farmed,[40] casting doubt on the notion that there are simply too many people for scarce resources.

Many long-time observers of Africa's agricultural development tell us that the real reasons for Africa's food problems are no mystery.[41] Africa's food potential has been distorted and thwarted as follows:

- The colonial land grab that continued into the modern era displaced peoples and the production of foodstuffs from good lands toward marginal ones, giving rise to a pattern by which good land is mostly dedicated to the production of cash crops for export or is even unused by its owners.[42] Furthermore, colonizers and, subsequently, national and international agencies have discredited peasant producers' often sophisticated knowledge of ecologically appropriate farming systems. Promoting 'modern,' often imported, and ecologically destructive technologies,[43] they have cut Africa's food producers out of economic decisions most affecting their very survival.

- Public resources, including research and agricultural credit, have been channeled to export crops to the virtual exclusion of peasant-produced food crops such as millet, sorghum, and root crops. In the 1980s increased pressure to export to pay interest on foreign debt further reinforced this imbalance.[44]

- Women are principal food producers in many parts of Africa, yet both colonial policy and, all too often, ill-conceived foreign aid and investment projects have placed decisions over land use and credit in the domain of men. In many cases that has meant preferential treatment for cash crops over food crops, skewing land-use and investment patterns toward cash crops.[45]

- Aid policies unaccountable to African peasant producers and pastoralists have generally bypassed their needs in favor of expensive, large-scale projects. Africa has historically received less aid for agriculture than any other continent, and only a fraction of it has reached rain-fed agriculture, on which the bulk of grain production depends.[46] Most of the aid has backed irrigated, export-oriented, elite-controlled production.

- Because of external as well as domestic factors, African governments have often maintained cheap food policies whereby peasants are paid so poorly for their crops that they have little incentive to produce, especially for official market channels.[47] The factors responsible for these policies have included developed-country dumping of food surpluses in African markets at artificially low prices, developed-country interest in cheap wages to guarantee profitable export production, middle-class African consumer demand for affordable meat and dairy products produced with cheap grain, and government concerns about urban political support and potential unrest.[48] The net effect has been to both depress local food production and divert it toward informal, and therefore unrecorded, markets.

- Until recently many African governments also overvalued their currencies, making imported food artificially cheap and undercutting local producers of millet, sorghum, and cassava. Although recent policy changes have devalued currencies, which might make locally produced food more attractive, accompanying free-trade policies have brought increased imports of cheap food from northern countries, largely canceling any positive effect.[49]

- Urban tastes have increasingly shifted to imported grain, particularly wheat, which few countries in Africa can grow economically. Thirty years ago, only a small minority of urban dwellers in sub-Saharan Africa ate wheat. Today bread is a staple for many urbanites, and bread and other wheat products account for about a third of all the region's grain imports.[50] U.S. food aid and advertising by multinational corporations ("He'll be smart. He'll go far. He'll eat bread.")[51] have played parts in molding African tastes to what the developed countries have to sell.[52]

Beneath the 'scarcity diagnosis' of Africa's food situation lie many human-made (often Western-influenced) and therefore reversible causes. Even Africa's high birth rates are not independent variables but are determined by social realities that shape people's reproductive choices.

A Future of Scarcity?

A centuries-old debate has recently heated up: Just how close are we to the earth's limits?

Major studies have arrived at widely varying conclusions as to the earth's potential to support future populations. In a 1995 book Professor Joel Cohen of Rockefeller University surveyed estimates put forth over four centuries.[53] Always a slippery concept,[54] estimates of the earth's 'carrying capacity,' or the number of people who could be supported, have varied from a low of one billion in a 1970 study to a high of 1,022 billion put forth in 1967. Among studies published between 1990 and 1994, the range was from "much less than our current population of 5.5 billion," according to Paul Ehrlich and others, to a high of 44 billion estimated by a Dutch research team, with most estimates falling into the ten to fourteen billion range.[55] By contrast, the 1996 United Nations forecast, generally considered to be the best future population projection, predicts that the world population will peak at 9.36 billion in the year 2050 and stabilize thereafter[56] (projections of the maximum future population have been coming down over the past few years). This is well within what most experts view as the capacity of the earth.

In view of today's abundant food supplies, as well as the potential food supplies, we question the more pessimistic predictions of demographic catastrophe. Only fifty years ago, China pundits predicted that that famine-ridden nation could never feed its population. Today more than twice as many people eat—and fairly adequately[57]—on only one-fourth the cropland per person used in the United States.[58]

Not that anyone should take the more pessimistic predictions lightly; they underscore the reality of the inevitably finite resource base entrusted to us. They should therefore reinforce our sense of urgency to address the root causes of resource misuse, resource degradation, and rapid population growth.

Lessons from Home

Finally, in probing the connection between hunger and scarcity, we should never overlook the lessons here at home. More than thirty million Americans cannot afford a healthy diet; 8.5 percent of U.S. children are hungry, and 20.1 percent are at risk of hunger.[59] But who would argue that not enough food is produced? Surely not U.S. farmers; overproduction is their most persistent headache. Nor the U.S. government, which maintains huge storehouses of cheese, milk, and butter. In 1995, U.S. aid shipments abroad of surplus food included more than 3 million metric tons of cereals and cereal products,[60] about two-thirds consisting of wheat and flour. That's enough flour to bake about six hundred loaves of bread per year for every hungry child in the United States.[61]

Here at home, just as in the Third World, hunger is an outrage precisely because it is profoundly needless. Behind the headlines, the television images, the superficial clichés, we can learn to see that hunger is real; scarcity is not.

Only when we free ourselves from the myth of scarcity can we begin to look for hunger's real causes.

The Green Revolution Is the Answer

MYTH: The miracle seeds of the Green Revolution increase grain yields and therefore are a key to ending world hunger. Higher yields mean more income for poor farmers, helping them climb out of poverty, and more food means less hunger. While the Green Revolution may have missed poorer areas, with more marginal lands, we can learn valuable lessons from that experience to help launch a 'Second' Green Revolution to defeat hunger once and for all.

OUR RESPONSE: People have been improving seeds through experimentation since the beginning of agriculture,[62] but the term 'Green Revolution' was coined in the 1960s to highlight a particularly striking breakthrough. In test plots in northwest Mexico, improved varieties of wheat dramatically increased yields. Much of the reason these 'modern varieties' produced more than traditional varieties was that they were more responsive to controlled irrigation and to petrochemical fertilizers, allowing for much more efficient conversion of industrial inputs into food. With a big boost from the International Agricultural Research Centers created by the Rockefeller and Ford Foundations,[63] the 'miracle' seeds quickly spread to Asia, and soon new strains of rice and corn were developed as well.[64]

By the 1970s, the term *revolution* was well deserved, for the new seeds—accompanied by chemical fertilizers, pesticides, and, for the most part, irrigation—had replaced the traditional farming practices of millions of Third World farmers. By the 1990s almost seventy-five percent of Asian rice areas were sown with these new varieties.[65] The same was true for almost half of the wheat planted in Africa and more than half of that in Latin America and Asia,[66] and about seventy percent of the world's corn as well.[67] Overall, it was estimated that forty percent of all farmers in the Third World were using Green Revolution seeds,[68] with the greatest use found in Asia, followed by Latin America. The Green Revolution, however, had made fewer inroads in Africa.[69]

Clearly, the production advances of the Green Revolution are no myth. Thanks to the new seeds, tens of millions of extra tons of grain a year are being harvested, Green Revolution promoters tell us. And we may be on the brink of a second Green Revolution based on further advances in biotechnology.[70]

But has the Green Revolution actually proven itself a successful strategy for ending hunger? Here the debate heats up. Let us capsulize two dominant sides in this debate.

Proponents claim that by increasing grain production, the Green Revolution has alleviated hunger, or at least has prevented it from becoming even worse as populations keep growing. Traditional agriculture just couldn't meet the demands of today's burgeoning populations. Moreover, they assert, dealing with the root causes of poverty that contribute to hunger takes a very long time and people are starving now. So we must do what we can—increase production. The Green Revolution buys the time desperately needed by Third World countries to deal with the underlying social causes of poverty and to cut birth rates. In any case, outsiders—like the scientists and policy advisors behind the Green Revolution—can't tell a poor country to reform its economic and political systems, but they can contribute invaluable expertise in food production.

The above view was rarely questioned when we began our work over twenty-five years ago. It was the official wisdom. But in response to many independent analyses, including the work of our institute and the track record of the Green Revolution itself, this 'wisdom' has been increasingly challenged.[71]

Those of us challenging the Green Revolution strategy know that production will have to increase if populations continue to grow. But we've also seen that focusing narrowly on increasing production—as the Green Revolution does—cannot alleviate hunger because it fails to alter the tightly concentrated distribution of economic power, especially access to land and purchasing power. If you don't have land on which to grow food or the money to buy it, you go hungry no matter how dramatically technology pushes up food production.

Introducing any new agricultural technology into a social system stacked in favor of the rich and against the poor—without addressing the social questions of access to the technology's benefits—will, over time, lead to an even greater concentration of the rewards from agriculture, as is happening in the United States.

Because the Green Revolution approach does nothing to address the insecurity that lies at the root of high birth rates—and can even heighten that insecurity—it cannot buy time while population growth slows. Finally, a narrow focus on production ultimately defeats itself, as it destroys the very resource base on which agriculture depends.

We've come to see that without a strategy for change that addresses the powerlessness of the poor, the tragic result will be more food and yet more hunger.

This debate is no esoteric squabble among development experts—it cuts to the core of our understanding of development and therefore deserves careful probing. So first we will explore why a narrow production strategy like the Green Revolution is bound to fail to end hunger, drawing on the experience of the past four decades. Then, recognizing that this debate itself overlooks critical considerations, we will step outside its confines to ask what approach could offer genuine food security.

More Food and Yet More Hunger?

In 1985, the head of the international body overseeing Green Revolution research, S. Shahid Husain, declared that the poor are the beneficiaries of the new seeds' output. He even went so far as to claim that "added emphasis on poverty alleviation is not necessary" because increasing production itself has a major impact on the poor.[72] Husain's statement must have embarrassed many promoters of the Green Revolution, because by the 1980s few of even its avid defenders would make such a sweeping claim.

In fact, some within the World Bank—which finances the research network Husain chaired—concluded in a major 1986 study of world hunger that a rapid increase in food production does not necessarily result in food security, that is, less hunger. Current hunger can be alleviated only by "redistributing purchasing power and resources toward those who are undernourished," the study said.[73] In a nutshell—if the poor don't have the money to buy food, increased production is not going to help them. At last! was our response—for this fundamental insight was a starting point of our institute's analysis of hunger more than two decades ago.[74]

Despite three decades of rapidly expanding global food supplies, there are still an estimated 786 million hungry people in the world in the 1990s.[75] And where are these 786 million hungry people?

Since the early 1980s, media representations of famines in Africa have awakened Westerners to hunger there, but Africa represents less than one-quarter of the hunger in the world today. We are made blind to the day-in-day-out hunger suffered by hundreds of millions more.

By the mid-1980s, newspaper headlines were applauding the Asian success stories—India and Indonesia, we were told, had become "self-sufficient in food" or even "food exporters."[76] But Asia, precisely where Green Revolution seeds have contributed to the greatest production success,[77] is the home of roughly two-thirds of the undernourished in the entire world.[78]

Serious questions are raised when we look at the number of hungry people in the world in 1970 and in 1990, spanning the two decades of major Green Revolution advances. At first glance it looks as though great progress was made, with food production up and hunger down. The total food available per person in the world rose by eleven percent over those two decades,[79] while the estimated number of hungry people fell from 942 million to 786 million,[80] a sixteen percent drop. This was apparent progress, for which those behind the Green Revolution were understandably happy to take the credit.[81]

But these figures merit a closer look. If you eliminate China from the analysis, the number of hungry people in the rest of the world actually *increased* by more than eleven percent, from 536 to 597 million.[82] In South America, for example, while per capita food supplies rose almost eight percent, the number of hungry people also went up, by nineteen percent. It is essential to be clear on one point: It is not increased population that made for more hungry people—total food available per person actually increased—but rather the failure to address unequal access to food and food-producing resources.

In South Asia there was nine percent more food per person by 1990, but there were also nine percent more hungry people.[83] The remarkable difference in China, where the number of hungry *dropped* from 406 million to 189 million,[84] almost begs the question: which has been more effective at reducing hunger, the Green Revolution or the Chinese Revolution?

The volume of output alone tells us little about hunger. Whether the Green Revolution or any other strategy to boost food production will alleviate hunger depends on the economic, political, and cultural rules people make. Those rules determine who benefits as a *supplier* of the increased production (whose land and crops prosper and for whose profit) and who benefits as a *consumer* of the increased production (who gets the food and at what price).

According to *Business Week* magazine, "Even though Indian granaries are overflowing now," thanks to the success of the Green Revolution in raising wheat and rice yields, "5,000 children die each day of malnutrition. One-third of India's 900 million people are poverty-stricken." Since the poor can't afford to buy what is produced, "the government is left trying to store millions of tons of food. Some is rotting, and there is concern that rotten grain will find its way to public markets." The article concludes that the Green Revolution may have reduced India's grain imports substantially, but it did not have a similar impact on hunger.[85]

Early Criticisms of the Green Revolution

The impact of the Green Revolution on the rural poor has been controversial, with persuasive supporters and strong critics of this development strategy. While the two sides still disagree with each other on the fundamental issues, each has, over the years, accepted some points made by the other.[86] Yet the balance of evidence weighs in against the Green Revolution approach as a tool to alleviate hunger and poverty.

Early supporters of the Green Revolution varieties claimed that boosting production would raise the incomes of farmers, thus raising them out of poverty. Yet in the early decades of the Green Revolution a disturbing trend was revealed, which fueled criticisms: Larger farmers were much quicker to adopt the new varieties than poorer ones, and the landless could, of course, do nothing with new seeds, lacking the land in which to plant them. The larger, wealthier farmers were producing more than ever, and the increased production brought grain prices down, putting the squeeze on smaller, poorer farmers. The poor were often left in a weaker competitive position than before the new seeds had arrived. Critics of the Green Revolution argued that technologies that required purchased inputs—improved seeds, fertilizers, and pesticides—would inherently favor those with money over the poor, who would eventually lose their land and be forced to migrate to burgeoning urban shantytowns.[87]

Other early criticisms leveled at the Green Revolution were that it focused on just two or three crops, ignoring the rich diversity of foods grown by small farmers such as pulses and legumes, which are nutritionally key for lower-income families; that the technologies worked only on good-quality farmland with irrigation and were inappropriate to the diverse and marginal lands farmed by the poor;[88] and that agrochemicals were dangerous to the health of farmers and to the environment.[89]

Green Revolution Supporters Respond

Yet as the years passed, many poor farmers did eventually adopt modern varieties and begin to use chemical fertilizers and pesticides as well, sometimes even earning higher incomes as a result of yield increases.[90] As to why it took longer for small farmers to adopt new technologies, research by Green Revolution supporters revealed that the institutions that give credit, technical assistance, and marketing support to farmers are biased against the poor. This, rather than anything intrinsic about seeds, fertilizer, or pesticides, the supporters now say, explains the slower adoption by the poor. They argue that seeds, pesticides, and fertilizers do not discriminate between richer and poorer farmers because they are 'divisible' into tiny amounts—you can buy just a few ounces if that is all you can afford—unlike a tractor, for example, which is totally out of reach for a very small farmer.[91]

In the late 1980s and early 1990s the supporters tried to take back the initiative from the critics of the Green Revolution,[92] arguing for a redoubling of efforts to address the anti-poor biases of institutions, so that the benefits of the new technologies could be better shared by the poor.[93] In the past, they conceded, government institutions—ministries of agriculture and extension services—failed to effectively deliver Green Revolution technological packages to small farmers in marginal areas, while nongovernmental organizations (NGOs, called nonprofits here in the United States) showed a greater capacity to reach and work with the poor.

The principal supporters of the Green Revolution now argue that foreign aid monies be partially redirected from government institutions toward NGOs to assure that the second Green Revolution reaches those left out by the first.[94] The idea is a sort of affirmative action for improved seeds, chemical fertilizers, and pesticides. Supporters further point out that research has now been initiated on a whole range of food crops grown by poor farmers and on how to adapt the basic technological packages for use on marginal lands.[95]

What Critics Have Said All Along

Prominent Green Revolution critic Keith Griffin finds it odd that blame has now been placed on societal structures. He argues that "the purpose of the Green Revolution was precisely to circumvent the need for institutional change. Technical progress was to be regarded as an alternative to land reforms and institutional transformation—the Green Revolution was to substitute for the red—and it is misleading twenty years later to claim that the fault lies entirely with inappropriate institutions and policies."[96]

The critics of the Green Revolution have argued all along that institutions that are biased against the poor are a significant part of the problem. They might also agree that NGOs should be part of an alternative. But critics stop short of absolving the technology and blaming only the institutions. One cannot separate a technological package made up of expensive inputs—agrochemicals—that farmers must purchase with their limited money, which are manufactured by global corporations with vested interests in increasing sales, from the institutions that promote and perpetuate those technologies. It is well documented that, from the beginning, chemical companies and development institutions collaborated in the promotion of Green Revolution technologies.[97]

Today, two leading pesticide companies figure among the funders of the "2020 Vision" program of the International Food Policy Research Institute (IFPRI), which serves as the ideological leader in the push for the second Green Revolution.[98] The International Fertilizer Industry Association—representing manufacturers—works closely with IFPRI, the World Bank, and the Food and Agriculture Organization of the UN (FAO), to promote increased fertilizer use as part of the package.[99]

Furthermore, while Green Revolution proponents argue for change from above—mandated by international lenders and aid agencies—critics argue that real change, truly benefiting the poor, is possible only if it is driven from below, i.e., by the demands of the poor themselves.[100] The critics of the Green Revolution are very concerned about the repetition and intensification of mistaken policies. Several kinds of evidence underlie this concern.

It is only in areas that had relatively low levels of inequality before the Green Revolution arrived that inequity has not totally derailed potentially poverty-alleviating effects of the new technologies. The least disruptive impacts have come in those areas of Asia with centuries-long traditions of irrigated rice. In those areas a long history of communally managed traditional irrigation systems has left a legacy of relative equality in land holdings, and it is there that we see the most uniform adoption of the new seeds.[101]

By comparison, other crops and other areas, like corn in Latin America, are highly inequitable, and the entry of relatively costly Green Revolution practices into communities has often sharpened the differences, leading to increased poverty and even conflict. The 1994 Zapatista rebellion in Chiapas, Mexico, was partially driven by a growing gap between rich and poor farmers, produced by the introduction of Green Revolution technologies, principally herbicide and chemical fertilizer.[102]

In a 1995 study reviewing every research report published on the Green Revolution over a thirty-year period—more than three hundred in all—the author showed that eighty percent of those with conclusions on equity found that inequality increased, including even seventy percent of the studies that focused on India and the Philippines.[103]

This is not a minor problem. If hunger and poverty existed principally because the poor didn't know how to grow corn, or rice, then one might be willing to accept greater inequality as the cost of teaching them how to do so with 'modern' technology. This argument is further bolstered if the 'rising tide' of higher yields 'raises all ships,' that is, if the poor obtain higher yields too, even if they didn't benefit as much as the wealthy, precisely what a few pro-Green Revolution studies show.[104]

But in fact people are *not* hungry and poor because they do not know how to grow corn. They are not even hungry because not enough corn or rice is grown in the world; there is plenty for all, today. Rather, they are hungry and poor because of *inequalities*—inequalities in access to land, to jobs, to income and other resources, and to political power. We cannot strike at the roots of hunger and poverty with approaches that accentuate their very basis in inequality.

The Green Revolution and the Landless

In theory, the Green Revolution was to alleviate hunger by helping poor farmers produce more food for themselves and generate more income from their land. But the new seeds' potential to relieve poverty and hunger by making farmers more prosperous depends—to start with—on what portion of the poor are farmers. This seems pretty elementary, but it is often overlooked—despite the fact that more than half a billion rural people in the Third World are either landless or have too little land to feed their households.[105] Very often the landless are the most hungry, and in many countries they make up the majority of the poorest in rural areas.

Given a choice between redistributing land and boosting crop yields, those with little or no land clearly would gain the most from redistribution. Yet proponents of the Green Revolution correctly argue that the landless may also gain if production increases generate more off-farm jobs or lead to greater demand for farmworkers and therefore higher wages. Investigators dispute to what extent either has in fact happened. While the findings are mixed, a number of studies have reported improvements in both employment and wages in some Green Revolution areas.[106] But research also shows that employment and wage gains are frequently offset by the tendency for successful farmers to use herbicides and/or to mechanize, thus reducing demands for hand weeding and labor-intensive land preparation techniques. Furthermore, the Green Revolution has by and large bypassed those marginal areas where rural poverty—and the landless—are most concentrated.

British economist Michael Lipton, in the most comprehensive and even-handed treatment of the Green Revolution to date, concluded: "We have learned that employment gains per hectare, created by modern varieties, fall off as better off farmers seek labor-saving ways to weed and thresh… More important, poverty—and modern varieties' impact on it—is not a problem mainly for farm households in [Green Revolution] lead areas, but for farm households outside them."[107]

Within Green Revolution areas, whether the landless benefit hinges in large part on how those profiting from the new seeds use their new wealth. Do they replace workers with machines, thereby reducing the number of jobs? Even the staunchest supporters of the Green Revolution admit this is a problem,[108] though they rightly point out that the massive import of tractors by Third World countries often preceded the introduction of new varieties.[109]

Subsidies to large, better-off farmers and overvalued exchange rates have long encouraged the adoption of labor-saving technologies, ironically, in parts of the world where labor is most abundant and capital most scarce. Lipton calls this mechanization a "socially-inefficient response, made privately profitable for big farmers only by their success in lobbying for subsidies on fuel, credit, and tractors to displace labor."[110] Of course, large growers want to mechanize not for higher yields but to dispense with 'labor' problems like unions, demands for minimum wages, and uppity tenant farmers.[111]

Much depends on the political organization of the landless. In the Indian state of Kerala, where agricultural workers are well organized, real wages of farmworkers rose, in contrast with many other parts of India.[112] In fact, Kerala is the state with the largest drop over the last thirty years in the percentage of the population living in poverty, far outstripping Green Revolution lead areas like Punjab and Uttar Pradesh.[113]

The Intertwined Fates of Rich and Poor

In most of the Indian subcontinent, it is "generally true" that "the poor benefit from the Green Revolution but not equally or proportionately because of the several advantages enjoyed by the rich and the handicaps suffered by the poor," argues one of the architects of India's Green Revolution, D. P. Singh. The gulf between the two, he concludes, "widens in absolute as well as relative terms."[114]

What are those handicaps suffered by the poor? Most of all, the poor lack clout. They can't command the subsidies and other government favors accruing to the rich. Neither can the poor count on police or legal protection when wealthier landowners abuse their rights to resources.

The poor also pay more and get less. Poor farmers can't afford to buy fertilizer and other inputs in volume; big growers can get discounts for large purchases. Poor farmers can't hold out for the best price for their crops, as can larger farmers whose circumstances are far less desperate.[115]

In much of the world, water is the limiting factor in farming success, and irrigation is often out of the reach of the poor. Canal irrigation favors those near the top of the flow. Tubewells, often promoted by development agencies, favor the bigger operators, who can better afford the initial investment and have lower costs per unit.[116]

Credit is also critical. It is common for small farmers to depend on local moneylenders and pay interest rates several times as high as wealthier farmers. Government subsidized credit overwhelmingly benefits the big farmers.[117]

So what? some will say. If the poor are using the improved seeds, haven't they gained, even if not as much as the better off?

But poor and rich participate in a single social dynamic—their fates are inevitably intertwined. A study of two rice-growing villages in the Philippines dramatizes this reality. In both villages, large and small farmers alike adopted the new seeds. In the village where landholdings were relatively equal and a tradition of community solidarity existed, the new technology did not polarize the community by disproportionately benefiting the better off. But in the village dominated by a few large landowners, their greater returns from the Green Revolution allowed them to advance at the expense of the small farmers. After ten years, the large farms in the village had grown in size by over fifty percent.[118] Land absorbed by the rich meant less for the poor.

Where a critic might want to blame the Green Revolution for the greater misery of poor farmers, the real fault, as some proponents have argued, lies in a social order permitting a tight grip on resources by only a few families. But contrary to what these proponents might like, that social order is a reality, one that the Green Revolution, as a technology-based, production-oriented approach, fails to, and in fact cannot, address.

Who Survives the Farm Squeeze?

With the Green Revolution, farming becomes petro-dependent. Some of the more recently developed seeds may produce higher yields without manufactured inputs,[119] but the best results require the right amounts of chemical fertilizer, pesticides, and water.[120]

So as the new seeds spread, petrochemicals become part of farming. In India, adoption of the new seeds has been accompanied by a sixfold rise in fertilizer use per acre. Yet the quantity of agricultural production per ton of fertilizer used in India dropped by two-thirds during the Green Revolution years.[121] In fact, over the past thirty years the annual growth of fertilizer use on Asian rice has been from three to forty times faster than the growth of rice yields.[122]

Because farming methods that depend heavily on chemical fertilizers do not maintain the soil's natural fertility, and because pesticides generate resistant pests, farmers need ever more fertilizers and pesticides just to achieve the same results.[123] At the same time, machines—though not *required* by the new seeds—enter the fields, as those profiting fear labor organizing and use their new wealth to buy tractors and other machines.

This incremental shift we call the industrialization of farming. What are its consequences?

Once on the path of industrial agriculture, farming costs more. It can be more profitable, of course, but only if the prices farmers get for their crops stay ahead of the costs of petrochemicals and machinery. Green Revolution proponents claim increases in net incomes from farms of all sizes once farmers adopt the more responsive seeds. But recent studies also show another trend: Outlays for fertilizers and pesticides may be going up faster than yields, suggesting that Green Revolution farmers are now facing what U.S. farmers have experienced for decades—a cost-price squeeze.[124]

In Central Luzon, Philippines, rice yield increased thirteen percent during the 1980s but at the cost of a twenty-one percent increase in fertilizer use. In the Central Plains yields went up only 6.5 percent, while fertilizer use rose twenty-four percent and pesticides jumped by fifty-three percent. In West Java a twenty-three percent yield increase was virtually canceled by sixty-five and sixty-nine percent increases in fertilizers and pesticides, respectively.[125]

To anyone following farm news here at home, these reports have a painfully familiar ring—and why wouldn't they? After all, the United States—not Mexico—is the true birthplace of the Green Revolution. Improved seeds combined with chemical fertilizers and pesticides have pushed corn yields up nearly threefold since 1950, with smaller but still significant gains for wheat, rice, and soybeans.[126] As larger harvests have pushed down the prices farmers get for their crops while the costs of farming have shot up, farmers' profit margins have been drastically narrowed since World War II. By the early 1990s, production costs had risen from about half of gross farm income to over eighty percent.[127]

So who survives today? Two very different groups: those few farmers who chose not to buy into industrialized agriculture, and those able to keep expanding their acreage to make up for their lower per-acre profit.

Among this second select group are the top 1.2 percent of farms by income, those with $500,000 or more in yearly sales, dubbed superfarms by the U.S. Department of Agriculture. In 1969, the superfarms earned sixteen percent of net farm income, and by the late 1980s they garnered nearly forty percent.[128]

Superfarms triumph, not because they are more efficient food producers,[129] or because the Green Revolution technology itself favored them, but because of advantages that accrue to wealth and size.[130] They have the capital to invest and the volume necessary to stay afloat even if profits per unit shrink. They have the political clout to shape tax policies in their favor. Over time, why should we expect the result of the cost-price squeeze to be any different in the Third World?

In the United States, we've seen the number of farms drop by two-thirds since World War II,[131] and average farm size has more than doubled.[132] The gutting of rural communities, the creation of inner-city slums, and the exacerbation of unemployment all followed in the wake of this vast migration from the land. Think what the equivalent rural exodus means in the Third World, where the proportion of jobless people is already double or triple our own.

Not Ecologically Sustainable

There is growing evidence that Green Revolution-style farming is not ecologically sustainable, even for large farmers.[133] Furthermore, dependence on expensive purchased inputs can increase indebtedness and the precarious nature of small-farmer life.[134]

In the 1990s Green Revolution researchers themselves sounded the alarm about a disturbing trend that had only just come to light. After dramatic increases in the early stages of the technological transformation, yields began falling in a number of Green Revolution areas. In Central Luzon, Philippines, rice yields grew steadily during the 1970s, peaked in the early 1980s, and have been dropping gradually ever since.[135] Long-term experiments conducted by the International Rice Research Institute (IRRI) in both Central Luzon and Laguna province confirm these results.[136] Similar patterns have now been observed for rice-wheat systems in India and Nepal.[137] The causes of this phenomenon have to do with forms of long-term soil degradation that are still poorly understood by scientists.[138]

An Indian farmer told *Business Week* his story: "Dyal Singh knows that the soil on his 3.3-hectare [eight-acre] farm in Punjab is becoming less fertile. So far, it hasn't hurt his harvest of wheat and corn. 'There will be a great problem after five or ten years,' says the 63-year-old Sikh farmer. Years of using high-yield seeds that require heavy irrigation and chemical fertilizers have taken their toll on much of India's farmland... So far, six percent of agricultural land has been rendered useless."[139]

Where yields are not actually declining, the rate of growth is slowing rapidly or leveling off, as has now been documented in China, North Korea, Indonesia, Myanmar, the Philippines, Thailand, Pakistan, and Sri Lanka.[140]

Pests and pesticides have been another source of problems for the Green Revolution. IRRI released the first Green Revolution rice variety, called IR8, in 1966. With irrigation, chemical fertilizers, and pesticides, IR8 produced more than twice as much per acre as older varieties in traditional cropping systems. But with hindsight we can see that IR8, and other new varieties, had serious ecological drawbacks. Because of their high degree of genetic uniformity, they had a narrower base of resistance to insect pests than did the more diverse traditional varieties. And their shorter stature—an attribute deliberately bred for in order to support more grain per plant—made them poor competitors against weeds, increasing the need for labor for weeding or for herbicides.[141]

Within a few years of releasing IR8, reports started trickling in of major insect, crop disease, and weed problems. Relatively minor pests exploded to epidemic proportions. Tungro virus, transmitted by an insect called the green leafhopper, destroyed tens of thousands of acres of rice in the Philippines in 1971 and 1972. The brown planthopper, a minor pest in the 1960s, devastated rice across Southeast Asia in the 1970s. Massive spraying of insecticides proved useless in fighting the outbreak.[142]

In late 1973 IRRI released IR26, a rice variety containing a resistance gene for brown planthopper. Shortly thereafter, the institute released IR28, IR29, and IR30, other varieties with the same gene—all in the Laguna region of the Philippines. For two years it appeared that the solution had been found. But in 1975 a new strain of the brown planthopper, labeled biotype 2, broke out in the Laguna area. It could successfully attack all of the supposedly resistant varieties because they carried the identical gene against planthoppers. Biotype 2 soon spread to other parts of the Philippines and to Indonesia, where it wiped out 600,000 acres in one fell swoop.[143]

In 1975 IRRI released IR32, with a resistance gene for biotype 2. Once again it appeared that the solution had been found. But three years later biotype 3 appeared and attacked IR32. Farmers stepped up insecticide spraying, with very serious health effects for themselves,[144] but utterly failed to control the planthopper, which was resistant to the chemicals. Experiments at IRRI showed that early application of insecticides actually led to thirty- to forty-fold increases in planthopper numbers afterwards—a phenomenon dubbed the 'pesticide treadmill,' as it led farmers to spray over and over again. Dr. Peter Kenmore, then in the Philippines working on his doctoral dissertation for the University of California at Berkeley, discovered that the pesticides eliminated other insects that normally preyed on the planthopper, thus releasing it from natural population controls.[145]

This story has a relatively happy ending, however. Thanks in large part to Dr. Kenmore's later work at the Food and Agriculture Organization of the UN, and to an anthropologist named Dr. Grace Goodell, a new approach to pest management was pioneered in the Philippines. Recognizing that the top-down approach, by which international scientists developed technology packages at research stations, just wasn't producing the desired results, Dr. Kenmore turned to the farmers themselves. Over the next few years, farmers were organized into 'field schools,' where they learned pest biology and acquired the skills needed to manage pests based on observation and minimal pesticide use.[146]

Pest management based on farmer field schools has now spread from the Philippines to Indonesia, Malaysia, Thailand, Sri Lanka, Bangladesh, India, China, and Vietnam. By 1993 18,000 extension agents and 500,000 farmers participated in field schools, achieving an average fifty percent reduction in insecticide use, equivalent to $325 million per year.[147]

While this is a very positive step, questions remain. The scale of farmer field school success is enormous yet pales when measured against the estimated 120 million rice farmers in Asia. "Are the benefits of extending the model worth the cost?" asks Green Revolution advocate Dr. Prabhu Pingali. He questions whether a couple of simple rules suitable for dissemination via mass media, like "Don't spray in the first forty days," might not be more cost effective than the field schools.[148]

This approach raises a fundamental question about the directions of technological change in agriculture. Green Revolution critics see the pesticide treadmill in rice as symptomatic of the top-down research and extension process that characterizes the international research centers, and they see the farmer field schools as proof that empowering farmers is far more important in the larger scheme of things.[149] A couple of simple rules may resolve today's problem, but they can leave farmers dependent on outside experts when the next problem arises.

On the other hand, empowering farmers with the self-confidence to analyze and solve their own problems, while seeking outside advice only on an as-needed basis, creates the basis for long-term sustainability and self-reliance. Furthermore, the cost of organizing farmer field schools is only high as long as the job is done by international centers with staff earning international salaries. The very successful farmer-to-farmer program in Central America has been run largely by farmer organizations with minimal outside funding.[150]

The larger question is what the farmer field school experience tells us about the Green Revolution itself. Economist Pingali argues that it proves the "system works," that it is able to innovate new methodologies when faced with crisis.[151] Critics feel that it is more evidence that the top-down, capital-intensive approach is fundamentally wrong.

A partially successful solution to one problem does little to avoid further problems. The newest crisis to emerge in Asian rice production is that of weeds—even as insecticide use drops, herbicide use is skyrocketing.[152] As research centers release higher-yielding varieties that are weaker competitors against other plants, weed problems grow. Wealthier farmers use herbicides, fine tuning of irrigation water, and mechanization to deal with these problems, while poorer farmers are increasingly at a disadvantage.[153]

Critics argue for alternative, more ecologically sound methods of food production. Farmers' organizations must be empowered and supported in their efforts to develop alternatives tailored to the needs of the rural poor in diverse circumstances.[154] NGOs can and should assist in this task.

Big Winners Off the Farm

We've tried to clarify which farmers come out ahead in the Green Revolution, but the biggest winners may not be farmers at all. "Neither the farmer nor the landlord reaped the benefits of the Green Revolution," writes researcher Hiromitsu Umehara about the Philippines. "The real beneficiaries were the suppliers of farm inputs, farm work contractors, private moneylenders and banks." In his village study in Central Luzon, their share of the value of the rice harvest rose from one-fifth to more than one-half in only nine years.[155]

In the United States, the same pattern holds. Not only has the share of the consumer food dollar going to the farmer dropped from thirty-eight to seventeen percent since 1940, but of that seventeen percent the farmer must pay out an ever bigger share to banks and corporate suppliers.[156]

The pressure toward the industrialization of agriculture—leading to ever bigger and fewer farms—may be no more inherent in the new seeds themselves than in any farm technology that costs money. But any strategy to increase production that does not directly address this underlying dynamic will ultimately contribute to the displacement of rural people, and thus, especially in the Third World, to greater poverty and hunger.

The Green Revolution: The Price of Dependency

The more dire consequences of displacing people from the land is one obvious difference between the industrialization process now under way in the Third World and that well advanced in U.S. agriculture. But there's a less visible difference. Here the manufacturers of farm inputs are largely based in this economy. So as corporations selling inputs to farmers capture a bigger share of the national food dollar, at least it ends up in part creating jobs and returning profits here.

But in countries in the Third World that must import most of their fertilizers, pesticides, irrigation equipment, and machines, benefits leave the country altogether. In India, the cost of importing fertilizer rose six hundred percent between the late 1960s and 1980.[157] And India has exceptional industrial capacity and is unusually well endowed with its own fertilizer-making resources. Even where multinational firms set up plants in Third World countries to produce, for example, fertilizers and pesticides, profits are controlled by the parent company.[158]

With the agriculture of Third World countries increasingly dependent on imports that must be purchased with scarce foreign exchange, rural poverty then becomes vulnerable to fluctuations in exchange rates, dollar reserves, and inflation. In the late 1980s and early 1990s India implemented free-market economic reforms including cuts in subsidized agricultural inputs and currency devaluations, which caused inflation. The result was a dramatic upsurge in rural poverty.[159]

The future breakthroughs of the Green Revolution promise to make farmers and entire nations even more dependent on a handful of corporate suppliers. The first stage relied on improving seeds by breeding for desired qualities. These seeds were for the most part developed within the public domain in research institutes funded by governments and international lending agencies.[160]

The next stage of the revolution is biotechnology, especially gene splicing from one species to another to create 'transgenic' varieties. Its techniques can be applied to virtually any crop and even to livestock. These products are being patented by major chemical and pharmaceutical companies, which have already acquired a major share of the seed industry.[161]

Among the consequences may be even greater dependency of Third World farmers on imported inputs. For example, new seeds from Monsanto and AgrEvo are genetically engineered to work only with certain herbicides, and a farmer must purchase the whole package. What will farmers' cost savings be if they buy the genetically engineered seed—and then have to buy still another product?[162]

Even those who see great potential benefits for Third World agriculture in biotechnology breakthroughs, such as the Agricultural and Rural Development Department of the World Bank, are concerned about what one might call research dependency.[163] The widespread patenting of biotechnology processes and products will make it more difficult—and prohibitively expensive—for Third World scientists to adapt biotechnologies to the needs of their countries. Moreover, poor farmers just don't make an attractive market for the private sector.[164]

There are considerable ecological and environmental risks associated with transgenic crops. Among these is the possible 'escape' of herbicide tolerance genes to wild relatives of crops, perhaps creating 'super weeds' that are resistant to control.[165] Many concerns have been raised about the transfer of a gene from the bacteria *Bacillus thuringiensis*, known as Bt, to crop plants. Bt is currently widely used as a safe alternative for controlling insect pests by organic farmers and others who want to reduce their reliance on chemical pesticides. Its recent release by Monsanto in commercial crop varieties means that insects will be exposed to it much more regularly, probably leading to accelerated resistance and the loss of Bt as a useful tool in alternative or organic pest management.[166] Finally, the patenting of crop genes means that farmers in the future may be obliged to pay royalties to foreign companies on varieties bred by their ancestors.[167]

The Green Revolution: Some Lessons

Having seen food production advance while hunger widens, we are now prepared to ask, under what conditions are greater harvests doomed to failure in eliminating hunger?

First, where farmland is bought and sold like any other commodity and society allows the unlimited accumulation of farmland by a few, superfarms replace family farms and all society suffers.

Second, where the main producers of food—small farmers and farmworkers—lack bargaining power relative to suppliers of farm inputs and food marketers, producers get a shrinking share of the rewards from farming.

Third, where dominant technology destroys the very basis for future production by degrading the soil and generating pest and weed problems, it becomes increasingly difficult and costly to sustain yields.

Under these conditions, mountains of additional food could not eliminate hunger, as hunger in America should never let us forget. Fortunately, there are alternatives and substantial evidence that they work.

Toward an Agriculture We Can Live With

The post-World War II Green Revolution represents not just the breeding of more responsive seeds and a (flawed) strategy for ending hunger; it is a mechanical way of viewing agriculture. In the Green Revolution–*cum*–industrial agriculture framework, farming means extracting maximum output from the land in the shortest possible time. It is a 'mining' operation. In this process, humanity seeks to defeat its competition—nonfood plants, insects, and disease. So agriculture becomes a battlefield, in a war we believe we're winning as long as growth in food production stays ahead of population.[168]

In this war, weapons are chosen through a cost-calculus derived from the marketplace, from the apparently extrahuman economic laws of supply and demand. If the market tells farmers that more pesticides and chemical fertilizers will produce enough to cover their costs, then, of course, they use them.

But there's one big problem: Intent on winning the war, we fail to perceive how the food-producing resources our very security rests on are being diminished and destroyed. The potential of food resources is diminished as prime land gets degraded through desertification, soil erosion, and pesticide contamination. But the list goes on:

- Groundwater is being rapidly depleted.
- Overuse and poor drainage are causing the salinization of water used in agriculture.
- Prime farmland is being gobbled up by urban sprawl.
- The world's plant genetic resources, essential for developing new seed varieties, are shrinking. In India, which had thirty thousand wild varieties of rice only half a century ago, no more than fifty will likely remain in fifteen years.[169] (The industrial countries are heavily dependent on the Third World's genetic diversity. According to one analysis, genes from local and wild varieties have contributed an estimated $66 billion to the U.S. economy, more than the combined debts

of Mexico and the Philippines. Most of this material once belonged to sovereign states and local peoples in the Third World, the vast majority of whom have never been compensated.)[170]

- Fossil fuels, on which the industrial model of agriculture rests, are being depleted, and they are nonrenewable resources.

This is only a partial list. Yet none of these threats is addressed within the Green Revolution framework. We have to ask if there are alternative approaches up to the challenge of feeding today's growing populations.

Over thousands of years, in many areas of the world, agricultural systems have evolved along principles that are fundamentally different from those of industrial agriculture. Productivity is an important goal, but not above stability and sustainability. Today, the emerging field of agroecology[171]—built on the ecological principles of diversity, interdependence, and synergy—is applying modern science to improve rather than displace traditional farming wisdom.[172]

Industrial agriculture is simple; its tools are powerful. Agroecology is complex; its tools are subtle. Industrial agriculture is costly in both money and energy. Agroecology is cheap in dollars and fossil fuel energy, but in knowledge, labor, and diversity of plant and animal life it is rich.[173]

Traditional rice farming in Asia, for example, produced ten times more energy in food than was expended to grow it, but today's Green Revolution rice production cuts that net output in half, according to Cambridge University geographer Tim Bayliss-Smith. The gain drops to zero, he reports, in a fully industrialized system, such as that of the United States.[174]

Why such a difference? Instead of continuous production of one crop, agroecology relies on intercropping, crop rotations, and the mixing of plant and animal production—all time-honored practices of farmers throughout the world. With intercropping, several crops grow simultaneously in the same field. Rotating cereals with legumes (fixing nitrogen in the soil for use by other plants) and interplanting low-growing legumes with a cereal or in stubble help to maintain soil fertility without costly purchased fertilizers. Mixing annual and perennial crops better uses the soil's lower strata and helps prevent the downward leaching of nutrients.[175]

Ecologist John Vandermeer of the University of Michigan at Ann Arbor explains the scientific basis of intercropping.[176] Because different plants have different needs and different timings of those needs, intercropping takes better advantage of available light, water, and nutrients so more total growth takes place. Instead of depleting the soil, as does monocropping of row crops (with soil between the rows exposed to rain and wind erosion), intercropping increases the organic matter content of soils, thereby promoting better tillage and higher yields. It also insures against disaster, since the more plant varieties, the less chance of all failing simultaneously.

Integrating crops and animals on the same farm allows the return of organic matter to the fields. Using some animals—ducks or geese in rice farming—can reduce weeds without herbicides. Animals also provide emergency income and food, adding overall stability to the farmstead. Limiting pest damage by crop rotation and intentional diversity, along with careful timing of planting and harvesting, can maximize yields without the heavy doses of pesticides that threaten farm families' and consumers' health.

Miguel Altieri of the University of California at Berkeley highlights three types of traditional agriculture from which much can be learned:

- paddy rice culture, which can produce edible aquatic weeds and fish as well as rice
- shifting cultivation, involving complex combinations of annual crops, perennial tree crops, and natural forest regrowth
- raised-bed agriculture, the ancient Aztec method of constructing islands of rich soil scraped from swamps and shallow lakes

Striking biodiversity and a relatively closed nutrient cycle typifies each of these farming systems. Throughout the tropics, farming systems involving both crops and trees commonly contain over one hundred plant species used as construction materials, firewood, tools, medicinal plants, and livestock feed, as well as human food.[177]

Agroecology does not mean going backward—it means applying modern biological science to improve rather than displace traditional agriculture. This can be done only by empowering farmers themselves to take the lead, using scientists trained in agroecology as consultants.[178]

But why, then, has traditional agriculture been perceived simply as an obstacle to development?

If It Could Have Worked, It Would Have

In a heated response to our criticisms of the Green Revolution, one of its most distinguished defenders once told us that the notion of an alternative drawn from traditional agriculture is hopelessly naive. Obviously, if traditional agriculture could have fed the Third World's growing populations, it would have; today's terrible hunger is proof of its failure.

Did traditional agriculture really fail, or was it destroyed by the Green Revolution and other forces?[179] Such disdain for traditional farming practices clearly bolsters the role of corporate suppliers of manufactured inputs. Calgene is one such supplier, a U.S.-based biotechnology firm proud of its seeds with 'built-in management,' such as resistance to herbicides. Norman Goldfarb, Calgene's chief executive, suggested that such seeds would be particularly relevant in Africa. "In Africa, there are a lot of unsophisticated farmers," he said. "You can't even expect them to drive a tractor straight; you might ask them to put the seed on the field evenly."[180]

Goldfarb's arrogance reflects more than simple ignorance. It typifies a widespread blindness to traditional agriculture's potential. Because its principles profoundly contradict the rules that guide industrial agriculture—particularly, allowing market values almost exclusively to dictate use of resources—traditional agriculture has been undervalued. Criteria derived from industrial agriculture are simply inadequate for assessing traditional approaches.

In measuring success, industrial agriculture asks, How much of the main commercial crop is produced this year per acre and per labor hour? But in traditional agriculture, where intercropping of several crops is more common, such a single measure grossly underestimates production. And judging productivity by how few people can produce a given quantity of food—the industrial yardstick—is hardly appropriate in societies where many people are out of work. Finally, traditional agriculture asks how much can be harvested not just this year but indefinitely into the future.

To move beyond industrial agriculture thus involves rethinking how we judge performance. What happens when we evaluate an entire farming system—including its year-to-year stability, its sustainability, and the productivity of its diverse elements—instead of just this year's output of the top cash crop?

In agroecological farming, two or more crops in the same field produce a total output that would require as much as three times more land if the crops were cultivated separately.[181] As many as twenty-two crops in the same field is not unknown in traditional systems.

Such systems hold special importance for Africa, where fragile soils have been seriously abused for many decades but considerable knowledge of alternatives remains.[182] In West Africa's Senegal, for example, where soils have been depleted by cash crops, one study suggests that the traditional mix of millet, cattle raising, and a park-like cover of acacia trees could support a population almost double the present density, already considered high. Acacia trees yield nitrogen the soil so badly needs and high-protein pods for animal feed. The tree's drought-tolerant tap root goes down almost one hundred feet, and, conveniently, the leaves fall just before the short crop-growing season, so the trees do not compete with millet for sunlight, moisture, and nutrients.[183]

Ironically, Africa is now seen as the main target for the second Green Revolution.[184] A 'Washington Consensus' of the World Bank, the U.S. Agency for International Development, IFPRI, FAO, and others has coalesced around the idea that what Africa needs is more new seeds, agrochemicals, biotechnology, and free trade.[185] For the previously stated reasons, we think that would be a disaster for the poor and the hungry in Africa.

The alternative exists, not just for Africa, but for everywhere. It is to create a viable and productive small-farm agriculture based on land reform, and using the principles of agroecology.[186] That is the only model with the potential to end rural poverty, feed everyone, and protect the environment and the productivity of the land for future generations.[187]

Successful Examples

That sounds good, but has it ever worked? From the United States to India, alternative agriculture is proving itself viable. In the United States, a landmark study by the prestigious National Research

Council found that "alternative farmers often produce high per acre yields with significant reductions in costs per unit of crop harvested," despite the dispiriting fact that "many federal policies discourage adoption of alternative practices." The Council concluded that: "Federal commodity programs must be restructured to help farmers realize the full benefits of the productivity gains possible through alternative practices."[188]

In South India, a 1993 study was carried out to compare 'ecological farms' with matched 'conventional,' or chemical-intensive farms. The study's author found that the ecological farms were just as productive and profitable as the chemical ones. He concluded that if extrapolated nationally, ecological farming would have "no negative impact on food security" and would reduce soil erosion and the depletion of soil fertility while greatly lessening dependence on external inputs.[189]

But Cuba is where alternative agriculture has been put to its greatest test.[190] Changes underway in that island nation since the collapse of trade with the former socialist bloc provide evidence that the alternative approach can work on a large scale. Before 1989 Cuba was a model Green Revolution–style farm economy, based on enormous production units, using vast quantities of imported chemicals and machinery to produce export crops, while over half of the island's food was imported.[191] Although the government's commitment to equity, as well as favorable terms of trade offered by Eastern Europe, meant that Cubans were not undernourished, the underlying vulnerability of this style of farming was exposed when the collapse of the socialist bloc was added to the already existing and soon to be tightened U.S. trade embargo.

Cuba was plunged into the worst food crisis in its history, with consumption of calories and protein dropping by perhaps as much as thirty percent. Nevertheless, by 1997 Cubans were eating almost as well as they did before 1989, yet comparatively little food and agrochemicals were being imported.[192] What happened?

Faced with the impossibility of importing either food or agrochemical inputs, Cuba turned inward to create a more self-reliant agriculture based on higher crop prices to farmers, agroecological technology, smaller production units, and urban agriculture.

The combination of a trade embargo, food shortages, and the opening of farmers' markets meant that farmers began to receive much better prices for their products.[193] Given this incentive to produce, they did so, even in the absence of Green Revolution–style inputs. They were given a huge boost by the reorientation of government education and research and extension toward alternative methods, as well as the rediscovery of traditional farming techniques.

As small farmers and cooperatives responded by increasing production, while large-scale state farms stagnated and faced plunging yields, the government initiated the newest phase of revolutionary land reform, parceling out the state farms to their former employees as smaller-scale production units. Finally, the government mobilized support for a growing urban agriculture movement—small-scale organic farming on vacant lots—which, together with the other changes, transformed Cuban cities and urban diets in just a few years.[194]

The Cuban experience tells us that we *can* feed a nation's people with a small-farm model based on agroecological technology, and in so doing we can become more self-reliant in food production. A key lesson is that when farmers receive fairer prices, they produce—with or without Green Revolution inputs. If these expensive and noxious inputs are unnecessary, then we can dispense with them.

A Metaphor

One of the arguments used by proponents of the Green Revolution says that outsiders are not the ones to instigate the political and economic reforms essential to ending hunger. All that concerned foreigners can really offer is expert technical help—like the Green Revolution—to boost production.

There are many ways U.S. citizens—whether we like it or not—mightily affect the lives of people within the Third World, often blocking the very changes needed to alleviate hunger. As outsiders, we can help reverse the nature of our influence. But another response emerges: Should not the many negative consequences of our 'expert' advice render us more humble in assuming that our development model is superior to the values and knowledge that have been developed in many Third World societies over hundreds, even thousands, of years?

The Green Revolution–*cum*–industrial model single-mindedly asks, How can we get more out of the land? And to anyone raising questions, the response is, How heartless you are—without the Green Revolution many more people would be dying of hunger!

Perhaps a metaphor will clarify our answer. Imagine for a moment our global food resources as a large house gradually burning to the ground. The fire destroying the house represents all the ways in which our present food-producing resources are being degraded and diminished. How does the Green Revolution respond to the catastrophe? It rushes into the burning dwelling to rescue as many people as possible. Its defenders declare, "Look, the Green Revolution works—it saves lives."

There's only one problem. The house is still burning! The fire is consuming all who remain inside and making the house uninhabitable for those who will need its protection in the future. And plenty of evidence suggests that the rescue team itself—the Green Revolution—inadvertently adds oxygen to the flames as it smashes down the doors to save as many victims as possible. The 'oxygen' is its single-minded pursuit of production, which contributes to the soil degradation, erosion, pesticide abuse, and so on that helped start the fire in the first place.

Clearly, congratulating the rescue team for saving lives is beside the point. We must rebuild the dwelling and make it fireproof.

As we put out the flames and begin to rebuild, we must never forget that increased production can go hand in hand with greater hunger. So even if the house were made sound, keeping its doors open for all depends on social forces, not technical ones. By social forces we mean the rules people make through custom, laws, and—too often—brute force that govern the life-and-death question of who eats and who doesn't. The new, fireproof dwelling can offer genuine security only as people change those rules, as they make the claim to food, the right to life itself, effectively universal.

The model of industrial agriculture that the Green Revolution carries with it is constructed by the rules of the market and the unlimited accumulation of productive resources. Ending hunger does not necessitate throwing out the market or property ownership. It will require, however, transforming such rules from dogma into mere economic devices serving the entire community, no longer a privileged few. Here are some questions that can free us from dogma and allow us to begin addressing the underlying forces generating hunger:

- How can claims to land and other food-producing resources and claims to income to buy food be made equitable?
- How can poor farmers—the majority of the world's food producers—augment production and maintain and improve soil fertility without increasing their dependency on costly technologies?
- How can decisions leading to the destruction of food-producing resources be brought under democratic direction so that the destruction can be reversed?

Only by facing such questions squarely can we assume our rightful responsibility—no longer abdicating our morality to supposedly automatic laws of the marketplace. Once we are moving in this direction, farming practices could be based on a more sophisticated calculation of cost effectiveness. Melding traditional wisdom and growing scientific appreciation of our complex biological interdependence with plant and animal life, we could then finally achieve food security for everyone and responsibly safeguard resources needed by future generations.

After more than three decades of the Green Revolution we stand at a crossroads. Do we stake the food security of the poor on a second, 'new and improved' Green Revolution; or do we change directions, opting for a more agroecological path, in which grassroots movements supported by NGOs and enlightened governments play a role in helping empower the poor to develop their own alternatives?

1 Uvin, Peter. "The State of World Hunger," in *The Hunger Report: 1995*, eds. Ellen Messer and Peter Uvin (Amsterdam: Gordon and Breach Publishers, 1996), pp. 1–17, table 1.6; estimates vary.

2 United Nations Children's Emergency Fund (UNICEF). *The State of the World's Children 1993* (Oxford, UK: Oxford University Press and UNICEF, 1993), statistical note, unnumbered. See also Richard A. Hoehn's introduction in *Hunger 1997: What Governments Can Do* (Silver Spring, MD: Bread for the World Institute, 1996).

3 Clements, Charles. *Witness to War* (New York: Bantam, 1984), pg. 104.

4 For discussion of the nuances of gender and food production, see Judith Carney, "Contracting a Food Staple in the Gambia," chapter 5 in *Living Under Contract: Contract Farming and Agrarian Transformation in Sub-Saharan Africa,* eds. Peter D. Little and Michael J. Watts, (Madison, WI: University of Wisconsin Press, 1994); Beverly Grier, "Pawns, Porters and Petty Traders: Women in the Transition to Export

Agriculture in Ghana," African Studies Center, Boston University, *Working Papers in African Studies,* no. 144, 1989; Gita Sen and Caren Grown, *Development, Crises and Alternative Visions: Third World Women's Perspectives* (New York: Monthly Review Press, 1986); and Carmen Diana Deere and Magdalena León, eds., *Rural Women and State Policy: Feminist Perspectives on Latin American Agricultural Development* (Boulder, CO: Westview Press, 1987).

5 Rehman, Sobhan. *Agrarian Reform and Social Transformation* (London, UK: Zed Books, 1993), table 1.

6 Langevin, Mark S. and Peter Rosset. "Land Reform from Below: The Landless Workers Movement in Brazil," *Food First Backgrounder* (Oakland, CA: Institute for Food and Development Policy, 1997) no. 3, vol. 4.

7 United Nations Conference on Trade and Development. *Handbook of International Trade and Development Statistics 1994* (New York and Geneva: United Nations, 1995), table A2.

8 Conroy, Michael E., and Douglas L. Murray and Peter M. Rosset. *A Cautionary Tale: Failed U.S. Development Policy in Central America* (Boulder, CO: Lynne Rienner/Food First Development Studies, 1996), figure 4.4.

9 For examples see *Structural Adjustment and the Spreading Crisis in Latin America* (Washington, DC: Development Gap for Alternative Policies, 1995); and Walden Bello with Shea Cunningham and Bill Rau, *Dark Victory: The United States and Global Poverty* (Oakland, CA: Food First Books, 1998).

10 Calculated from Food and Agriculture Organization, *1992 FAO Production Yearbook,* vol. 46 (Rome: Food and Agriculture Organization, 1993). Thirty-eight percent of the world's grain supply is now fed to livestock (*World Resources 1996–97,* [New York: Oxford University Press, 1996] table 10.3). Most of the land and other resources now used to produce feed grain could be used to grow grain and other foods for human consumption. Feed grains are grown because better-off consumers prefer livestock products, making feed grains more profitable than food grains. While daily calorie requirements vary greatly and are notoriously difficult to estimate, it seems reasonable to use the FAO figure of 2,450 per day for the average person (see P.R. Payne, "Measuring Malnutrition," *IDS Bulletin* 21, no. 3, July 1990). Calories adequate to cover energy needs are generally sufficient to meet protein needs, except for people (especially young children) subsisting on low-protein roots, tubers, and plantains.

11 Food and Agriculture Organization. *FAO Production Yearbook 1995,* vol. 49 (Rome: Food and Agriculture Organization, 1996).

12 Calculated from *FAO Time Series for State of Food and Agriculture,* (Rome: Food and Agriculture Organization, 1994).

13 Mitchell, Donald O. and Merlinda D. Ingco. *The World Food Outlook* (Washington, DC: World Bank, 1993).

14 Ingco, Merlinda D. and Donald O. Mitchell and Alex F. McCalla. *Global Food Supply Prospects,* World Bank Technical Paper no. 353, 1996.

15 Food and Agriculture Organization. *FAO Production Yearbook* 1966, 1974, 1984, 1995 (Rome: Food and Agriculture Organization), table 9. See Philip Raikes, *Modernising Hunger: Famine, Food Surplus & Farm Policy in the EEC & Africa* (London, UK:

Catholic Institute for International Relations, 1988). Raikes argues that the FAO statistics seriously underestimate African production because, among other things, often only products sold through official marketing channels are counted. He suggests that there has been a decline in the proportion of production sold through such channels, as farmers sell elsewhere in response to artificially low prices.

16 Smith, Lisa C. Science, Engineering and Diplomacy Fellow, American Association for the Advancement of Science. *The FAO Measure of Chronic Undernourishment: What Is It Really Measuring?* (Washington, DC: U.S. Agency for International Development, Office of Population, Health and Human Development, June 1997), pp. 6–7.

17 Poleman, Thomas T. "Quantifying the Nutrition Situation in Developing Countries," *Food Research Institute Studies* 18, no. 1: 9. This article is a good discussion of the multifaceted problems of most agricultural and nutritional statistics. See also Donald McGranahan et al., *Measurement and Socioeconomic Development* (Geneva: United Nations Research Institute for Social Development, 1985).

18 *Handbook of International Trade and Development Statistics* (New York and Geneva: United Nations Conference on Trade and Development, 1995), table 3.2. Data for food items include beverages, tobacco, and edible oil seeds.

19 Calculated from Food and Agriculture Organization, *1994 FAO Trade Yearbook*, vol. 48 (Rome: Food and Agriculture Organization, 1994), tables 7, 8. Classification of lowest-income countries according to the *World Development Report, 1992* (Washington, DC: World Bank, 1993).

20. According to the FAO, twenty-one percent of Indians, or about 185 million people, are undernourished (*Mapping Undernutrition: An Ongoing Process*, poster, [Rome: Food and Agriculture Organization, 1996]). However, in *The FAO Measure* cited above, Lisa Smith argues that this number is a gross underestimation, as it is calculated not from real surveys but from projections based on food production and imports, and thus does not reliably take into account food distribution and access. A better index of the prevalence of hunger, she suggests, is the percent of children who are malnourished, which is calculated from surveys and represents the outcome of food availability *and* access. The FAO estimate for percent of children under five who are malnourished in India is 61 percent, the second highest in the world (*The Sixth World Food Survey*, [Rome: Food and Agriculture Organization, 1996], appendix 2, table 8), suggesting that the total number of hungry people is probably far higher than the twenty-one percent or 185 million figures reported.

21 FAOSTAT Database (Rome: Food and Agriculture Organization, 1990–97).

22 Boyce, James K. "Agricultural Growth in Bangladesh, 1949–50 to 1980–81: A Review of the Evidence," *Economic and Political Weekly* 20, no. 13 (March 30, 1985): pp. A31–A43.

23 FAOSTAT Database.

24 World Bank. *Bangladesh: Economic and Social Development Prospects*, report no. 5409 (Washington, DC: World Bank, 1985), pg. 18.

25 Food and Agriculture Organization. *FAO Production Yearbook 1995*, vol. 49 (Rome: Food and Agriculture Organization, 1996), table 17.

26 Jones, Steve. "Agrarian Structure and Agricultural Innovation in Bangladesh: Panimara Village, Dhaka District," in *Understanding Green Revolutions,* ed. Tim Bayliss-Smith and Sudhir Wanmali (New York: Cambridge University Press, 1984), pg. 194.

27 United Nations Development Program. *Human Development Report 1994* (New York: United Nations Development Program, 1994), table 2.

28 Food and Agriculture Organization. *FAO Trade Yearbook 1995*, vol. 49 (Rome: Food and Agriculture Organization, 1996) 13. See also *Human Development Report 1994*, pg. 134.

29 *The FAO Measure of Chronic Undernourishment: What Is It Really Measuring?* appendix A, table A1. See also *Food Security in Africa,* (Washington, DC: U.S. General Accounting Office, 1996), GAO testimony, statement of Harold J. Johnson, associate director, International Relations and Trade Issues, National Security and International Affairs Division.

30 Pinstrup-Andersen, Per. "World Trends and Future Food Security," *Food Policy Report* (Washington, DC: International Food Policy Research Institute, 1994), pp. 9–10.

31 Calculated from Food and Agriculture Organization, *1994 FAO Trade Yearbook*, table 8. When all agricultural products are taken into account, the figure rises to twenty countries (Ibid., table 7). Note that the increase of imports in a specific country may reflect an increase in food aid received by it. Calculations of imports include food aid, according to James Hill, senior economist at the North American UN FAO Liaison Office, Washington, DC, interviewed by Joseph Collins in May 1986.

32 Calculated from *1984 FAO Trade Yearbook.* For Chad, Niger, Mauritania, Mali, Burkina Faso, and Senegal, years of net exports were 1980, 1982, and 1983. Years of net imports were 1981 and 1984 due to exceptionally high imports into Senegal and Niger those two years.

33 Reports from two of the Sahelian countries, Niger and Burkina Faso (formerly Upper Volta), estimate that traders smuggle out as much as half the grain produced to sell elsewhere to customers able to pay more. (Interview with chief economist, U.S. Agency for International Development Mission, Ouagadougou, Upper Volta on January 17, 1977).

34 During the comparatively less severe drought of 1982–85, the value of food exported by these countries was three-fourths that of food imported, Chad, Mali, and Niger being net exporters (calculated from *FAO Trade Yearbook 1985*, table 7).

35 *FAO Production Yearbook* 1966, 1974, 1984, 1990, tables 9 and 10.

36 Calculated from FAOSTAT Database, data on production, consumption, availability, and imports of cereal products in 1995, September 17, 1997.

37 Berry, Sara S. "The Food Crisis and Agrarian Change in Africa: A Review Essay," *African Studies Review 27*, no. 2 (June 1984): pp. 59–97. See also Jane Guyer, "Women's Role in Development," in *Strategies for African Development,* ed. Robert J. Berg and Jennifer Seymour Witaker (Berkeley: University of California Press, 1986), pp. 393–396. Also see *Modernising Hunger.*

38 *The FAO Measure of Chronic Undernourishment: What Is It Really Measuring?*, pp. 15–16. The much reported but, she argues, inaccurate FAO estimate of chronic undernourishment for sub-Saharan Africa is forty-three percent, while it is only twenty-two percent for South Asia. Yet the percent of children under five who are malnourished, a better index of hunger is fifty-three percent in South Asia, almost double that of sub-Saharan Africa (thirty percent).

39 Plucknett, Donald L. "Prospects of Meeting Future Food Needs through New Technology," in *Population and Food in the Early Twenty-First Century: Meeting Future Food Demand of an Increasing Population,* ed. Nurul Islam (Washington, DC: International Food Policy Research Institute, 1995).

40 Heilig, Gerhard K. "How Many People Can Be Fed on Earth?" in *The Future Population of the World: What Can We Assume Today?* ed. Wolfgang Lutz (Laxenburg, Austria: International Institute for Applied Systems Analysis, 1996).

41 See Stephen K. Commins et al., eds., *Africa's Agrarian Crisis: The Roots of Famine* (Boulder, CO: Lynne Rienner, 1985); Nigel Twose, *Fighting the Famine* (San Francisco: Food First Books, 1985); Michael Watts, "Entitlements or Empowerment? Famine and Starvation in Africa," *Review of African Political Economy,* no. 51 (1991): pp. 9–26.

42 Harrison, Paul. *The Greening of Africa: Breaking Through in the Battle for Land and Food* (Nairobi: Academy Science Publishers, 1996). See also Eric R. Wolf, *Europe and the People Without History* (Berkeley: University of California Press, 1982). See also Barbara Dinham and Colin Hines, *Agribusiness in Africa: A Study of the Impact of Big Business on Africa's Food and Agricultural Production* (Trenton, NJ: Africa World Press, 1984).

43 Richards, Paul. "Ecological Change and the Politics of African Land Use," *African Studies Review* 26, no. 2 (June 1983). See also Paul Richards, *Indigenous Agricultural Revolution* (Boulder, CO: Westview Press, 1985).

44 See Bill Rau, *Feast to Famine* (Washington, DC: Africa Faith and Justice Network, 1985), esp. chapter 6. See also Bonnie K. Campbell, "Inside the Miracle: Cotton in the Ivory Coast," in *The Politics of Agriculture in Tropical Africa,* ed. Jonathan Barker (Beverly Hills: Sage, 1984), pp. 154–168.

45 World Resources Institute. *World Resources, 1994–1995* (New York: Oxford University Press, 1994), pp. 48–49. For examples of the nuances of gender, policy, investment, and food production, see Judith Carney, "Contracting a Food Staple in the Gambia," in *Living under Contract: Contract Farming and Agrarian Transformation in Sub-Saharan Africa,* ed. Peter D. Little and Michael J. Watts (Madison, WI: University of Wisconsin Press, 1994); Beverly Grier, "Pawns, Porters and Petty Traders: Women in the Transition to Export Agriculture in Ghana," African Studies Center, Boston University, *Working Papers in African Studies,* no. 144, 1989.

46 Independent Commission on International Humanitarian Issues. *Famine: A Manmade Disaster?* (New York: Vintage Books, 1985), pp. 85–89.

47 See *Africa, Make or Break: Action for Recovery* (London, UK: Oxfam UK, 1993); and John Mihevc, *The Market Tells Them So: The World Bank and Economic Fundamentalism in Africa* (Penang, Malaysia, and Accra, Ghana: Third World Network, 1995).

48 *The Market Tells Them So;* and Manfred Bienefeld, *Structural Adjustment and Rural Labor Markets in Tanzania* (Geneva: International Labor Organization, 1991).

49 *The Market Tells Them So,* op. cit.

50 Calculated from FAOSTAT Database.

51 A typical ad appearing in an African newspaper in the early 1970s, cited in Jean-Yves Carfantan and Charles Condamines, *Vaincre la Faim, C'est Possible* (Paris: L'Harmattan, 1976), pg. 63.

52 Andrae, Gunilla and Björn Beckman. *The Wheat Trap: Bread and Underdevelopment in Nigeria* (London, UK: Zed Books, 1985).

53 Cohen, Joel E. *How Many People Can the Earth Support?* (New York: W. W. Norton, 1995), esp. appendix 3.

54 Wisner, Ben. "The Limitations of 'Carrying Capacity,'" *Political Environments* (Winter/Spring 1996): pg. 1, 3–6.

55 *How Many People Can the Earth Support?* See appendix 3.

56 UN Department for Economic and Social Information and Policy Analysis, Population Division. *World Population Prospects: The 1996 Revision. Annex 1: Demographic Indicators,* table A2.

57 According to the *1994 Human Development Report,* by the United Nations Development Program (New York: Oxford University Press, 1994), the average per capita calorie supply for China in the early 1990s was twelve percent above minimum requirements. See also Elizabeth Croll, *The Family Rice Bowl: Food and the Domestic Economy in China* (Geneva: UN Research Institute for Social Development, 1982). Though Lester Brown of the WorldWatch Institute recently created a media frenzy when he argued that China's future grain consumption would grossly deplete the food available to the rest of the world (Lester R. Brown, *Who Will Feed China? A Wake-Up Call for a Small Planet* [New York: W. W. Norton, 1995]), calmer heads responded by recalculating his data, as well as furnishing new data that clearly demonstrates China's ability to produce sufficient grain for its future needs. See Vaclav Smil, *Who Will Feed China? Concerns and Prospects for the Next Generation* (Fourth Annual Hopper Lecture, University of Guelph, 1996); Jikun Huang et al., *China's Food Economy to the Twenty-First Century: Supply, Demand, and Trade* (Washington, DC: International Food Policy Research Institute, Food, Agriculture and the Environment Discussion Paper no. 19, 1997); Feng Lu, *Grain versus Food: A Hidden Issue in China's Food Policy Debate* (China Center for Economic Research, Peking University, Working Paper no. E1996003); Xiaoguang Kang, *Excerpts of "How Chinese Feed Themselves"—Reply to Lester R. Brown's "Who Will Feed China?"* (Chinese Academy of Sciences, Research Center for Eco-Environmental Sciences, Department for the Study of National Conditions, 1997).

58 In 1994, the United States had 1.6 ha/per capita of cropland and China had 0.4 ha/per capita (calculated from FAOSTAT Database).

59 *Hunger, 1997: What Governments Can Do,* Seventh Annual Report on the State of World Hunger (Silver Spring, MD: Bread for the World Institute, 1996), pp. 114–15, tables 5, 6.

60 World Food Program, INTERFAIS Internet information system, September 19, 1997.

61 Calculations based on the 1994 figure for hungry children under twelve years of age. *Hunger, 1997*, pg. 114, table 5.

62 See Yrju Halla and Richard Levins, *Humanity and Nature: Ecology, Science and Society* (London, UK: Pluto Press, 1992); Christopher J. Baker, "Frogs and Farmers: The Green Revolution in India, and Its Murky Past" in *Understanding Green Revolutions,* ed. Tim Bayliss-Smith and Sudhir Wanmali (New York: Cambridge University Press, 1984), pg. 40. The first major Green Revolution in the Indian Punjab took place over a hundred years ago when Punjabi farmers adopted three new varieties of sugar cane in a single generation, with striking results.

63 Jennings, Bruce H. *Foundations of International Agricultural Research: Science and Politics in Mexican Agriculture* (Boulder, CO: Westview Press, 1988); John H. Perkins, "The Rockefeller Foundation and the Green Revolution, 1941–1956," *Agriculture and Human Values* 7, no. 3/4 (1990): pp. 6–18.

64 For an analysis of the role of internationally funded agricultural research in this process, see Shripad D. Deo and Louis E. Swanson, "Structure of Agricultural Research in the Third World" in *Agroecology,* ed. C. Ronald Carroll, John H. Vandermeer, and Peter M. Rosset (New York: McGraw-Hill, 1990).

65 Pingali, P.L. and M. Hossain and R.V. Gerpacio. *Asian Rice Bowls: The Returning Crisis* (Wallingford, UK: CAB International, 1997), pg. 4.

66 Oram, Peter A. and Behjat Hojjati. "The Growth Potential of Existing Agricultural Technology," in *Population and Food in the Early Twenty-First Century: Meeting Future Food Demand of an Increasing Population,* ed. Nurul Islam (Washington, DC: International Food Policy Research Institute, 1995), pp. 167–189, table 7.8.

67 Dowsell, Christopher R. and R. L. Paliwal and Ronald P. Cantrell. *Maize in the Third World* (Boulder, CO: Westview Press, 1996), pg. 137.

68 Lipton, Michael and Richard Longhurst. *New Seeds and Poor People* (Baltimore, MD: Johns Hopkins University Press, 1989), pg. 3.

69 Food and Agriculture Organization. "Lessons from the Green Revolution: Towards a New Green Revolution" in *World Food Summit: Technical Background Documents,* vol. 2 (Rome: Food and Agriculture Organization, 1996).

70 Agriculture and Rural Development Department, World Bank, "Agricultural Biotechnology: The Next 'Green Revolution'?" *World Bank Technical Papers* no. 133, 1991.

71 See part four of Frances Moore Lappé and Joseph Collins, *Food First: Beyond the Myth of Scarcity* (New York: Ballantine Books, 1977); Susan George, *How the Other Half Dies* (London: Penguin, 1976); George Kent, *The Political Economy of Hunger: The Silent Holocaust* (New York: Praeger, 1984); Keith Griffin, *The Political Economy of Agrarian Change: An Essay on the Green Revolution,* 2nd ed. (London, UK: Macmillan, 1979); Keith Griffin and Jeffrey James, *Transition to Egalitarian Development: Economic Policies for Structural Change in the Third World* (New York: St. Martin's Press, 1981); Andrew Pearse, *Seeds of Plenty, Seeds of Want: Social and Economic Implications of the Green Revolution* (New York: Oxford University Press, 1980); Vandana Shiva, *The Violence of the Green Revolution: Third World Agriculture, Ecology and Politics* (Penang, Malaysia: Third World Network, 1991).

72 Husain was chairman of the Consultative Group on International Agricultural Research, quoted in *Bank's World* vol. 4 (1985) no. 12:1.

73 World Bank. *Poverty and Hunger: Issues and Options for Food Security in Developing Countries* (Washington, DC: World Bank, 1986), pg. 49.

74 *Food First*, pg. 121.

75 "The State of World Hunger," pp. 1–17, table 1.6.

76 Lewis, Paul. "The Green Revolution Bears Fruit," *The New York Times*, June 2, 1985.

77 *Poverty and Hunger*, pg. 1.

78 "The State of World Hunger," table 1.6.

79 Calculated from table 3.2 in Nikos Alexandratos, "The Outlook for World Food and Agriculture to Year 2010" in Islam, *Population and Food in the Early Twenty-First Century*, pp. 25–48.

80 Ibid.

81 See Gregg Easterbrook, "Forgotten Benefactor of Humanity," *The Atlantic Monthly* 279, no. 1 (January 1997): pp. 75–82.

82 "The State of World Hunger," calculated from table 1.6. China reduced the number of hungry people by 53 percent, while the number rose a combined 27 percent in sub-Saharan Africa, South Asia, and South America.

83 Ibid.

84 Ibid. In all fairness, we should point out that the number of hungry people also fell in East Asia and in the Near East, though the numbers are nowhere as dramatic as in China. It is also worth remembering that Chinese policy toward hunger was undoubtedly strongly influenced by the terrible famine that struck that country in the early 1960s.

85 Pluenneke, John E. and Sharon Moshavi. "A Revolution Comes Home to Roost… Leaving Hunger in the Midst of Plenty," *Business Week*, November 6, 1994.

86 Balanced historical treatments of the Green Revolution debate can be found in Frederick H. Buttel and Laura T. Raynolds, "Population Growth, Agrarian Structure, Food Production, and Food Distribution in the Third World" in *Food and Natural Resources*, ed. David Pimentel and Carl W. Hall (New York: Academic Press, 1989), pp. 341–351; and Anthony Bebbington and Graham Thiele, *Non-Governmental Organizations and the State in Latin America: Rethinking Roles in Sustainable Agricultural Development* (London, UK: Routledge, 1993), chapter 4.

87 See Clifton R. Wharton, Jr., "The Green Revolution: Cornucopia or Pandora's Box," *Foreign Affairs* 47 (1969): pp. 464–476; and K. Griffin, *The Political Economy of Agrarian Change: An Essay on the Green Revolution* (London, UK: Macmillan, 1974).

88 Green Revolution supporter Walter Falcon cautioned in 1970 that Green Revolution success in some regions was counterbalanced by poor results in marginal areas, and that the new varieties and associated technologies were accompanied by increasing pest, crop disease, and weed problems (he felt that better pesticide programs were the solution). While he argued that these would prove to be 'short-run issues,' they remain with us three decades later. See Walter P. Falcon, "The Green Revolution:

Generations of Problems," *American Journal of Agricultural Economics* 52 (1970): pp. 698–710; and Magnus Jirström, *In the Wake of the Green Revolution: Environmental and Socio-Economic Consequences of Intensive Rice Agriculture—The Problems of Weeds in Muda, Malaysia* (Lund, Sweden: Lund University Press, 1996), pg. 35.

89 See M. Perelman, *Farming for Profit in a Hungry World* (Montclair, NJ: Allanheld, Osmun, 1977); W. Ophuls, *Ecology and the Politics of Scarcity* (San Francisco: Freeman, 1977); P.R. Mooney, *Seeds of the Earth* (Ottawa: Inter Pares, 1979).

90 *New Seeds and Poor People,* pg. 118.

91 *Non-Governmental Organizations and the State,* pg. 64; J. Rigg, "The New Rice Technology and Agrarian Change: Guilt by Association?" *Progress in Human Geography* 13, no. 2: pp. 374–399.

92 "Population Growth, Agrarian Structure," pp. 345–347, summarize what they call the "formidable counterattack" by defenders of the Green Revolution. See also "Lessons from the Green Revolution."

93 "Lessons from the Green Revolution"; *Non-Governmental Organizations and the State,* pp. 64–67.

94 Ibid.

95 "Population Growth, Agrarian Structure," pp. 345–347; Peter Hazell and James L. Garrett, "Reducing Poverty and Protecting the Environment: The Overlooked Potential of Less-Favored Lands," International Food Policy Research Institute, *2020 Vision Brief,* no. 39, 1996; "Lessons from the Green Revolution."

96 Griffin, Keith. *Alternative Strategies for Economic Development* (New York: St. Martin's Press, 1989), pg. 147.

97 See *Farming for Profit* for a historical treatment of how the agrochemical industry has worked with development agencies to promote their products through the Green Revolution.

98 International Food Policy Research Institute. "Donors to the 2020 Vision Initiative" in *A 2020 Vision for Food, Agriculture and the Environment: The Vision, Challenge and Recommended Action* (Washington, DC: International Food Policy Research Institute, 1995), pg. 51.

99 Letter to Ambassador Robert O. Blake, chairman, Committee on Agricultural Sustainability for Developing Countries, from Jack Whelan, External Relations, International Fertilizer Industry Association, dated May 7, 1996, a copy of which was leaked to the authors.

100 See Robert Chambers, "Farmer-First: A Practical Paradigm for the Third Agriculture" in *Agroecology and Small Farm Development,* ed. Miguel A. Altieri and Susanna B. Hecht (Ann Arbor: CRC Press, 1990), pp. 237–244; and Eric Holt-Gimenez, "The Campesino a Campesino Movement: Farmer-led, Sustainable Agriculture in Central America and Mexico," *Food First Development Report,* no. 10, (Oakland: Institute for Food and Development Policy, 1996).

101 "Population Growth, Agrarian Structure," pg. 345, note 7; see also F. Bray, *The Rice Economies* (Oxford, UK: Blackwell, 1986).

102 See George A. Collier with Elizabeth Lowery Quaratiello, *Basta! Land and the Zapatista Rebellion in Chiapas* (Oakland: Food First Books, 1994); and Peter Rosset with Shea Cunningham, "Chiapas: Social and Agricultural Roots of Conflict," *Global Pesticide Campaigner* 4, no. 2 (1994): pp. 1, 8–9, 16. For a foreshadowing of these problems, written before the uprising, see George A. Collier, "Seeking Food and Seeking Money: Changing Productive Relations in a Highland Mexican Community," United Nations Research Institute for Social Development *Discussion Paper,* no. 10, 1990.

103 Freebairn, Donald K. "Did the Green Revolution Concentrate Incomes? A Quantitative Study of Research Reports," *World Development* 23, no. 2 (1995): pp. 265–279. Despite the 'maturation' of the Green Revolution in recent decades, later studies were just as likely to find heightened inequality as were studies carried out in the early years. Three 'landmark' books that attempt to refute the argument that inequality has increased have been published in the 1990s by Green Revolution supporters. One of the most widely cited is a multi-investigator, village-level study provocatively named *The Green Revolution Reconsidered: The Impacts of High-Yielding Rice Varieties in South India,* edited by P.B.R. Hazell and C. Ramasamy (Baltimore, MD and Washington, DC: Johns Hopkins University Press and the International Food Policy Research Institute, 1991). The editors conclude that the "Green Revolution had a favorable impact" in their study villages, including "increasing aggregate output," and "across-the-board gains in income, employment, and the quality of diet" (pg. 251). Nevertheless, the team's anthropologist, John Harriss, observed a "remarkable stability" of social class relations over the course of the study, despite a good deal of social mobility, "rather more of it downward than upward" (pg. 73). Rita Sharma and Thomas T. Poleman, in *The New Economics of India's Green Revolution: Income and Employment Diffusion in Uttar Pradesh* (Ithaca, NY: Cornell University Press, 1993), take a look at the off-farm economic impacts of the Green Revolution. Massive investment in agriculture during the Green Revolution has generated a diversity of off-farm employment opportunities related to the storage, transport, processing, and marketing of farm production. They conclude that "the Green Revolution can be a double-barreled blessing to India," providing both increased production and off-farm employment (pg. 254). The weakness of their argument is that *any* agricultural growth, regardless of whether it is based on the Green Revolution, would have positive effects on the rest of the economy. The authors did not consider whether alternative development strategies could have generated more and/or better off-farm effects.

We argue for a different kind of rural development, based on social equity, which if the experiences of South Korea, Japan, and Taiwan are worth anything, can generate far more positive impacts on the off-farm economy than did the Green Revolution in India. Cristina C. David and Keijiro Otsuka, the editors of *Modern Rice Technology and Income Distribution in Asia* (Boulder, CO and Manila: Lynne Rienner and International Rice Research Institute, 1994), take on the issue of regional equity. They ask: "What effect has technological change in the favorable rice-growing areas had on the welfare of people in the unfavorable areas bypassed by the new technology?" (pg. 7). Their rather lukewarm conclusion is that Green Revolution adoption "is limited to irrigated and favorable rainfed environments, and thus the yield gap between favorable and unfavorable rice-growing areas has widened." Lest this conclusion generate concern, they offer that "when the indirect effects through labor, land, and product market adjustments are accounted for, differential adoption of modern varieties across production environments *does not significantly worsen income distribution*" (pg. 427, emphasis added). The problem with this argument is that we need to significantly *improve* income distribution if we are to achieve the kind of broad-based development needed to attack the structural roots of hunger and poverty.

104 Summarized in *Non-Governmental Organizations and the State in Latin America,* chapter 4.

105 Prosterman, Roy L. and Mary N. Temple and Timothy M. Hanstad, eds. *Agrarian Reform and Grassroots Development: Ten Case Studies* (Boulder, CO: Lynne Rienner, 1990), pg. 1.

106 See Michael Lipton, "Inter-Farm, Inter-Regional and Farm Non-Farm Income Distribution: The Impact of the New Cereal Varieties," *World Development* 6, no. 3 (1978): 319–337; and R. Barker and V.G. Cordova, "Labor Utilization in Rice Production" in *Economic Consequences of the New Rice Technology,* ed. R. Baker and Y. Hayami (Los Banos, Philippines: International Rice Research Institute, 1978). In a more recent study by P.B.R. Hazell et al., "Economic Changes Among Village Households," pp. 29–56 in *The Green Revolution Reconsidered,* the authors found higher wages in new off-farm employment, though their "results suggest that the green revolution did little to increase total crop employment" (pg. 37). In the editors' conclusions to the entire volume (chapter 11), they admit that they may not have detected an increase in the absolute poverty of the landless because "of migration from the villages to towns" (pg. 252). In other words, the landless left the area if they could not find work, fueling rural-urban migration.

107 *New Seeds and Poor People,* pp. 401, 415. See also Michael Lipton, "Successes in Anti-Poverty," International Labor Office, Development and Technical Cooperation Department, *Issues in Development Discussion Paper,* no. 8, (Geneva: International Labor Office, 1996), pp. 66–71.

108 See Y. Hayami and V.W. Ruttan, *Agricultural Development: An International Perspective* (Baltimore: Johns Hopkins University Press, 1985), pg. 341.

109 Tractor numbers in the third world have, however, continued to grow at a rapid clip, increasing by 68 percent between 1980 and 1994 (*FAO Production Yearbook 1990*, table 118 and 1995, table 106).

110 "Successes in Anti-Poverty," pg. 63.

111 *Food First,* pp. 174–177; for a case in the United States see John H. Vandermeer, "Mechanized Agriculture and Social Welfare: The Tomato Harvester in Ohio," *Agriculture and Human Values* 3, no. 3 (Summer 1986): pp. 21–25; and Peter M. Rosset and John H. Vandermeer, "The Confrontation Between Processors and Farm Workers in the Midwest Tomato Industry and the Role of the Agricultural Research and Extension Establishment," *Agriculture and Human Values* 3, no. 3 (Summer 1986): pp. 26–32.

112 Khan, Azizur Rahman and Eddy Lee. *Poverty in Rural Asia,* International Labor Organization and the Asian Employment Program (Bangkok: International Labor

Office, 1984), pp. 126–130. See also Richard W. Franke and Barbara H. Chasin, *Kerala: Radical Reform as Development in an Indian State*, 2nd edition (Oakland: Food First Books, 1994).

113 Ravallion, Martin. "Poverty and Growth: Lessons from 40 Years of Data on India's Poor," Development Economics Vice Presidency of the World Bank, *DEC Notes Research Findings No. 20*, 1996, figure 3 (Washington, DC: World Bank, 1996).

114 Singh, D.P. "The Impact of the Green Revolution," *Agricultural Situation in India 13*, no. 8 (1980): pg. 323.

115 *New Seeds and Poor People*, pp. 128–133.

116 Ibid., pp. 121–125.

117 Zeller, Manfred and Gertrud Schrieder, Joachim von Braun, and Franz Heidhues. "Rural Finance and Food Security for the Poor: Implications for Research and Policy," *Food Policy Review* 4 (Washington, DC: International Food Policy Research Institute, 1997), chapter 3. For an overview of how the availability of subsidized credit grew in the 1970s and shrank in the 1980s, see Willem C. Beets, *Raising and Sustaining Productivity of SmallHolder Farming Systems in the Tropics* (Alkmaar, Holland: AbBé Publishing, 1990), pp. 594–598.

118 Hayami, Yujiro and Masao Kikuchi. "Directions of Agrarian Change: A View from Villages in the Philippines" in *Agricultural Change and Rural Poverty*, ed. John W. Mellor and Gunvant M. Desai (Baltimore, MD: Johns Hopkins University Press, 1985), 132 ff.

119 *New Seeds and Poor People*, pp. 42–51.

120 Ibid., chapter 2.

121 In 1970 India used an average of 12.7 kg/ha of fertilizer; by 1995 the figure stood at 76.6. The ratio of agricultural production to fertilizer use was calculated by dividing the FAO Agricultural Production Index by the number of millions of metric tons of fertilizer used in India, giving a ratio of 24.4 in 1970 and 8.15 in 1995. Data from *Global Data Manager 3.0*, CD-ROM (Philadelphia: World Game Institute, 1996).

122 *Asian Rice Bowls*, table 4.5.

123 McGuinness, Hugh. *Living Soils: Sustainable Alternatives to Chemical Fertilizers for Developing Countries* (Yonkers, NY: Consumer Policy Institute, 1993), discusses the hidden costs of fertilizer use.

124 See John Harriss, "What Happened to the Green Revolution in South India? Economic Trends, Household Mobility and the Politics of an 'Awkward Class,'" IFPRI/TNAU Workshop on Growth Linkages, International Food Policy Research Institute, Washington, DC 1986, pp. 29–30. A very rough comparison of net farmer revenues from Philippine rice production in 1971 and 1986 can be made by comparing tables 3.1 and 3.7 in *Asian Rice Bowls*, suggesting that net farmer profits per hectare may have fallen from U.S. $481 to $296, a thirty-eight percent drop.

125 Pingali, P. L. "Diversifying Asian Rice Farming Systems: A Deterministic Paradigm" in *Trends in Agricultural Diversification: Regional Perspectives*, paper no. 180, ed. S. Barghouti, L. Garbux, and D. Umali (Washington, DC: World Bank, 1992).

126 "Agroecology versus Input Substitution," figure 3a–d.

127 Ibid., calculated from figure 2.

128 Krebs, A.V. *The Corporate Reapers: The Book of Agribusiness* (Washington, DC: Essential Books, 1992), pg. 29.

129 For extensive discussion of this point, see Marty Strange, *Family Farming: A New Economic Vision* (San Francisco: Food First Books, 1988).

130 Not only can the biggest farms take advantage of bulk discounts in purchasing and premium prices for large volume sales, but they also benefit disproportionately from contracts with processors and from tax policies favoring large capital investments.

131 "Agroecology versus Input Substitution," figure 1.

132 Lobao, Linda M. *Locality and Inequality: Farm and Industry Structure and Socioeconomic Conditions* (Albany, NY: State University of New York Press, 1990), table 2.1.

133 *Asian Rice Bowls,* chapters 4 and 5.

134 See Clara Ines Nicholls and Miguel A. Altieri, "Conventional Agricultural Development Models and the Persistence of the Pesticide Treadmill in Latin America," *International Journal of Sustainable Development and World Ecology* 4 (1997): pp. 93–111; "Agroecology versus Input Substitution," pp. 283–295; Douglas L. Murray, *Cultivating Crisis: The Human Cost of Pesticides in Latin America* (Austin: University of Texas Press, 1994); McGuinness, *Living Soils,* chapters 1, 2, and 3.

135 Cassman, K.G. and P.L. Pingali. "Extrapolating Trends from Long-Term Experiments to Farmers' Fields: The Case of Irrigated Rice Systems in Asia" in *Agricultural Sustainability: Economic, Environmental and Statistical Considerations,* ed. Vic Barnett, Roger Payne, and Roy Steiner (London, UK: John Wiley & Sons, 1995), 1: pp. 67–68.

136 Ibid., figures 5.6 and 5.7; see also *Asian Rice Bowls,* figures 4.1 and 4.2.

137 Cassman, K.G. and R.R. Harwood, "The Nature of Agricultural Systems: Food Security and Environmental Balance," *Food Policy* 20, no. 5 (1995): pp. 439–454, pp. 447–448; K.K.M. Nambiar, "Long-Term Experiments on Major Cropping Systems in India" in *Agricultural Sustainability,* ed. Vic Barnett et al., 133–170; D. Byerlee, "Technical Change, Productivity and Sustainability in Irrigated Cropping Systems of South Asia: Emerging Issues in the Post-Grain Era," *Journal of International Development* 4, no. 5: pp. 477–496.

138 "The Nature of Agricultural Systems;" "Extrapolating Trends;" and *Asian Rice Bowls.*

139 "A Revolution Comes Home to Roost."

140 *Asian Rice Bowls,* pg. 26.

141 Chapter 8 in Michael Hansen, *Escape from the Pesticide Treadmill: Alternatives to Pesticides in Developing Countries* (Mount Vernon, NY: Institute for Consumer Policy Research, 1987), pg. 134.

142 Ibid., pp. 137–138.

143 Ibid., pg. 138.

144 See *Asian Rice Bowls,* pp. 110–118, for a discussion of health problems associated with pesticide use in rice.

145 *Escape from the Pesticide Treadmill*, pg. 139.

146 Ibid., pp. 143–148; PANUPS, "Farmer First: Field Schools Key to IPM Success," Pesticide Action Network North America Updates Service, August 16, 1994, San Francisco, CA: www.panna.org/panna; Peter E. Kenmore, *Indonesia's Integrated Pest Management: A Model for Asia* (Manila: FAO Intercountry IPC Rice Programme, 1991).

147 "Farmer First," op. cit.

148 *Asian Rice Bowls, 267.*

149 See Monica Moore, *Redefining Integrated Pest Management: Farmer Empowerment and Pesticide Use Reduction in the Context of Sustainable Agriculture* (San Francisco: Pesticide Action Network, 1995).

150 "The Campesino a Campesino Movement," op. cit.

151 Dr. Prabhu Pingali, comments made at discussion workshops held as part of "The Keystone Center Workshop on Critical Variables and Long-Term Projections for Sustainable Global Food Security," Warrentown, VA, March 10–13, 1997.

152 Naylor, Rosamond. "Herbicide Use in Asian Rice Production," *World Development* 22, no. 1 (1994): pp. 55–70.

153 *In the Wake of the Green Revolution*, pp. 226–229. This 1996 study of the phenomenon concludes that "intensive Green Revolution agriculture is associated with a set of sustainability problems having effects which do not seem to be scale-neutral." The author, a Green Revolution supporter himself, warns that "there is little room for complacency about the distributional impacts of current technologies… On the contrary, the present spread and adoption of labor-displacing technologies such as direct-seeding and herbicides may, unless the circumstances are right, pose a major threat to the less well-to-do."

154 See "Farmer-First" and "The Campesino a Campesino Movement."

155 Umehara, Hiromitsu. "Green Revolution for Whom?" in *Second View from the Paddy*, ed. Antonio J. Ledsma, S. J. et al. (Manila: Institute of Philippine Culture, Ateneo de Manila University), pg. 37.

156 Strange, Marty. "Family Farming: Faded Memory or Future Hope?" *Food First Action Alert*, (San Francisco: Institute for Food and Development Policy, 1989).

157 Doyle, Jack. "The Agricultural Fix," *Multinational Monitor* 7, no. 4 (1986): pg. 3.

158 See *Farming for Profit*, for some history on this point.

159 "Poverty and Growth," pg. 2 and figure 1.

160 Buttel, Frederick H. and Randolph Barker. "Emerging Agricultural Technologies, Public Policy, and Implications for Third World Agriculture," *American Journal of Agricultural Economics* 67, no. 5 (1985): pp. 1170–1175; and Jack R. Kloppenburg, *First the Seed: The Political Economy of Plant Biotechnology, 1492–2000* (New York: Cambridge University Press, 1988).

161 Kenney, Martin and Frederick Buttel, "Biotechnology: Prospects and Dilemmas for Third World Development," *Development and Change* 16 (1985): 61–91; Henk Hobbelink, *Biotechnology and the Future of World Agriculture: The Fourth Resource* (London, UK: Zed Books, 1991); Vandana Shiva, *Monocultures of the Mind:*

Perspectives on Biodiversity and Biotechnology (London, UK: Zed Books and Third World Network, 1993).

162 Dawkins, Kristin. *Gene Wars: The Politics of Biotechnology* (New York: Seven Stories Press, 1997), pg. 31.

163 "Agricultural Biotechnology: The Next 'Green Revolution'?"

164 "Biotechnology," pg. 68. It is possible that some patented genes will be provided to third world countries on a donated or at least 'favorable' license basis, according to Klaus M. Leisinger of the Ciba-Geigy Foundation for Cooperation in Development, a charitable foundation established by one of the world's largest pesticide and biotechnology companies (Klaus M. Leisinger, "Sociopolitical Effects of New Biotechnologies in Developing Countries," *2020 Vision Brief* No. 35, [Washington, DC: International Food Policy Research Institute, 1996]). Dr. Michael Hansen of the Consumer Policy Institute in New York, an expert on the biotechnology industry, told the authors of this book that if the industry decides that poor farmers in certain countries do not have sufficient resources to constitute a profitable market, some companies will provide genes free of charge purely for their advertising and public relations value.

165 Rissler, Jane and Margaret Mellon. *Perils Amid the Promise: Ecological Risks of Transgenic Crops in a Global Market* (Cambridge, MA: Union of Concerned Scientists, 1993).

166 Hileman, Bette. "Views Differ Sharply Over Benefits, Risks of Agricultural Biotechnology," *Chemical & Engineering News,* August 21, 1995 (http://pubs.acs.org).

167 *Monocultures of the Mind,* op. cit.

168 For an excellent philosophical discussion of the history of this mindset, see Edmund P. Russell III, "'Speaking of Annihilation': Mobilizing for War Against Human and Insect Enemies, 1914–1945," *Journal of American History* 82, no. 4: pp. 1505–1529.

169 *Monocultures of the Mind,* pg. 67.

170 Ibid., pg. 80.

171 Altieri, Miguel A. *Agroecology: The Science of Sustainable Agriculture,* second edition (Boulder, CO: Westview Press, 1995); Jules N. Pretty, *Regenerating Agriculture: Policies and Practices for Sustainability and Self-Reliance* (London, UK: Earthscan, 1995).

172 Altieri, Miguel A. "Why Study Traditional Agriculture?" chapter 20 in Carroll et al., *Agroecology.*

173 "Agroecology versus Input Substitution."

174 Bayliss-Smith, Tim P. "Energy Flows and Agrarian Change in Karnataka: The Green Revolution at Micro-scale," in Bayliss-Smith and Wanmali, *Understanding Green Revolutions,* pp. 169–170. While types of energy are not strictly comparable, such comparisons are meaningful in designing agricultural systems appropriate to farmers with varying access to energy sources.

175 Vandermeer, John. *The Ecology of Intercropping* (Cambridge, UK: Cambridge University Press, 1989); Donald Q. Innis, *Intercropping and the Scientific Basis of Traditional Agriculture* (London, UK: Intermediate Technology Publications, 1997).

176 *The Ecology of Intercropping,* op. cit.

177 Altieri, Miguel A. and M. Kat Anderson. "An Ecological Basis of the Development of Alternative Agricultural Systems for Small Farmers in the Third World, *American Journal of Alternative Agriculture* 1, no. 1 (1986): pp. 33–34; *Agroecology.*

178 "Farmer-First" and "The Campesino a Campesino Movement."

179 *The Violence of the Green Revolution,* op. cit.

180 "The Agricultural Fix."

181 *Intercropping,* op. cit.

182 For a useful discussion, see Paul Richards, "Ecological Change and the Politics of African Land Use," *African Studies Review 26,* no. 2 (1983). See also Kurt G. Steiner, *Intercropping in Tropical Smallholder Agriculture with Special Reference to West Africa* (Eschborn, Germany: GTZ, 1984); Lloyd Timberlake, *Africa in Crisis: The Causes, the Cures of Environmental Bankruptcy* (London, UK: Earthscan, 1985); Sustainable Agriculture Networking and Extension, *An Agroecology Reader for Africa* (New York: UNDP-SANE, 1995).

183 Freeman, Peter H. and Tomas B. Fricke. "Traditional Agriculture in Sahelia: A Successful Way to Live," *The Ecologist 13,* no. 6 (1983): pp. 210–212.

184 "Lessons from the Green Revolution," op. cit.

185 "Challenging the 'New Green Revolution," *Food First News and Views 19,* no. 65, (Oakland, CA: Institute for Food and Development Policy, 1997); for an example of this model in Ethiopia, see the Letter to the Editor in the same issue, by Dr. Mario Pareja, food and livelihood security coordinator, CARE–East Africa.

186 Dudley, Nigel and John Madeley, and Sus Stolton, eds. *Land Is Life: Land Reform and Sustainable Agriculture* (London, UK: Intermediate Technology Publications, 1992).

187 Some proponents have recently made the claim that the Green Revolution offers the best way to protect the environment (see Dennis Avery, *Saving the Planet with Pesticides and Plastic: The Environmental Triumph of High-Yield Farming,* [Indianapolis, IN: Hudson Institute, 1995]). They argue that by boosting yields on favorable lands it will be unnecessary to farm less favorable ones, which are more likely to be more important wildlife refuges or pristine forests, and can thus be saved. This argument is specious for several reasons: It recognizes only one way to boost production (their way), when in fact there are many; it fails to take into account the environmental destruction wrought by industrial-style farming; and it also fails to consider the greater compatibility with the environment offered by alternative farming methods. For a report that rebuts this argument, see Tracy Irwin Hewitt and Katherine R. Smith, *Intensive Agriculture and Environmental Quality: Examining the Newest Agricultural Myth* (Greenbelt, MD: Henry Wallace Institute for Alternative Agriculture, 1995).

188 National Research Council. *Alternative Agriculture* (Washington, DC: National Academy Press, 1989), pp. 8, 10, 17.

189 van der Werf, Erik. "Agronomic and Economic Potential of Sustainable Agriculture in South India," *American Journal of Alternative Agriculture* 8, no. 4 (1993): pp. 185–191.

190 Collier, Robert. "Cuba Turns to Mother Earth: With Fertilizers and Fuel Scarce, Organic Farming Is In," *San Francisco Chronicle,* February 21, 1998, pp. A1, A6.

191 Rosset, Peter and Medea Benjamin. *The Greening of the Revolution: Cuba's Experiment with Organic Agriculture* (Melbourne, Australia: Ocean Press, 1994).

192 Rosset, Peter. "Alternative Agriculture and Crisis in Cuba," *Technology and Society* 16, no. 2 (1997): 19–25.

193 The Cuban government tried a short-lived experiment with farmers' markets in the 1980s, which were subsequently closed because the leadership felt that middlemen were taking a large share of the profits. For a thorough discussion of these and other food issues from the 1980s, see Medea Benjamin, Joseph Collins, and Michael Scott, *No Free Lunch: Food and Revolution in Cuba Today* (San Francisco: Food First Books, 1989).

194 Companioni, N. and A. A. Rodríguez Nodals, Mariam Carrión, Rosa M. Alonso, Yanet Ojeda, and Ana María Viscaíno, "La Agricultura Urbana en Cuba: Su Participación en la Seguridad Alimentaria," pp. 9–13 in Asociación Cubana de Agricultura Urbana (ACAO), *III Encuentro Nacional de Agricultura Orgánica 14 al 16 de mayo de 1997*, Universidad Central de las Villas, Villa Clara, Cuba. Conferencias (Havana, Cuba: ACAO, 1997). For an excellent discussion of the enormous potential that urban agriculture has worldwide, see Jac Smit, Annu Ratta, and Joe Nasr, *Urban Agriculture: Food, Jobs and Sustainable Cities* (New York: UNDP, 1996).

From *Food First: Beyond the Myth of Scarcity* by Frances Moore Lappé and Joseph Collins

Why Can't People Feed Themselves?

QUESTION: You have said that the hunger problem is not the result of overpopulation. But you haven't yet answered the most basic and simple question of all: Why can't people feed themselves? As Senator Daniel P. Moynihan put it bluntly, when addressing himself to the Third World, "Food growing is the first thing you do when you come down out of the trees. The question is, how come the United States can grow food and you can't?"

OUR RESPONSE: In the very first speech I, Frances, ever gave after writing *Diet for a Small Planet*, I tried to take my audience along the path that I had taken in attempting to understand why so many are hungry in this world. Here is the gist of the talk that was a turning point in my life.

> When I started, I saw a world divided into two parts: a *minority* of nations that had 'taken off' through their agricultural and industrial revolutions to reach a level of unparalleled material abundance and a *majority* that remained behind in a primitive, traditional, undeveloped state. This lagging behind of the majority of the world's peoples must be due, I thought, to some internal deficiency or even to several of them. It seemed obvious that the underdeveloped countries must be deficient in natural resources—particularly good land and climate—and in cultural development, including modern attitudes conducive to work and progress.
>
> But when looking for the historical roots of the predicament, I learned that my picture of these two separate worlds was quite false. My 'two separate worlds' were really just different sides of the same coin. One side was on top largely because the other side was on the bottom. Could this be true? How were these separate worlds related?

Colonialism appeared to me to be the link. Colonialism destroyed the cultural patterns of production and exchange by which traditional societies in 'underdeveloped' countries previously had met the needs of the people. Many precolonial social structures, while dominated by exploitative elites, had evolved a system of mutual obligations among the classes that helped to ensure at least a minimal diet for all. A friend of mine once said, "Precolonial village existence in subsistence agriculture was a limited life indeed, but it's certainly not Calcutta." The misery of starvation in the streets of Calcutta can only be understood as the end-point of a long historical process—one that has destroyed a traditional social system.

'Underdeveloped,' instead of being an adjective that evokes the picture of a static society, became for me a verb (to 'underdevelop') meaning the *process* by which the minority of the world has transformed—indeed often robbed and degraded—the majority.

That was 1972. I clearly recall my thoughts on my return home. I had stated publicly for the first time a world view that had taken me years of study to grasp. The sense of relief was tremendous. For me the breakthrough lay in realizing that today's 'hunger crisis' could not be described in static, descriptive terms. Hunger and underdevelopment must always be though of as a *process.*

To answer the question, "why hunger?" it is counterproductive to simply *describe* the conditions in an underdeveloped country today. For these conditions, whether they be the degree of malnutrition, the levels of agricultural production, or even the country's ecological endowment, are not static facts—they are not 'givens.' They are rather the *results* of an ongoing historical process. As we dug ever deeper into that historical process for the preparation of this book, we began to discover the existence of scarcity—creating mechanisms that we had only vaguely intuited before.

We have gotten great satisfaction from probing into the past since we recognized it is the only way to approach a solution to hunger today. We have come to see that it is the *force* creating the condition, not the condition itself, that must be the target of change. Otherwise we might change the condition today, only to find tomorrow that it has been recreated—with a vengeance.

Asking the question "Why can't people feed themselves?" carries a sense of bewilderment that there are so many people in the world not able to feed themselves adequately. What astonished us, however, is that there are not *more* people in the world who are hungry—considering the weight of the centuries of effort by the few to undermine the capacity of the majority to feed themselves. No, we are not crying "conspiracy!" If these forces were entirely conspiratorial, they would be easier to detect and many more people would by now have risen up to resist. We are talking about something more subtle and insidious; a heritage of a colonial order in which people with the advantage of considerable power sought their own self-interest, often arrogantly believing they were acting in the interest of the people whose lives they were destroying.

The Colonial Mind

The colonizer viewed agriculture in the subjugated lands as primitive and backward. Yet such a view contrasts sharply with documents from the colonial period now coming to light. For example, A. J. Voelker, a British agricultural scientist assigned to India during the 1890s, wrote:

> Nowhere would one find better instances of keeping land scrupulously clean from weeds, of ingenuity in device of water-raising appliances, of knowledge of soils and their capabilities, as well as of the exact time to sow and reap, as one would find in Indian agriculture. It is wonderful, too, how much is known of rotation, the system of 'mixed crops' and of fallowing… I, at least, have never seen a more perfect picture of cultivation.[1]

Nonetheless, viewing the agriculture of the vanquished as primitive and backward reinforced the colonizer's rationale for destroying it. To the colonizers of Africa, Asia, and Latin America, agriculture became merely a means to extract wealth—much as gold from a mine—on behalf of the colonizing power. Agriculture was no longer seen as a source of food for the local population, nor even as their livelihood. Indeed, the English economist John Stuart Mill reasoned that colonies should not be thought of as civilizations or countries at all but as 'agricultural establishments' whose sole purpose was to supply the "larger community to which they belong." The colonized society's agriculture was only a subdivision of the agricultural system of the metropolitan country. As Mill acknowledged, "Our West India colonies, for example, cannot be regarded as countries... The West Indies are the place where England *finds it convenient* to carry on the production of sugar, coffee and a few other tropical commodities."[2]

Prior to European intervention, Africans practiced a diversified agriculture that included the introduction of new food plants of Asian or American origin. But colonial rule simplified this diversified production to single cash crops—often to the exclusion of staple foods—and in the process sowed the seeds of famine.[3] Rice farming once had been common in Gambia. But with colonial rule so much of the best land was taken over by peanuts (grown for the European market) that rice had to be imported to counter the mounting prospect of famine. Central Ghana, once famous for its yams and other foodstuffs, was forced to concentrate solely on cocoa. Most of the Gold Coast thus became dependent on cocoa. Liberia was turned into a virtual plantation subsidiary of Firestone Tire and Rubber. Food production in Dahomey and southeast Nigeria was all but abandoned in favor of palm oil; Tanganyika (now Tanzania) was forced to focus on sisal and Uganda on cotton.

The same happened in Indochina. About the time of the American Civil War the French decided that the Mekong Delta in Vietnam would be ideal for producing rice for export. Through a production system based on enriching the large landowners, Vietnam became the world's third largest exporter of rice by the 1930s; yet many landless Vietnamese went hungry.[4]

Rather than helping the peasants, colonialism's public works programs only reinforced export crop production. British irrigation works built in nineteenth-century India did help increase production, but the expansion was for spring export crops at the expense of millets and legumes grown in the fall as the basic local food crops.

Because people living on the land do not easily go against their natural and adaptive drive to grow food for themselves, colonial powers had to force the production of cash crops. The first strategy was to use physical or economic force to get the local population to grow cash crops instead of food on their own plots and then turn them over to the colonizer, generally for export. The second strategy was the direct takeover of the land by large-scale plantations growing crops for export.

Isn't the Backwardness of Small Farmers to Blame?

QUESTION: You seem to think the small farmer is the savior of the hungry world. But isn't one basic reason for low production levels in the poor countries that so much land is in farms too small to be efficient? Aren't most small farmers too backward and tradition-bound to respond to development programs?

OUR RESPONSE: Whether a small farm is necessarily less efficient and less productive than a large operation is of critical importance in assessing the production potential of the underdeveloped countries, since about eighty percent of all farms are of less than twelve acres of land.

So we asked, does smallness equal low levels of production? To answer that question we looked at studies from all over the world and everywhere the verdict is the same: Contrary to our previous assumption, the small farmer in most cases produces more per unit of land than the large farmer. Here are just a few examples:

- The value of output per acre in India is more than one-third higher on the smallest farms than on the larger farms.[5]
- In Thailand plots of two to four acres produce almost sixty percent more rice per acre than farms of 140 acres and more.[6]
- In Taiwan net income per acre of farms with less than one and a quarter acres is nearly twice that of farms over five acres.[7]
- The World Bank has reported on an analysis of the differences in the value of output on large and small farms in Argentina, Brazil, Chile, Colombia, Ecuador, and Guatemala. The conclusion? The small farms were three to fourteen times more productive per acre than the large farms.[8]

Such comparisons go a long way toward explaining the low productivity of agriculture in underdeveloped countries when you bear in mind that, according to a study of eighty-three countries, only three percent of all the landowners control a staggering eighty percent of all farmland.[9] The point is that the largest landholders control most of the farmland, yet studies from all over the world show that they are the least productive.

To explain the higher productivity of the small farmer, one need not romanticize the peasant. Peasant farmers get more out of their land precisely because they need to survive on the meager resources allowed to them. Studies show that small landholders plant more carefully than a machine would, mix and rotate complementary crops, choose a combination of cultivation and livestock that is labor-intensive and, above all, work their perceptibly limited resources (especially themselves) to the fullest. Farming for the peasant family is not an abstract calculation of profit to be weighed against other investments. It is a matter of life and death.

What About Small Farmer Efficiency in the United States?

Every single study ever made by the United States Department of Agriculture (USDA) has found that the most efficient farm, measured in terms of cost per unit of output, is the mechanized one- or two-farmer unit, not the largest operations. Any savings associated with sheer size (and there are remarkably few) are quickly offset by the higher management, supervisory, and labor costs of large farms.[10]

To test this finding we looked at net income per acre by farm size in the United States from 1960 through 1973. We found that in all those fourteen years there were only two in which the biggest farms realized a net income per acre greater than the family farm.[11] The very pattern of greater productivity by small farmers that struck us in our study of underdeveloped countries is found right here in America.

Why Don't Small Farmers Produce More?

QUESTION: You say that small landholders produce more per acre than the large landholders. But are not the yields of small peasant holdings in the Third World only a fraction of the yields of the farms in the industrial countries? Why don't they produce more?

OUR RESPONSE: Compared to large landholders, small peasant producers do not have equal access to credit and agricultural inputs, such as water, fertilizer, and tools. As Green Revolution pioneer Norman Borlaug himself expressed it:

> I have a lot of respect for the small farmer... Almost invariably when you look at what he's doing with his land, you find he's producing the maximum under the situation he has to work with. The thing is that he usually doesn't have much to work with.[12]

The Small Pay More

Small farmers often cannot get ahead because their initiatives are actively obstructed by the landed elite, who are threatened by any advance that would make the village's small farmers less dependent on them.

Moreover, necessities such as fertilizer and water do not reach small farmers because they have neither the cash nor the credit to buy them. Quite often loans from government agencies stipulate a minimum holding that cuts out the small farmer. In Pakistan, to get a loan for a tubewell from the Agricultural Development Bank, a peasant must have at least 12.5 acres. This single stipulation excludes over eighty percent of Pakistan's farmers.[13] One estimate is that only about five percent of Africa's farmers have access to institutional credit—and it is not hard to guess which five percent![14]

Sudhir Sen, Indian economist and commentator on the Green Revolution, has estimated that roughly one-half of India's small farmers lack any recorded right to the land, without which they are unable to obtain crop loans from credit institutions.[15] (Even where there are exceptions, the tenant is still penalized. In Tamil Nadu, India, the tenant is allowed only 60 percent of the amount of credit per acre advanced to landowners.[16]) Perhaps, even more important, small farmers are reluctant to use their land for loan collateral anyway. Poor farmers quite sensibly decide that they do not want to risk losing their land.

Largely excluded from institutional credit, small landholders are left dependent on private moneylenders and merchants who charge usurious rates of interest. We have seen estimates of interest rates ranging from fifty to two hundred percent! In one area of the Philippines fifteen percent of the borrowers paid an interest rate of over 200 percent while twenty percent of the borrowers paid only sixteen percent. Moreover, merchant-creditors can increase the interest by underpricing farm products used to repay loans and overpricing the goods that debtors buy from them.[17] By contrast, the large operator may pay no interest, or even come out ahead by borrowing money. When the nominal rates of interest on credit available to large operators from commercial institutions are adjusted for inflation, the real rate of interest is often negative.[18]

Obligations to Moneylenders and Landlords

Debt bondage keeps many peasant farmers in a form of perpetual vassalage. As agricultural economist Keith Griffin so aptly puts it, "The *campesinos* of Latin America have suffered not from insecurity of tenure but from excessively secure tenure." Debt bondage, he points out, has been used to tie peasants to the land to assure landowners that labor would be available, particularly in labor-scarce economies in Latin America.[19] What is the impact on production? Inevitably, motivation to increase production is stifled because the trapped peasants know that higher yields will never benefit them, only the landowner or moneylender. "The constantly indebted peasant is virtually bound by contract to sell his produce at prices set by the private moneylender cum-trader, as no effective marketing cooperatives exist to safeguard his interests," explain Erich and Charlotte Jacoby in their classic *Man and Land.*[20]

Debt bondage can mean that the peasant farmer must work off the debt by tending the fields of the creditor. The peasant's own plot then suffers neglect. Unable to work his land adequately, the peat farmer often has no choice but to give it up.

Consider sharecroppers who represent a significant portion of the rural population in many underdeveloped countries Although in many cases they must provide all of the inputs, they get only a portion of the crop. Why then make the investments necessary to increase production? In Bangladesh, we learned that while owner-cultivators need the prospect of a two-to-one advantage in order to take the risk of adopting a new technology, sharecroppers need the prospect of a four-to-one advantage since they get only half the crop.[21]

Insecure tenancies result in soil depletion. Tenants in constant indebtedness and unsure of whether or not they will be on the same plot next year, can hardly be expected to protect the soil fertility by rotating crops and leaving fields fallow.

Without a certain minimum landholding, security of tenure, credit at a reasonable rate, and control over what is produced, farmers make the realistic assessment that it is not in their best interest to invest to increase production or to take steps to preserve the soil fertility. It is not the alleged 'backwardness' of the peasant farmers that keeps them from buying fertilizer and other modern inputs, but hard economic sense.

Is Small Always Beautiful?

QUESTION: If small farmers have proven to be more productive, should not the primary concern of development be to channel more credit, equipment, seeds, fertilizers and irrigation to them?

OUR RESPONSE: First of all, to focus on the small farmer is to miss entirely a large portion—in many countries eighty to ninety percent—of the rural labor force. A recent study from Cornell University concludes that the landless and near-landless constitute a majority of the rural labor force in Asia, approaching ninety percent in Java, Bangladesh, and Pakistan. In Latin America the landless and near-landless make up a majority in every country studied, exceeding eighty percent in Bolivia, El Salvador, Guatemala, and the Dominican Republic.[22]

Nor should one make the mistake of believing that the small farm is inherently more productive than the large. We have found that the size of the parcel of land matters less than the relationship of the people to it.

We have seen that small farms can be very productive—as in Japan—where the people working the land know that the productivity will benefit them. And we have seen exactly the opposite: small farms with low productivity when credit, debt, and tenancy arrangements deny those who work the fields the fruits of their labors.

Likewise with large farming units. They can be productive where those working the land know that their labor will benefit them. Thai

Binh in North Vietnam is one example. Since 1965 a single cooperative involving 4000 people produces rice, small animals, such as ducks and geese, as well as fish in over 100 acres of village-controlled fish ponds. Harvesting two or even three rice crops annually, Thai Binh can produce almost eighty percent more than the annual production of the less-than-five-acre plot characteristic of India, for example.[23] But large units are not necessarily productive. We have just documented the inefficiencies of many privately owned, large landholder operations. Exchange these private landowners for antidemocratic bureaucrats and productivity will still remain low, as developments in Soviet agriculture have amply demonstrated.

1 Sinha, Radha. *Food and Poverty* (New York: Holmes and Meier, 1976), pg. 26.

2 Mill, John Stuart. *Political Economy*. Book 3, chapter 25.

3 Feldman, Peter and David Lawrence. "Social and Economic Implications of the Large-Scale Introduction of New Varieties of Foodgrains," *Africa Report*, preliminary draft (Geneva, Switzerland: UNRISD, 1975), pp. 107-108.

4 Owens, Edgar. *The Right Side of History*. Unpublished manuscript, 1975.

5 Owens, Edgar and Robert Shaw. *Development Reconsidered: Bridging the Gap Between Government and People* (Lexington, MA: DC Health, 1972), pg. 20.

6 World Bank. *The Assault on World Poverty—Problems of Rural Development, Education, and Health* (Baltimore: Johns Hopkins University Press, 1975), pg. 215.

7 *Development Reconsidered*, op. cit.

8 *The Assault on World Poverty*, pp. 215-216.

9 Food and Agricultural Organization. *Report on the 1960 World Census of Agriculture*, (Rome: Food and Agricultural Organization, 1971).

10 U.S. Department of Agriculture. "The One-Man Farm," prepared by Warren Bailey, USDA/ERS-519 (Washington, DC: Government Printing Office, 1973). See also Angus McDonald, "The Family Farm is the Most Efficient Unit of Production," in Peter Barnes, ed., *The People's Land* (Emmaus, PA: Rodale Press, 1975).

11 Calculated from U.S. Department of Agriculture, *Statistical Bulletin*, no. 547, Farm Income Statistics, Table 3D, USDA/ERS (Washington, DC: Government Printing Office, 1975), pg. 60; and "The Balance Sheet of the Farming Sector by Value of Sales Class, 1960-1973," supplement no.1, *Agricultural Information Bulletin*, no. 376, Table 2, USDA/ERS (Washington, DC: Government Printing Office, 1975), pg.30.

12 *The New York Times*, November 5, 1974, pg. 14.

13 Griffin, Keith. *The Political Economy of Agrarian Change* (Cambridge, MA: Harvard University Press, 1974), pg. 27.

14 *The Assault on World Poverty*, pg. 105.

15 Sen, Sudhir. Reaping the Green Revolution (Maryknoll, NY: Orbis, 1975), pg. 11.

16 Pearse, Andrew. "Social and Economic Implications of the Large-Scale Introduction of the New Varieties of Foodgrains," Part 2 (Geneva, Switzerland: UNDP/UNRISD, 1975), pp. 8-9.

17 *Political Economy*, pg. 28.

18 Griffin, Keith. *Land Concentration and Rural Poverty* (New York: Macmillan, 1976), pg. 122.

19 Griffin, Keith and Azizur Rahman Khan, eds. *Poverty and Landlessness in Rural Asia: A Study by the World Employment Programme* (International Labor Office, 1976), pp. 1-31.

20 Jacoby, Erich and Charlotte Jacoby. *Man and Land* (New York: Alfred A. Knopf, 1971), pg. 79.

21 Brammer, Hugh. FAO, Bangladesh. Interviewed by Joseph Collins, January 1978.

22 Esman, Milton J. *Landlessness and Nearlandlessness in Developing Countries,* Center for International Studies (Ithaca, NY: Cornell University, 1978.

23 Gough, Kathleen. "The 'Green Revolution' in South India and North Vietnam," Social Scientist, Kerala, India, August 1977, no. 61. See also Gough, *Ten Times More Beautiful* (New York: Monthly Review Press, 1978).

From *Taking Population Seriously* by Frances Moore Lappé and Rachel Schurman

---

Children: Poor People's Source of Power

Living at the economic margin, many poor parents perceive their children's labor as necessary to augment meager family income. By working in the fields and around the home, children also free up adults and elder siblings to earn outside income.

One study on the island of Java, Indonesia found children to be an extremely important asset in the rural economy. As early as age seven, a boy assumes responsibility for his family's chicken and ducks. At nine, he can care for goats and cattle, cut fodder, and harvest and transplant rice. And as early as twelve, he can work for wages. And by his fifteenth birthday, a Javanese boy has, through his labor, repaid the entire investment his family has made in him.[1] Similarly in Bangladesh, by age six a son provides labor and/or income for his family. By twelve, he contributes more than he consumes.[2]

The labor of girls is equally important. In Tanzania, one study found that girls between the ages of five and nine spend an average of three and a half hours a day working on economic activities and in the home.[3] In Peru, as in many other Third World countries, girls of six and seven can be seen carting younger brothers and sisters around while their mothers are selling wares in the marketplace. And on tea plantations in India, girls as young as twelve rise at 4 AM to help their mothers prepare breakfast and then spend up to ten hours a day in the fields picking tea during harvest time.[4]

John Caldwell, a foremost theorist in population studies, criticizes his colleagues for underrating the importance of children to the village economy:

> The analysis of the value of children in the village household economy often completely fails to understand the subsistence nature of village services... [In Western] societies, most...expenditure is on goods...which merely occupy space while providing services. Vacuum cleaners pick up the dust; mowers keep down the grass; washing machines clean up the dishes; hot water systems provide hot water... In a society where nearly all these consumer durables are vastly expensive and difficult or impossible to maintain, the services are largely provided by either cheap labor or more frequently by family subsistence labor.. *In terms of the availability of cheap or ostensibly free labor, the farmer would be irrational who forced himself and his family to produce ever more crops for sale to buy expensive gadgets to supply the services he could get directly [from his children].*[5]

Children aren't assets only for people living in the countryside, Caldwell points out. Among the Yoruba in Nigeria, urban white collar families rely on many children to enhance their position through 'sibling assistance chains.' As one child grows up and completes school, he or she (most often he!) can help younger siblings climb up the educational ladder. Each successive child is thus in a position to get a higher-paying job. The family gets added income, status, and security.[6]

Moreover, the 'lottery mentality' is associated with poverty everywhere. With no reliable channels for advancement in sight, Third World parents can always hope that the next child will be the one clever and bright enough to get an education and land a city job, despite the odds. In many countries, income from just one such job in the city can support a whole family in the countryside.

In nearly all Third World societies, those rendered powerless by unjust economic structures also know that without children to care for them in old age, they will have nothing.[7] According to a 1984 World Bank report, eighty to ninety percent of the people surveyed in Indonesia, South Korea, Thailand, and Turkey expect to rely on their children for support in their old age.[8]

In addition to providing old age security children represent insurance against risk for many rural families, argues Mead Cain, a researcher at the Center for Population Studies in New York. Drawing on extensive experience in South Asia, Cain notes that children's financial help provide "an important means of insuring against property loss" to families for whom a bad crop year or unexpected expense can spell certain catastrophe, forcing a distress sale of their land. It is only when other forms of economic security become available to impoverished families, posits Cain, will the transition to smaller families be instigated.[9]

Of course, the value of children to their parents cannot be measured just in hours of labor or extra income. The intangibles may be just as important. Influence within the community is one. In community affairs, bigger families can carry more weight. And for most parents, children offer incomparable satisfaction, fulfillment, and joy. For the poor—whose lives are marred by much more grief and sacrifice than is true of the better-off—the role children play in fulfilling these very real human needs cannot be underestimated.

The question of children's economic contributions to the family is not a settled one among population researchers, however. Recently some have begun to argue that the economic advantages children offer may be diminishing in ways that will reduce fertility. As primary and secondary education becomes more widespread, for instance, children have less time to contribute to their families.[10] Also, as more people lose or are unable to acquire land and migrate to the cities, they will take wage-paying jobs. For wage laborers children come to represent a cost rather than just a benefit.

The Population Reference Bureau's Thomas Merrick uses this argument to help explain recent fertility declines in Mexico, Colombia, and Brazil. In each case, he argues, the economic squeeze reflected in declining real wages has made consumer aspirations more difficult to attain. In order to maintain their standard of living, however low, families react by limiting births.[11]

While the shift to wage labor and the simultaneous growth in Third World cities may change the economic value of children for *some* families, we wonder whether it will have as dramatic an effect on fertility as Merrick and others assume. Before children can become a net drain on their wage-earning parents, those parents must have jobs. Yet in Third World countries, vast numbers of urban people have not been able to find paid employment. So people find other ways to survive, and in many of these desperate strategies children are still assets. Children can engage in petty commerce, selling trinkets or snacks on the street. In the shantytowns of metropolitan Manila, whole families survive by collecting and selling scrap material from dumps. Children begging from tourists is another source of family income. And in the absence of alternatives, children also become prostitutes to support themselves and often their parents as well. Bangkok and Sao Paolo are both notorious for child prostitutes, many of whom are the sole supporters of their families.

When people are made powerless to provide for their children, children will continue to provide a primary way for parents, themselves, to survive.

When Many Babies Die

But to achieve any of the potential benefits of children, the Third World poor realize that they need to have many. Where food resources and health care are the monopoly of the better-off, as many as one in every four children dies before the age of five (see table 1). Seeing their babies dying or their children in such poor health that the threat of death is ever present, parents naturally are motivated to have more children. In India, of the six or seven births women average, only four can be expected to survive.[12] One Sudanese doctor angrily described the problem:

> In a country like ours where the infant mortality rate is 140 per 1,000 births, where infectious diseases kill so many children, where malnutrition affects about thirty to fifty percent of the people, where measles is a killer disease although it could be stopped by immunization, how can you tell us to stop having children? When a mother has twelve children, only three or four may live.[13]

Studies by the World Heath Organization confirm that both the actual death and the fear of death of a child will increase the fertility of a couple, regardless of income or family size.[14]

The positive impact of infant death on fertility rates is also a biological fact. Breastfeeding tends to prolong the period during which a woman cannot become pregnant, so that when her infant dies this infertile period is cut short. And this is no insignificant phenomenon. Lactation has been a primary means of birth spacing in many cultures, especially in Africa where babies have typically been breastfed for two years or longer.

Women: Powerlessness and High Birth Rates

High birth rates reflect not only the labor-income-security needs of the poor, but the disproportionate powerlessness of women. Excluded from many decisions that determine their role in the family, as well as in the society at large, many women have little opportunity for pursuits outside the home. Perpetual motherhood becomes their only 'choice.'

Perhaps the best proof that the powerlessness of women undergirds high fertility comes from extensive research on the effect of women's education. In one study after another, women's education turns out to be the single most consistent predictor of lower fertility. As women's schooling increases, fertility typically falls.

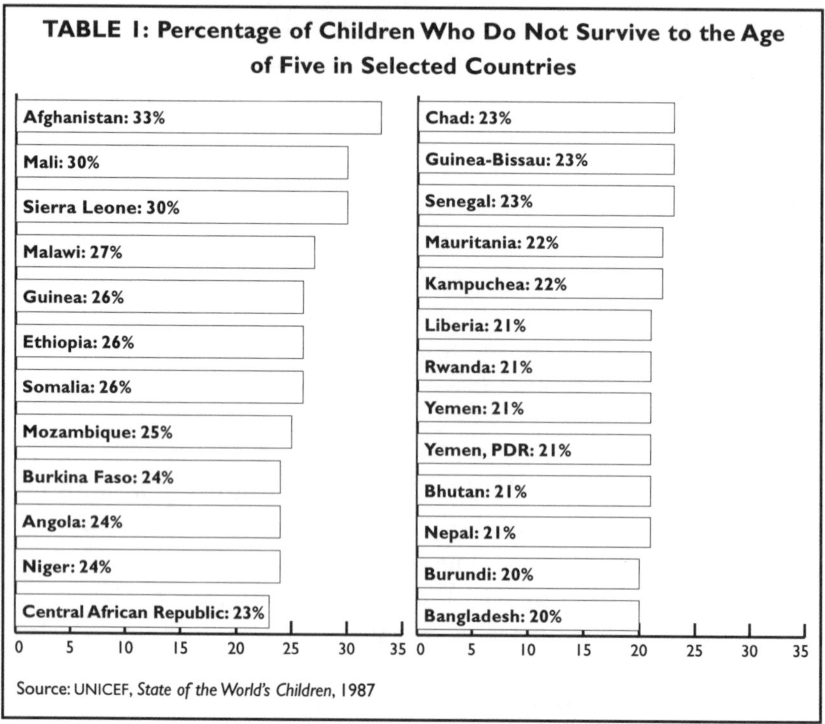

TABLE 1: Percentage of Children Who Do Not Survive to the Age of Five in Selected Countries

Source: UNICEF, *State of the World's Children*, 1987

Of course, few would interpret these findings literally—that women learn how to limit births. Rather, demographers surmise that the fact that women are getting educated reflects a multitude of changes in society that allow women greater power.

All this cuts to the core of the population issue because women's subordination to men within the family often translates into a direct loss of control over heir own fertility. It has been widely documented that after several births many Third World women want to avoid or delay pregnancy. But women simply do not have the power to act on their desire. As one doctor in a Mexican clinic explained:

> When a wife wants to… [try] to limit the number of mouths to feed in the family, he husband will become angry and even beat her. He thinks it is unacceptable that she is making a decision of her own. She is challenging his authority, his power over her—and thus the very nature of his virility.[15]

Women who do try to limit their pregnancies—either with or without the consent of their partners—often receive little or no help from the state. Poor women have particularly limited access to health services, including birth control devices. In desperation, many resort to illegal, and often fatal, abortions. A major cause of maternal death in the Third World,[16] complications from illegal abortions are estimated to kill over 200,000 women a year, most of them poor and illiterate Third World women.[17]

Where a woman's choices are severely limited—where women are discouraged from working outside the home—children often represent her only source of power. In Kenya, population researchers note that the low status of women pushes them into early marriage and frequent childbearing.[18] "If society impedes other avenues to power such as pursuit of economic activity,[19] point out two African scholars, "then women may compensate by having large numbers of children." Sally Mugabe, writing in *Popline*, underscores the point:

> For a [Zimbabwean] woman, bearing and rearing children is the primary source of status in the family and the community. The larger number of children a woman has, the higher the status she enjoys.[20]

Other cultural forces severely limit a woman's freedom to choose fewer births. The influence of the Catholic church is significant in many countries. This is particularly true in Latin America. In the book *Don't Be Afraid, Gringo* a Honduran peasant woman talks in intimate detail about the many forces depriving her of the power to provide for her family. She reflects on why Honduran women have so many children:

> Not many campesina women use birth control. They just keep having babies, babies, and more babies... I've thought a lot about why we have so many children... Part of the reason might be the Catholic church. Most of us are Catholics, and the church tells us that it's natural to have children and that going against nature is going against God.[21]

One can well imagine how difficult it is for Catholic women to use birth control if it means having to confess to a central authority figure in your community, the priest, that you have sinned. For many poor women, whose self-esteem is already low, challenging church authority can be virtually unthinkable.

The Fertility Consequences of Son Preference

Patriarchal family and community attitudes also pressure a woman to keep having children until she gives birth to a son, regardless of her own wishes or even possible jeopardy to her health. Male attitudes and power over women are critical. As we just noted, many women do not resist family planning, but as an Egyptian woman explains, it's the men "who are sometimes against it... They want children until they get a boy."[22] In India, a preference for sons on the part of both parents is so strong that amniocentisis is now being used in many areas to determine the sex of the fetus. According to population researcher Betsy Hartmann, Indian women found to be carrying females are often pressured to abort by husbands and in-laws.[23] A study in Bombay of 8,000 abortions following amniocentisis found that all but one of the aborted fetuses were female.[24]

Son preference is not only linked to enhanced social status; it often has financial implications as well. China is a good example. A daughter offers her parents much less security than a son. Upon marriage, she leaves to live in her husband's home, whereas a son's wife comes to live with his parents, providing security and companionship in their old age. In Bangladesh, where many women are subject to the Islamic custom of Purdah (forbidden to leave, much less work, outside the home), the incentive is strong to bear sons for future social and economic support.[25] Sons can also better protect rights to the land, especially important to widows.[26]

Not Only Women Are Made Powerless

While the power-structures perspective helps explain the high birth rates of women subordinated within the family and society, it recognizes that often the men who hold power over women are themselves part of a subordinate group—those with little or no claim to income-producing resources. This, too, has important implications for fertility.

As long as poor men are denied sources of self-esteem through productive work, and are denied access to the resources they need to act responsibly toward their families, it's likely they will cling even more tenaciously to their superior power vis-a-vis women. For many men, this may mean showing their virility through siring large numbers of children. Men who are forced to migrate for work, for example, may decide to start up a second family, further increasing the number of children.

In many cultures, men unable to bring in enough income to support dependents feel inadequate to maintain a permanent household. The sad irony is that self-blame for this failure, lowering self-esteem, can result in a behavior pattern of moving in and out of relationships and the fathering of even more children.

Summarizing the Power-Structures Perspective

In our view, this varied evidence—drawn by anthropologists and sociologists working in the Third World—about why the poor have many children, suggests that high fertility can best he understood as a response to antidemocratic structures of power within which people are often left with little choice but many births.

To recap, freedom of choice in fertility is nonexistent where:

- one's financial security depends entirely or largely on one's surviving children
- many births are necessary to ensure that even several children live to maturity
- health services, including birth control, are available almost exclusively to the better-off in urban areas, not to the poor
- a woman has no choice other than marriage and her only source of power is derived from her children, especially sons
- few opportunities for education and employment exist for women outside of homemaking

The power-structures analysis—particularly in recent years[27]—stresses the impact on fertility of women's subordination to men, a condition that contributes to the social pressure for many births. But it places this problem within the context of unjust economic structures that deny people the realistic alternatives to unlimited reproduction. Within such a framework, rapid population growth is seen to result largely from efforts by the poor to cope, given their powerlessness in he face of the concentrated economic strength of an elite.

The narrowly constricted power of Third World women can only be understood in light of relationships extending far beyond the family and even the community (see table 1). From the level of international trade and finance, down to jobs and income available to men as well as women, antidemocratic structures of decision making set limits on people's choices which ultimately influence their reproductive options.

In a report of this length, we can only offer a few examples to suggest how decisions at these many levels can affect fertility.

At the international level, consider the debt crisis. In the 1970s, Third World governments received large loans from banks in the industrial nations, investing the money in big-ticket projects—airports, arms, nuclear power plants, and so on—responding to the interests of their wealthiest citizens. In the 1980s, many of these loans came due, just as interest rates climbed and prices of raw material exports from the Third World hit a thirty-year low. As a result, between 1982 and 1987, the net transfer from poor countries to banks and governments in the rich countries totaled $140 billion, or the equivalent of *two* Marshall Plans.[28]

How did Third World countries come up with such sums? Health and welfare budgets and food subsidies got slashed first. And to earn foreign exchange land and credit increasingly went toward export crops. But reduced health care budgets means that more babies die and fewer resources are available for comprehensive family planning care. More resources devoted to crops for export means that locally food becomes more scarce and more expensive. Add to this cuts in government food subsidies. Understandably, nutrition and health worsen; death rates rise. The link between debt and heightened death rates is so clear that sociologists have quantified it: every additional ten dollars in interest paid per person per year by seventy-three poor countries means 142 days shorter life on average than would be true if life expectancy had continued to improve at pre-debt levels.[29]

The 'international debt crisis'—seemingly remote from intimate reproductive behavior—ends up affecting conditions of basic family security, health, and nutrition known to influence fertility. High growth rates can in part be understood by reference to such far-reaching decisions that end up shifting resources away from the poor. From this analysis, one can surmise that in a country like the debt-burdened Philippines, the disappointing stall in the decline in birth rates is in part due to the increasing insecurity of the poor whose lives have become even-less secure in the last decade.

Government policies directly affect the poor majority's access to land, and thus influence the peasant family's sense of security which plays a part in its child-bearing choices (see table 1). In many countries, including the Philippines, El Salvador, and Brazil, for example, agrarian policy is beholden to the most wealthy landowners. They have made sure to block reforms transferring land to the poorest peasants. In Brazil, for example, 224 large farms still control as much land as 1.7 million peasant families.[30] Given what we now know about how the insecurity of landlessness affects poor families' view of their need for children, it should come as no surprise that in such countries fertility remains high.

Honduras offers another illustration. After Haiti, it is the poorest country in the western hemisphere. In 1980, two-thirds of its national budget was devoted to economic and social programs and one-third to debt repayment and defense. But in the 1980s, Honduras became a central staging ground for U.S. military operations in Central America. Millions in U.S. military aid went to Honduras and the government was pressured to increase its own military expenditures. By 1984, Honduras' budget priorities were completely reversed: education, housing, health and other such programs received only a third of the budget. The rest went to debt and the military. Given the link between improved health and education and fertility, it is clear that the geopolitical strategy of a foreign power-diverting the Honduran government from social programs—is powerfully influencing its potential to reduce its high birth rate.[31]

We've only sketched some of the layers of decision making power shaping human reproductive life, but the reader might draw back with skepticism. Does not such a far-reaching approach confuse more than clarify—for could not virtually every economic, political, and cultural fact of life be squeezed into such a broad perspective?

Our response is that to achieve a holistic understanding one's view must necessarily be far reaching. But this does not mean that it is without coherency. The pivot on which our perspective turns is the concept of power, a concept that we have found woefully missing in the perspectives we earlier critiqued. Without such a concept, we believe it is impossible to understand the complex and interacting problems of poverty, hunger, and population, much less act effectively to address them.

1 "Children: A Cost to the Rich, A Benefit to the Poor," *The New Internationalist* (June, 1977): pp. 16-17, cited in Morley and Lovel, *My Name is Today,* pg. 34. For more detail on this particular study, see *Population and Development Review*, September, 1977.

2 Cain, M.T. "The Economic Activities of Children in a Village in Bangladesh," *Populations and Development Review* 3 (1977): pp. 201-228, cited in W. Murdoch, *The Poverty of Nations* (Baltimore: Johns Hopkins University Press, 1980), pg. 26.

3 Kamuzora, C. Lwechungura. "High Fidelity and the Demand for Labor in Peasant Economies: The Case if Bukoba District, Tanzania," *Development and Change 15,* no.1 (January, 1984): pp. 105-123.

4 "Hundred Dollar Slaves," *The New Internationalist* (October, 1986): pg. 9.

5 Caldwell, John C. *Theory of Fertility Decline* (New York: Academic Press, 1982), pg. 37, cited in Betsy Hartmann, *Reproductive Rights and Wrongs* (New York: Harper and Row, 1987), pg. 7.

6 *Theory of Fertility Decline,* op. cit.

7 Nag, M. and B. White and R. C. Peet. "An Anthropological Approach to the Study of the Economic Value of Children in Java and Nepal," *Current Anthropology* 19 (1978): pp. 293-306.

8 World Bank. *The World Development Report 1984* (New York: Oxford University Press, 1984), pg. 52.

9 Cain, Mead. "Fertility as an Adjustment to Risk," *Population and Development Review* 9, no. 4 (December 1983): pp. 688-701, especially pg. 699.

10 A recent study of Thailand found that large families are increasingly perceived as an economic burden in part because the cost of educating children has risen substantially. See Knodel et. al., "Fertility Transition in Thailand: A Qualitative Analysis," *Population and Development Review* 10, no. 2 (June 1984): pp. 297-328.

11 Merrick, Thomas. "Recent Fertility Declines in Brazil, Columbia, and Mexico," *World Bank Staff Working Paper* no. 692 (Washington, DC: World Bank, 1985)

12 India's Sixth Five Year Plan, pg. 374, cited in Sheila Zurbrigg, *Rakku's Story: Studies of Ill-Health and the Source of Change* (Madras, India: George Joseph Printing

Company, 1984), pg. 70.

13 Interview with two women doctors on family planning. *Connexions*, Summer/Fall, 1985, pg. 49.

14 *Poverty of Nations*, pg. 45.

15 Huston, Perdita. *Message from the Village* (New York: The Epoch B Foundation, 1978), pg. 119, cited in Hartmann, *Reproductive Rights and Wrongs*, pg. 48.

16 Jacobson, Jodi L. "Planning the Global Family," *Worldwatch Paper* no. 80, (Washington, DC: Worldwatch Institute, 1987), pg. 20.

17 Ibid., pg. 21.

18 Faruqee and Gulhati. *Rapid Population Growth in Sub-Saharan Africa, Issues and Policies*, World Bank Staff Working Paper no. 559 (Washington, DC: World Bank, 1983), pp. 48-52.

19 Ibid., pg. 54.

20 Mugabe, Sally. "High Fertility Hampers Women's Status," *Popline*, June, 1987, pg. 2. *Popline* is a publication of the World Population News Service.

21 Benjamin, Medea, ed. *Don't Be Afraid, Gringo: A Honduran Woman Speaks From the Heart* (San Francisco, CA: Food First Books, 1987).

22 *Message from the Village*, pg. 38.

23 *Reproductive Rights and Wrongs*, pp. 247-248.

24 Spretnak, Charlene. "The Population Bomb: An Explosive Issue for the Environmental Movement," *Utne Reader*, May/June, 1988, pp. 86-87.

25 World Resources Institute. *World Resources 1986* (New York: Basic Books, 1986), pg. 21.

26 Hartmann, Betsy. Personal correspondence, January, 1988.

27 Hartmann's *Reproductive Rights and Wrongs* is only one of the most recent expressions of this attempt to incorporate a gender-based analysis of power within a larger structural framework.

28 George, Susan. "Debt: The Profit of Doom," *Food First Action Alert* (San Francisco, CA: Institute for Food and Development Policy, 1988). See also George, *Fate Worse than Debt* (New York: Grove Press/Food First Books, 1988)

29 Sell, Ralph R. and Steven J. Kunitz. "The Debt Crisis and the End in Mortality Decline," *Studies in Comparative International Development*, 1987. Cited in George, *Fate Worse than Debt*, pg. 134.

30 Riding, Alan. "In Northeastern Brazil Poverty Cycle Goes On," *The New York Times*, May 3, 1988.

31 Danaher, Kevin and Phillip Berryman and Medea Benjamin. "Help or Hindrance: United States Economic Aid in Central America," *Food First Development Report* no. 1, (San Francisco, CA: Institute for Food and Development Policy, 1987), pp. 19-21.

From *Breakfast of Biodiversity: The Truth About Rain Forest Destruction* by John Vandermeer and Ivette Perfecto

Slicing up the Rain Forest on Your Breakfast Cereal

The eastern slopes of the Balva volcano catch water-laden trade winds from the Caribbean to create the climate of Costa Rica's eastern coast, location of some of the most beautiful rainforests in the world. Here you can experience that special feeling that inspires poets and explorers—from the myriad vegetative forms so evident even on first glance (figure 1) to the misty mornings that invoke mysterious feelings and bucolic images of paradise lost. The rain forest here, as elsewhere, is a collective human construct that sometimes serves as our mystical Garden of Eden, but is also a material collection of fabulous plants and animals, a natural construct of the high temperature and heavy rainfall of equatorial climates. The trade winds rise as they encounter the eastern seaboard and with their ascent they cool, condensing the water vapor they borrowed during their voyage across the Caribbean. The consequent rains collect in several basins and come together roughly at the town of Puerto Viejo, continuing northward to empty into the San Juan River, the border with Nicaragua. Thus is the region known as the Sarapiqui (*sah rah pick ee*, with an accent on the *ee*), site of the world's most famous rainforest conservation areas (figure 2).

Streaming into the area to partake of the breathtaking beauty of the natural world in this 'one of the last' pristine places in the world, are biologists, ecologists, and ecotourists, spending their grant money or retirement savings to visit the 'heritage of humanity.' It is hardly necessary to repeat the clichés any longer—tropical rainforests cover only seven percent of the earth's surface yet harbor at least fifty percent of the world's plant and animal species (the earth's biodiversity); they are the lungs of the world, eating away at the excessive carbon dioxide we have excreted from our industrial metabolism; they are the source of foods and pharmaceuticals, bananas and Brazil nuts, chocolate, cashews, coffee and cocaine, cortisone and quinine. They

Figure 1

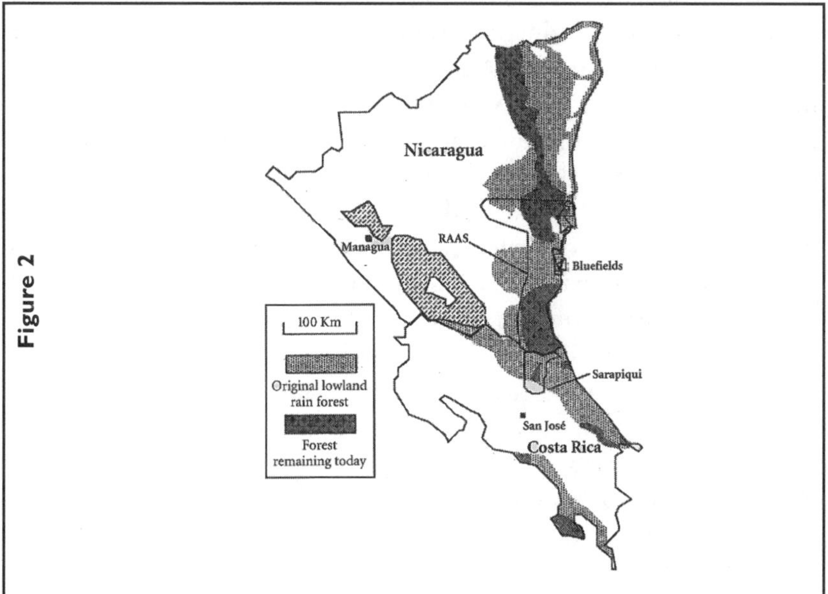

Figure 2

are also beautiful! The aesthetics of these forested lands cannot be overestimated, and the sense of wonder one experiences walking through this cradle of biodiversity cannot be expressed in words.

But as anyone visiting the Sarapiqui can readily see, all is not well in this Garden of Eden. Certainly, it remains beautiful inside of the conservation areas. The problem is outside those areas. And the problem is the same one Costa Rica has had ever since Minor C. Keith built his famous railroad, and helped found the United Fruit Company in 1898. The problem is the banana. Currently at least five major banana companies are converting vast acreage in the area to banana plantations, thus threatening both directly and indirectly the rain forests we so revere. Those same biologists, ecologists, and ecotourists, who love the rain forest when they're in Costa Rica, also love to slice bananas onto their cereal in the morning. And with our penchant for viewing the world in isolated little disconnected fragments, it is apparently difficult for us all to see the connection between the knife that slices the banana into our cereal bowl, and the chain saw that slices tree trunks onto the rain forest floor.

Not so long ago, environmental activists in the developed world became aware of the so-called 'hamburger connection.' Central American rain forests were being cut down at an alarming rate to make way for the production of low quality beef to supply the fast-food industry in the First World. Stop eating fast-food hamburgers, the argument went, and you would reduce the demand to cut down more forest. The banana expansion currently underway in Central America has been likened to this hamburger connection. But the whole argument surrounding the hamburger connection was flawed, and an attempt to construct the same argument for bananas would simply repeat that flaw. The expansion of bananas, like the expansion of pasture for beef production, is a tangled web of subtle connections. Tweak the web at one point and it reverberates all over, sometimes in unexpected ways. It's necessary to understand the nature of the connections in this 'web of causality.'

The transformation currently underway in the Sarapiqui is neither unprecedented nor unique, which makes it a useful example. Similar patterns occur throughout the tropics. Sometimes the pattern involves bananas, sometimes cattle, maybe citrus, African oil palm, or rubber trees—a variety of commodities, similar politically if quite distinct biologically. The pattern is a six stage process: 1. Visionary capitalists identify an economic opportunity for the market expansion of an agricultural product. In this case, the opportunity is the opening up of markets in Eastern Europe and the unification of

Western Europe, and the product is bananas. 2. They purchase (or steal, or bribe into a government concession) some land, including land that may contain rain forest, which is promptly cut down. 3. They import workers to produce the product (in this case workers come from all over Costa Rica and even from Nicaragua). 4. After a period of boom the product goes bust on the world market, which means scaling back production, which in turn means releasing a significant fraction of the work force. 5. The newly unemployed work force seeks and fails to find employment elsewhere, and must seek land to grow subsistence crops to tide themselves over until other work can be found. And finally, 6. The only place the now unemployed workers can find land no one will kick them off of is in the forest, which means yet more of the rain forest is converted to agriculture.

In this way Costa Rica, one of the world's showcases of conservation, is currently promoting a policy that actually encourages rain forest destruction. That is interesting by itself, but this specific example is not as important as the general idea it highlights. The crop in this instance happens to be bananas, but the general pattern is all too common.

Costa Ricans and Their Love/Hate of Bananas

An afternoon in Puerto Viejo, the little town located near the confluence of the rivers draining the Barva volcano, reveals what might surprise a European or North American tourist. Despite the fact that, given their history, Costa Ricans understandably love to hate bananas, it is difficult to find anyone in town who does not fully support the massive banana expansion that is currently underway. Furthermore, the government, both local and national, is encouraging the expansion with a vigor normally associated with a depressed Northern U.S. city courting a big assembly plant. (*You want no unions? You got it. You want tax breaks? Just say how much. You want license to pollute? Smoke your heart out. But please, just locate here.*) Costa Rica is as debt-laden as the rest of Latin America,[1] and needs all the money it can get just to service its debt. The expansion of bananas is one way to make money. Thus, despite the recognition that social and environmental problems will inevitably come along with the bananas, the vast majority of Costa Ricans, both in the Sarapiqui and elsewhere, welcome the current expansion. A small group of Costa Rican environmentalists are protesting, but they are overwhelmed by more powerful interests.

Of the approximately quarter of a million hectares[2] in the Sarapiqui valley, some 50,000 are devoted to biological preserves.[3] Another 100,000 hectares are in the legal hands of small peasant farming communities. The rest (approximately 100,000 hectares) is a mosaic of small farms, most without title to the land; secondary and old-growth forest; cattle pasture; and an occasional sizable ornamental-plant or fruit plantation.[4]

In the periphery of the valley lies an extensive banana plantation owned by the Standard Fruit Company, a major employer in the region for the past quarter century. The history of Standard Fruit provides an example of what might be expected on a larger scale in the future. Tales are common of pesticide abuse, waste-dumping into local waterways, deforestation, and the massive social problems normally associated with a frontier area. Best documented is the celebrated case in which Standard Fruit was accused of negligence in its use of DBCP (dibromochloropropane), a popular fungicide.[5] During the early 1970s more than 2,000 banana workers were rendered sterile by this poison. They are currently suing Standard Fruit in a United States court. This and other past records indicate that historically the banana companies have not accepted responsibility for the health and safety of their workers, the community, or the environment. With the current massive banana expansion there is no reason to assume that these adverse environmental, social, and health effects will not be repeated on an even larger scale.

History is perhaps even more ominous when we examine another long-term pattern evidenced by the Standard Fruit operation in the area. Standard Fruit employs workers who migrate to the Sarapiqui from other parts of Costa Rica. These workers are retained as long as the market for bananas is sufficiently robust, but are let go when sales slacken. The laid-off workers are mainly rural people, former peasants drawn into rural-wage labor. In past decades the ebb and flow of the banana business has created critical periods in which many workers were laid off and forced to fend for themselves. These layoffs were a natural product of the world economic system, due both to fluctuating banana prices, and to the very existence of a two-part economy—export bananas on the one hand and worker/farmer on the other. There is little employment opportunity in the area other than the banana companies, so when workers are laid off they must either migrate to the city, adding to the growing shanty towns, or they must turn to farming. In order to farm they have to find a homestead. Sometimes that small piece of land is in a rain forest. Other times it is a small corner of some large absentee landowner's cattle ranch, in which case, depending on complicated criteria, either the homesteading family is eventually forcibly evicted, or the state agrarian reform institute adjudicates a 'fair' purchase for the peasant family.

The past thirty or forty years have seen this arrangement persist, with rain forest cover in the region plummeting from almost ninety percent in 1950 to approximately twenty-five percent today. Only a small portion of the remaining twenty-five percent is not part of one of the four large biological reserves.

Loggers, Farmers, and Banana Companies—a Rich History[6]

This pattern, so readily observable today, is set in a rich ecological history beginning well before the current crisis. Early in the century, extensive river systems were used to transport both logs and bananas. Logging was a rather small-scale operation by modern standards, but had the effect of drawing workers to the area and creating pathways into the forest. Since only a handful of the many species of trees in the rain forest were actually valuable, it was necessary to scout out and then cut a path to the valuable trees, and after cutting them down to haul them out with teams of oxen or horses. By the 1930s, the land along almost all the rivers was deforested and planted with bananas, while the surrounding forest was riddled with trails made for dragging logs. The logging process intensified in the late 1940s and 1950s when machinery was brought into the area, and a complex network of logging roads crisscrossed what forests remained after the inroads already made by the banana plantations. Men who originally came to the area to work in the logging industry used these roads to gain access to logged areas and frequently established homesteads. Former banana workers did the same thing.

In the late 1940s everything changed in the Sarapiqui, as it did throughout the Atlantic coast of Costa Rica. Devastating fungal diseases routed the banana industry. The extensive plantations of the United Fruit Company and a variety of independent producers were decimated by this disease. No cure could be found and the company moved its entire operation to the other side of the mountains, where the disease had not been established. It was not until the mid 1950s that a genetic variety of banana that was resistant to the disease was developed, thus enabling the Standard Fruit Company to establish its plantations in the area in the late 1950s.

Bananas Today

Now the plot is thickening. In anticipation of an expected surge in the demand for bananas (the anticipated result of opening markets in Eastern Europe, and the economic unification of Western Europe), five or six major banana companies have been purchasing large expanses of land and expanding banana production accordingly. The area planted to banana rose from 20,000 hectares in 1985 to 32,000 hectares in 1991, and visits to the area in 1991 and 1992 revealed intense activity in setting up new banana plantations throughout the valley. As much as 45,000 hectares are expected to be in bananas by the end of 1995.[7] As had happened in the past, workers were drawn from all over the country. But breaking with past traditions, this time there apparently were not enough Costa Rican workers to do the necessary work, and workers were also attracted from Nicaragua, Panama, and even Honduras. It appears likely that within two years most if not all of the arable land not currently in either biological preserve or organized peasant agricultural communities, will become banana plantations.

A variety of factors make Costa Rica, and particularly the *Sarapiqui* basin, an especially attractive target of the banana companies. Notably, the infamous *Solidarista* movement has destroyed all union activity in the area. Some ten years ago, this church-based, U.S. supported, anti-union movement systematically moved into the Sarapiqui valley to replace all banana labor unions with a new concept for worker organization. *Solidarista* dogma outlaws strikes, does not recognize the right of workers to collectively bargain, and seeks to attract workers with frivolous benefits such as clubhouses and soccer fields. With massive funding from the Association for Free Labor Development, the international wing of the AFL-CIO long suspected to have CIA ties, democratic labor unions were systematically attacked throughout Costa Rica. The campaign was especially effective in the Sarapiqui where union membership now stands at zero and company officials proudly proclaim that no union people are able to find jobs.[8] A local political official told us in 1991 that the planned banana expansion would have been impossible without the existence of the *Solidarista* organizations.

A second important factor is the willingness of the Costa Rican government and its partner, the United States of America, to create infrastructural conditions, which favor the banana companies. Roads are being constructed, bridges are being built, hospitals and schools are planned, all for the purpose of creating an attractive infrastructure for the banana companies. The U.S. Army Corps of Engineers was enlisted in this effort In a 1992 program called 'Bridges for Peace,' Army Corps engineers built roads and bridges in the Sarapiqui. A cynical U.S. serviceman told us the program has been unofficially dubbed 'Bridges for Bananas,' as the construction so obviously focuses on improving infrastructure for the export of bananas. U.S. Army engineers built many of the roads and bridges that today carry the logs of the cut rain forest, and tomorrow will carry the harvest of the international banana companies. Indeed, with the infrastructure provided by U.S. taxpayers at the request of the Costa Rican government, from roads and bridges to the *Solidaristas*, from the 'converted' rain forest to new social infrastructure, investment opportunities look good—that is, if you are a banana company.

But the banana companies, mindful that their operations might attract outside attention, were prepared to pay 'expert' scientists to mollify the public. Corporación Bananera Nacional (CORBANA), was formed in 1990. Some twenty years earlier a smaller national effort, Asociación Bananera Nacional (ASBANA) had been formed by the Costa Rican government for the purpose of developing technical assistance for small producers of bananas in the country. Operating on a tiny budget, this small research operation persisted until two years ago, when the rush to privatization caught up with it and ASBANA changed to CORBANA and began to receive money directly from the banana companies. For every box of bananas exported, each company pays a fee to support the research efforts of CORBANA. Theoretically CORBANA conducts research aimed at making banana production more environmentally friendly. This research was to include proposed projects on using biotechnology to develop strains of bananas resistant to pests, development of organic fertilizers, and extensive surveys of fauna in the banana plantations (ostensibly to monitor the effects of the plantation on wildlife). However, in a visit to the CORBANA facility we observed very impressive projects on soils, plant diseases, parasites, and drainage, but none of the celebrated studies to promote environmental friendliness.

But we repeat and emphasize that the expansion of bananas is viewed as a positive event by nearly everyone in the Sarapiqui, and by most observers in the entire country. Local workers and peasants see jobs being created, local merchants see a potential surge in business, and local politicians see an increase in their power base. The Costa Rica government itself promotes the expansion since it sees the increased tax revenues as helping to pay debt service on its tremendous international debt. The accepted fact that almost all profit from banana farming will leave the country seems of little local concern. This is perhaps understandable given the state of the local and national economy. But less comprehensible is that segments of the international 'conservation' community have come on board, and either retain a calculated obliviousness to what is going on, or actively pursue a neutral position. Significant, yet weak opposition is coming from a small, loosely-structured local conservation movement, composed of Costa Ricans, and organized without the help of the international conservation community They are fighting what appear to be insurmountable odds.

Bananas Tomorrow

It is not difficult to predict what is likely to happen next. Unless history has significantly reversed free market laws, banana prices will fluctuate on the world market as they always have. Furthermore we can expect the banana companies to do what they are supposed to as good managers: reduce the cost of labor, which they will do at points of economic downturn by laying off workers, just as they always have in the past. But this time where will those former workers go? In the past there was always that mosaic of small farms, large cattle pastures, second growth, and primary forest between the organized agricultural settlements and the biological preserves. Such areas were normally considered externalities, sopping up the overflow of humanity that could not be accommodated in times of crisis. Now, however, that area will be taken up by banana plantations, and the only remaining area not already devoted to some form of agriculture (and therefore 'available'), will be within the four biological preserves in the area. It would take enormous naïveté to suppose that when their survival is at stake, these landless peasants will not begin cutting forest in the biological preserves.

This example illustrates the dynamics that occur, with different details of course, throughout Central America and in much of the rest of the world. Because of the nature of the world economic system, Costa Rica really has no choice but to promote the expansion of bananas. Costa Rica's international debt, accumulated because of its position in the world economy, and its need for the expansion of international capital, require that it seek tax revenues however it can. The banana companies themselves (at least one of which is Costa Rican) continue to play their historical role as international accumulators of capital, and temporary employers of peasants, thus maintaining the dysfunctional two-part economy. Peasants continue arriving from other parts of the country and now, even from Nicaragua, seeking jobs and the 'good life' and willing to accept minimal conditions—but since the *Solidarista* movement destroyed the unions, they are without significant political representation. The stage was initially set by the loggers with their systems of logging roads, and the first wave of banana plantations with their periodic layoffs that forced peasants into the forests. If the process continues with this basic overall structure, which we see no reason to doubt, there is little hope in the long run, even for those rain forests under protected status.

Current rumors in the area indicate that the long-running Standard Fruit Company operation in Rio Frio, will soon abandon all of their plantations near the Sarapiqui. Despite what some promoters claim, banana plantations do not last forever. A variety of ecological forces eventually catch up with such intensive production, and the plantations must be abandoned. Will the legacy of bananas leave the area with much degraded conditions of production, as has happened repeatedly in the past? Who will bear the costs of restoring conditions to their original state, if that will even be possible? And who will share concern with the thousands of rural people, deprived of their land and their livelihood with no place to go? Who will tell them that the rain forest is more important than feeding their families?

Viewing the Problem from Various Perspectives

In the midst of these dramatic events, an internationally recognized ecologist gave a public lecture at a local ecotourism center in the Sarapiqui, claiming that recent deforestation in the area was due to the inevitable march of Malthusian reality. He claimed that overpopulation was causing the destruction of the forest. In a sense, of course, such an observation is trivially true. There undoubtedly is an overpopulation of banana companies, an overpopulation of former banana workers looking for land, an overpopulation of adventurers seeking their fortunes in a new frontier zone, an overpopulation of greedy people and institutions, and even an overpopulation of ecotourists from Europe and the United States.

When this expert ecologist declared that a Malthusian crunch was the root of the problem, he was actually implying something rather different—that the pressure of having too many children, the birth rate of the population, is the real problem. This point of view implies that the main solution to the problem is birth control. It further implies that this is a sufficient solution, that it is useless to do anything other than promote birth control, and that as long as population densities remain as they are, the pattern of deforestation will continue.

An alternative viewpoint, expressed by a local conservationist, is that avaricious banana companies are cutting down rain forests because they are hungry for profits. They will stop at nothing to satisfy their need to accumulate ever greater quantities of capital and the forests will continue to disappear as long as banana companies are allowed to continue their greedy operations. This also is a distinct point of view. It implies that the only solution to the problem is to eliminate the capitalist. It further implies that this is a sufficient solution, that it is useless to do anything other than 'smash capitalism,' and that as long as the need to accumulate capital remains, the pattern of deforestation will continue.

These points of view are prisms through which the facts of the matter may be interpreted. They both encourage a single focus for solution: stop population growth or smash capitalism. We believe that they both are right in a very limited sense. But we also believe that they are both wrong in a broader, and more practical sense.

Ultimately each of these prisms focuses on a single thread in a fabric of causality. Eliminating one thread will not eliminate the problem. The problem is the fabric itself. The proper means of understanding the situation then, is to look at the complicated way that various forces are interdependent, especially focusing on the way countervailing tendencies are resolved. The approach we take may, at first glance, seem as narrow as the approaches of those who advocate population reduction, or smashing capitalism. We assert that food insecurity is the root cause of deforestation. It is a critical thread in the fabric of causality. We take this approach for two reasons. First, we wish to provide an antidote to the simplistic views that either overpopulation or avaricious capitalism cause deforestation. Second, we argue that given the ultimate goal of reworking the entire fabric of causality, the place to begin that process is with food security. We do not argue that providing peasants with food security will stop deforestation *per se*, but rather that beginning the political process of reorganizing socioeconomic-ecological systems by examining questions of food security, will force both analysis *and* practice into the realms ultimately necessary for the resolution of this issue. When neo-Malthusians suggest there are too many people for the land base, the food security position reveals several important particulars: that peasants seek land to feed their families, not because there are too many of them or too little land (at least right now), but because available land is occupied by other activities. Our orientation will also reveal that the techniques for sustainable agriculture in that zone have been replaced with destructive, chemically-based ones, and further, that the legal status of most peasants is 'landless' even when they clearly occupy a piece of land. When radicals purport that avaricious capitalism causes deforestation, the food security position shows that the evolution of modern agriculture has created international structures that force even progressive governments like Cuba to invite those greedy capitalists into their economies. The international order, which causes food insecurity in the developing world, is implicated in a chain of events that ultimately leads to the transformation from workers to peasants who must seek out rain forest land to farm in order to provide food for their families.

We do not wish to leave the impression that food insecurity is just another mechanistic cause to which the problem of rain forest destruction may be reduced. It is clearly not. But as a mode of analysis, examining food insecurity will cause us to deal with the entire complex web of ecological, sociological, economic, and political issues on which the poisonous spider of rain forest destruction crouches.

Two Models for Saving the Rain Forests

Current events in the Sarapiqui region are not unique. Tropical rain forest areas around the globe are experiencing similar complex socioeconomic forces that threaten to continue or even accelerate the destruction of this most diverse of all ecosystems. In all of these areas there has been some reaction from local and international concerns. Unfortunately, much of this response is misdirected because it is based on a distorted image of the facts, and on an implicit ideology—what can be called the mainstream environmental movement approach—which allows only a narrow range of possible courses. We feel there is an alternative philosophical approach, the political ecology strategy, which emphasizes basic issues of security: security of land ownership and the consequent ability to produce food for local consumption.

The mainstream environmental movement has raised large sums of money to purchase and protect islands of rain forest with little concern for what happens between those islands, either to the natural world or to the social world of the people who live there. We doubt this strategy has much chance of succeeding. It is likely that in the short term the landscape will be converted into isolated islands of tropical rain forest, surrounded by a sea of pesticide-drenched modern agriculture, underpaid rural workers, and masses of landless peasants looking for some way to support their families. The long term prospects, however, are worse, as the example of the Sarapiqui suggests.

Our alternative, the political ecology strategy, emphasizes the land and people *between* the islands of protected forest, and has greater credibility because of its willingness to see some of the interconnections in this complicated system. This point of view has been variously known as *ecological development, sustainable development,* or *eco-development,* though all of these terms have been cynically adopted by even the most environmentally destructive agencies.

Whatever we call it, this approach views the problem properly as a landscape problem with forests, forestry, agroforestry, and agriculture as interrelated land-use systems, and seeks to develop those land-use systems so that conditions of production according to the needs of the local population may be maintained. The political ecology strategy challenges non-sustainable development projects, such as modern banana plantations, and seeks to organize people to oppose ecologically and socially damaging development.

These two points of view lead to quite different projections of what the future rain forest areas of Central America might look like. If the mainstream position remains dominant, we expect to see, in the short term, a sea of devastation with islands of pristine rain forest, and in the long term, nothing but a sea of devastation. The political ecology point of view envisions a mosaic of land-use patterns: some protected natural forest, some extractive reserve, some sustainable timber harvest, some agroforestry, some sustainable agriculture, and, of course, human settlements. This mosaic would be sustainable over time.

But is this a practical vision in the real world? The decision to promote bananas in the Sarapiqui can hardly be faulted on 'modern' economic grounds. Sadly, if national and international commitment to the archaic economics of Adam Smith and the IMF persists, we fear continuing destruction of the rain forests and the deterioration of the lives of the people of Sarapiqui. The alternative requires a radical rethinking of what sorts of economic and political arrangements are to be tolerated, the sort of rethinking that can get you in trouble in Central America, the sort of rethinking that may even challenge the idea that it is our inalienable right to slice bananas onto our breakfast cereal.

Costa Rica, Bananas, and a General Pattern

The case study elaborated in this chapter is typical. Granted, there are cases in which rain forests are being cut with a profoundly different logic (several areas in Southeast Asia and much of the Amazon), but both historical and contemporary patterns the world over reflect the basic paradigms seen in this example. The details vary, but the underlying logic is remarkably consistent.

Costa Rica has been held up as one of the world's best examples of rain forest conservation. Its internationally recognized conservation ethic, its position of relative affluence, its democratic traditions, the remarkable importance of ecotourism to its national economy, its willingness to adopt virtually any and all programs of conservation promoted by western experts, make it the most likely place for the success of the traditional model of rain forest conservation. The fact that the model has been an utter failure in Costa Rica, where it had the greatest chance of success, calls the model itself into serious question.

Stopping individual logging companies and avaricious agroexporters can be only a small part of the solution; basic questions of land and food security are the most central component of any potentially effective political strategy. Such political strategies begin to look more like past political strategies, which helped stop the war in Vietnam and curtailed U.S. intervention in El Salvador and Nicaragua. Only by uniting with political forces that have similar fundamental goals can the future of the world's rain forests be brightened.

1 While Costa Rica's short term debt decreased from $575 million in 1980 to $341 million in 1992, probably due to the particular political situation of the 1980s, its long term debt actually increased from $2,112 million to $3,541 million during the same time period, and increase of sixty-eight percent. This is not as bad as other Latin American countries. El Salvador increased its debt by 208 percent during the same period, and Mexico managed a 101 percent increase. Compared to other Latin American countries, Costa Rica is doing better today than it was in 1980, but it is still one of the most debt-laden of them all—worse than Mexico. Only Panama, Argentina, Venezuela, and Nicaragua are worse off.

2 A hectare is equal to 2.47 acres.

3 The La Selva Research Station, Braulio Carrillo National Park, Tortugero National Park, and the corridor between La Selva and Braulio Carrillo.

4 Butterfield, R. "The Regional Context: Land Colonization and Conservation in Sarapiqui," and Montagnini, F. "Agricultural Systems in the La Selva Region," L.A. McDade, K.S. Bawa, H.A. Hespenheide, and G.S. Hartshorn, eds., *La Selva: Ecology and Natural History of a Neotropical Rainforest* (Chicago, IL: University of Chicago Press, 1994).

5 Thrupp, L.A. "The Human Guinea Pigs of Rio Frio: Standard Fruit Keeps Its Eye On the Bottom Line," *The Progressive*, April 1991.

6 Most of this history comes from conversations with the late Raphael Echeveria, a long-time local resident and former banana worker and timber cruiser. Information

is also derived from Danilo Brenes and Hector Gonzales. The history is effectively presented by Butterfield's "The Regional Context: Land Colonization and Conservation in Sarapiqui."

7 Lewis, S.A. "Banana Bonanza: Multinational Fruit Companies in Costa Rica," *The Ecologist*, issue 22, 1992.

8 In 1989 the president of the *Solidarista* organization at the Standard Fruit Company plantations in Rio Frio proudly told Vandermeer that "anyone trying to organize a union will be fired. We know who most of them are and we won't let them get jobs here in the first place." (Paraphrased)

Part Two

The American Connection

The Way to Our Dinner Table

For most Americans, food comes from the supermarket. We don't really know how it gets there—how it is produced, processed, and traded—nor do we much care. Occasionally, when there's a scare about pesticide residues or bacterial contaminants, we worry about whether to believe the assurances of agribusinesses and governments that the food on our table is truly safe. But by and large, what happens between the soil and the grocery store has not been our concern.

The great achievement of Food First, ever since Frances Moore Lappé's first best-selling book, *Diet for a Small Planet*, has been to convince Americans that it does matter. What happens along the complicated network of paths from field to table is of vital importance to our health, our pocketbooks, and our environment. The resources that are used to make our food—feed crops, pesticides, energy, water, and packaging—and the people who labor to make it—family farmers, farm workers, meat packers and truck drivers—are all linked together in the system which ultimately puts the steak on our plate. Our decisions about what we eat travel back through that system as signals and incentives which affect whether aquifers run dry, farm workers are poisoned, or family farmers go broke.

That system, as Lappé showed in *Diet for a Small Planet*, is incredibly wasteful. Modern livestock production as practiced in the United States and exported throughout the world, is based on cheap grains, cheap energy, cheap water, and valuing the earth cheaply. It uses grain, fuel, water, and soil to produce meat in a way that consumes far more resources than necessary. Carnivorous Americans are a much greater burden to the planet than their herbivorous cousins.

But Lappé saw that the wastefulness of our food system was far more than an argument for vegetarianism. She recognized that economies, not just eating habits, must be transformed, and that the only way to make that transformation is through political action. *Diet for a Small Planet* led to founding of Food First/Institute for Food and Development Policy and to the realization that, as in ecosystems, every part of the world food system is connected to each other.

Two vital links in that web are the export of pesticides from industrialized countries and their use on foods which are then exported back from the Third World. This is the *Circle of Poison.* Even pesticides that are banned as dangerous in the United States can end up on the vegetables Americans eat. But the concern goes far beyond the danger to consumers' health, for the circle causes poor farm workers and their families to be poisoned by the thousands each year. Despite pesticide companies' excuses, their exports do not help feed the hungry. On the contrary, by selecting for resistant pests and eliminating beneficial species, they actually make food production in Third World countries more difficult. As these same pesticide companies move into the seed business, they limit farmers' abilities to produce even more—even making it illegal for them to save their seed to plant the following year.

In the meat industry, corporate power is increasing at the expense of farmers, workers and consumers alike. Vertical integration, described by Cooper, Rosset, and Bryson in "Warning: Corporate Meat and Poultry May Be Hazardous to Workers, Farmers, the Environment, and Your Health," drives family farms out of business, endangers both workers and consumers, and damages our environment. Governments have allowed and even subsidized this trend.

Marty Strange's *Family Farming: A New Economic Trend* shows how the same economic trends which threaten field and factory workers in agriculture, are also driving family farms out of business. Saturated global markets and costly new technologies force farmers to become entrepreneurial competitors in a race which fewer and fewer can survive. The only winners are the giant agribusiness firms which can buy farmers' outputs more cheaply and sell them their inputs more dearly. As Strange makes clear, these fundamental problems were ignored by the traditional debate about farm support programs. Whether under Democratic price-subsidy programs in the 1980s or Republican free-market policies as in the 1996 farm bill (ironically called the 'Freedom to Farm' Act), the consequences have been the same: glutted markets and depressed prices. Working to raise the prices that farmers receive is not enough to save them from ruinous competition; only a change in the economic organization of agriculture can save the family farm from bankruptcy.

As the readings in this section illustrate, the world food problem is not just a question of starvation in poor countries. In the industrialized world, the system is failing us, threatening our health, our environment, our jobs, and our farms. Agribusiness extends all across the planet; the struggle against it must also be global.

From *Diet for a Small Planet* (20th Anniversary edition) by Frances Moore Lappé

Like Driving a Cadillac

A few months ago a Brazilian friend, Mauro, passed through town. As he sat down to eat at a friend's house, his friend lifted a sizzling piece of prime beef off the stove. "You're eating that today," Mauro remarked, "but you won't be in ten years. Would you drive a Cadillac? Ten years from now you'll realize that eating that chunk of meat is as crazy as driving a Cadillac."

Mauro is right: a grain-fed-meat-centered diet is like driving a Cadillac. Yet many Americans who have reluctantly given up their gas-guzzling cars would never think of questioning the resource costs of their grain-fed-meat diet. So let me try to give you some sense of the enormity of the resources flowing into livestock production in the United States. The consequences of a grain-fed-meat diet may be as severe as those of a nation of Cadillac drivers.

A detailed 1978 study sponsored by the Departments of Interior and Commerce produced startling figures showing that *the value of raw materials consumed to produce food from livestock is greater than the value of all oil, gas, and coal consumed in this country.*[1] Expressed another way, one-third of the value of all raw materials consumed for all purposes in the United States is consumed in livestock foods.[2]

How can this be?

The Protein Factory in Reverse

Excluding exports, about one-half of our harvested acreage goes to feed livestock. Over the last forty years the amount of grain, soybeans, and special feeds going to American livestock has doubled. Now approaching 200 million tons, it is equal in volume to all the grain that is now imported throughout the world.[3] Today our livestock consume ten times the grain that we Americans eat directly[4] and they outweigh the human population of our country four to one.[5]

These staggering estimates reflect the revolution that has taken place in meat and poultry production and consumption since about

1950. First, beef. Because cattle are ruminants, they don't need to consume protein sources like grain or soybeans to produce protein for us. Ruminants have the simplest nutritional requirements of any animal because of a unique fermentation 'vat' in front of their true stomach. This vat, the rumen, is a protein factory. With the help of billions of bacteria and protozoa, the rumen produces microbial protein, which then passes on to the true stomach, where it is treated just like any other protein. Not only does the rumen enable the ruminant to thrive without dietary protein, B vitamins, or essential fatty acids, it also enables the animal to digest large quantities of fibrous foodstuffs inedible by humans.[6]

The ruminant can recycle a wide variety of waste products into high-protein foods. Successful animal feeds have come from orange juice squeeze remainders in Florida, cocoa residue in Ghana, coffee processing residue in Britain, and bananas (too ripe to export) in the Caribbean. Ruminants will thrive on single-celled protein, such as bacteria or yeast produced in special factories, and they can utilize some of the cellulose in waste products such as wood pulp, newsprint, and bark. In Marin County, near my home in San Francisco, ranchers are feeding apple pulp and cottonseed to their cattle. Such is the 'hidden talent' of livestock.

Because of this 'hidden talent,' cattle have been prized for millennia as a means of transforming grazing land unsuited for cropping into a source of highly usable protein, meat. But in the last forty years we in the United States have turned that equation on its head. Instead of just protein factories, we have turned cattle into protein disposal systems, too.

Yes, our cattle still graze. From one-third to one-half of the continental land mass is used for grazing. But since the 1940s we have developed a system of feeding grain to cattle that is unique in human history. Instead of going from pasture to slaughter, most cattle in the United States now first pass through feedlots where they are each fed over 2,500 pounds of grain and soybean products (about twenty-two pounds a day) plus hormones and antibiotics.[7]

Before 1950 relatively few cattle were fed grain before slaughter,[8] but by the early 1970s about three-quarters were grain-fed.[9] During this time, the number of cattle more than doubled. And we now feed one-third more grain to produce each pound of beef than we did in the early 1960s.[10] With grain cheap, more animals have been fed to heavier weights, at which it takes increasingly more grain to put on each additional pound.

In addition to cattle, poultry have also become a big consumer of our harvested crops. Poultry can't eat grass. Unlike cows, they need a source of protein. But it doesn't have to be grain. Although prepared feed played an important role in the past, chickens also scratched the barnyard for seeds, worms, and bits of organic matter. They also got scraps from the kitchen. But after 1950, when poultry moved from the barnyard into huge factory-like compounds, production leaped more than threefold, and the volume of grain fed to poultry climbed almost as much.

Hogs, too, are big grain consumers in the United States, taking almost a third of the total fed to livestock. Many countries, however, raise hogs exclusively on waste products and on plants which humans don't eat. When Nobel Prize winner Norman Borlaug heard that China had 250 million pigs, about four times the number here, he could hardly believe it. What could they possibly eat? He went to China and saw "pretty scrawny pigs." Their growth was slow, but by the time they reached maturity they were decent-looking hogs, he admitted in awe. And all on cotton leaves, corn stalks, rice husks, water hyacinths, and peanut shells.[11] In the United States, hogs are now fed about as much grain as is fed to cattle.

All told, each grain-consuming animal 'unit' (as the Department of Agriculture calls our livestock) eats almost two and a half tons of grain, soy, and other feeds each year.[12]

What Do We Get Back?

For every sixteen pounds of grain and soy fed to beef cattle in the United States, we only get one pound back in meat on our plates.[13] The other fifteen pounds are inaccessible to us, either used by the animal to produce energy or to make some part of its own body that we do not eat (like hair or bones) or excreted.

To give you some basis for comparison, sixteen pounds of grain has twenty-one times more calories and eight times more protein—but only three times more fat—than a pound of hamburger.

Livestock other than cattle are markedly more efficient in converting grain to meat, as you can see in Figure 1; hogs consume six, turkeys four, and chickens three pounds of grain and soy to produce one pound of meat.[14] Milk production is even more efficient, with less than one pound of grain fed for every pint of milk produced. (This is partly because we don't have to grow a new cow every time we milk one.)

Now let us put these two factors together: the large quantities of humanly edible plants fed to animals and their inefficient conversion into meat for us to eat. Some very startling statistics result. If we exclude dairy cows, the average ratio of all U.S. livestock is seven pounds of grain and soy fed to produce one pound of edible food.[15] Of the 145 million tons of grain and soy fed to our beef cattle, poultry, and hogs in 1979, only 21 million tons were returned to us in meat, poultry, and eggs. *The rest, about 124 million tons of grain and soybeans, became inaccessible to human consumption.* (We also feed considerable quantities of wheat germ, milk products, and fishmeal to livestock, but here I am including only grain and soybeans.)

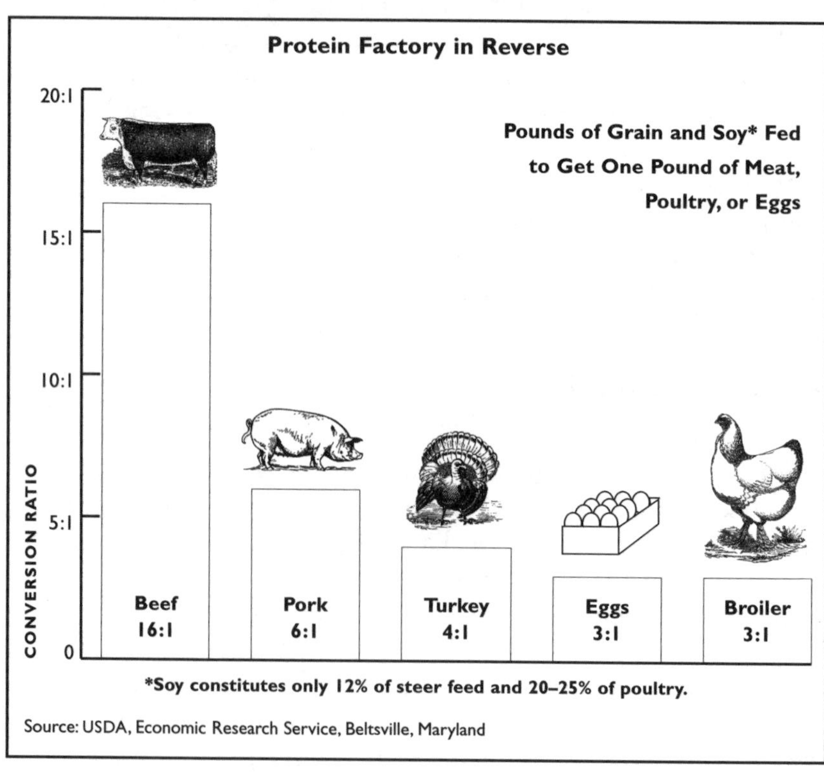

To put this enormous quantity in some perspective, consider that 120 million tons is worth over $20 billion. If cooked, it is the equivalent of one cup of grain for every single human being on earth every day for a year.[16]

Not surprisingly, *Diet for a Small Planet*'s description of the systemic waste in our nation's meat production put the livestock industry on the defensive. They even set a team of cooks to work to prove the recipes unpalatable! (Actually, they had to admit that they tasted pretty good.)

Some countered by arguing that you get *more* protein out of cattle than the humanly edible protein you put in! Most of these calculations use one simple technique to make cattle appear incredibly efficient: on the 'in' side of the equation they included only the grain and soy fed, but on the 'out' side they include the meat put on by the grain feeding *plus* all the meat the animal put on during the grazing period. Giving grain feeding credit for all of the meat in the animal is misleading, to say the least, since it accounts for only about forty percent. In my equation I have included only the meat put on the animal as a result of the grain and soy feeding. Obviously ail the other meat, put on by forage, would have been there for us anyway—just as it was before the feedlot system was developed. (My calculations are in note 13, so you can see exactly how I arrived at my estimate.)

The Feedlot Logic: More Grain, Lower Cost

On the surface it would seem that beef produced by feeding grain to livestock would be more expensive than beef produced solely on the range. For, after all, isn't grain more expensive than grass? To us it might be, but not to the cattle producer. As long as the cost of grain is cheap in relation to the price of meat, the lowest production costs per pound are achieved by putting the animal in the feedlot as soon as possible after weaning and feeding it as long as it continues to gain significant weight.[17] This is true in large part because an animal gains weight three times faster in the feedlot on a grain and high-protein feed diet than on the range.

As a byproduct, our beef has gotten fattier since the more grain fed, the more fat on the animal. American consumers have been told that our beef became fattier because we demanded it. Says the U.S. Department of Agriculture: "most cattle are fed today because U.S.

feed consumers have a preference for [grain]fed beef."[18] But the evidence is that our beef became fattier in spite of consumer preference, not because of it. A 1957 report in the *Journal of Animal Science* noted that the public prefers 'good' grade (less fatty) beef and would buy more of it if it were available.[19] And studies at Iowa State University indicate that the fat content of meat is not the key element in its taste anyway.[20] Nevertheless, more and more marbled 'choice' meat was produced, and 'good' lean meat became increasingly scarce as cattle were fed more grain. In 1957 less than half of marketed beef was graded 'choice;' ten years later 'choice' accounted for two-thirds of it.[21]

Many have misunderstood the economic logic of cattle feeding. Knowing that grain puts on fat and that our grading system rewards fatty meat with tantalizing names like 'choice' and 'prime,' people target the grading system as the reason so much grain goes to livestock. They assume that if we could just overhaul the grading system, grain going to livestock would drop significantly and our beef would be less fatty. (The grading system was altered in 1976, but it still rewards fattier meat with higher prices and more appealing-sounding labels.)

But what would happen if the grading system stopped rewarding fatty meat entirely? Would less fatty meat be produced? Would less grain be fed? Probably only marginally less. As long as grain is cheap in relation to the price of meat, it would still make economic sense for the producer to put the animal in the feedlot and feed it lots of grain. The irony is that, given our economic imperatives that produce cheap grain, most of the fat is an inevitable consequence of producing the cheapest possible meat. We got fatty meat not because we demanded fatty meat, but because fatty meat was the cheapest to produce. If we had demanded the same amount of leaner meat, meat prices would have been higher over the last thirty years.[22]

The Livestock Explosion and the Illusion of Cheap Grain

If we are feeding millions of tons of grain to livestock, it must be because it makes economic sense. Indeed, it does 'make sense' under the rules of our economy. But that fact might better be seen as the problem, rather than the explanation that should put our concerns to rest. We got hooked on grain-fed meat just as we got hooked on gas-guzzling automobiles. Big cars 'made sense' only when oil was cheap; grain-fed meat 'makes sense' only because the true costs of producing it are not counted.

But why is grain in America so cheap? If grain is cheap simply because there is so much of it and it will go to waste unless we feed it to livestock, doesn't grain-fed meat represent a sound use of our resources? Here we need to back up to another, more basic question: why is there so much grain in the first place?

In our production system each farmer must compete against every other farmer; the only way a farmer can compete is to produce more. Therefore, every farmer is motivated to use any new technology—higher yielding seeds, fertilizers, or machines—which will grow more and require less labor. In the last thirty years crop production has virtually doubled as farmers have adopted hybrid seeds and applied ever more fertilizer and pesticides. Since the 1940s fertilizer use has increased fivefold, and corn yields have tripled.

But this production imperative is ultimately self-defeating. As soon as one farmer adopts the more productive technology, all other farmers must do the same or go out of business. This is because those using the more productive technology can afford to sell their grain at a lower price, making up in volume what they lose in profit per bushel. That means constant downward pressure on the price of grain.

Since World War II real grain prices have sometimes fluctuated wildly, but the indisputable trend has been downward. The price of corn peaked at $6.43 per bushel in 1947 and fell to about $2.00 in 1967. In the early 1970s prices swung wildly up, but then fell to a low of $1.12 in 1977, or about one-sixth the price thirty years earlier. (All prices are in 1967 dollars.)[23]

This production imperative doesn't fully explain why production of feed doubled after 1950. In the 1950s the problem of agricultural surplus was seen as too much of certain crops, such as wheat, cotton, and tobacco; so government programs subsidized cutbacks of certain crops, but allowed farmers to expand their acreage in others, such as the feed crops barley, soybeans, and grain sorghum. In Texas, sorghum production leaped sevenfold after cotton acreage was limited by law in the 1950s.[24]

But neglected in this explanation of the low price of grain are the hidden production costs which we and future generations are subsidizing: the fossil fuels and water consumed, the groundwater mined, the precious topsoil lost, the fertilizer resources depleted, and the water polluted.

Fossil Fuel Costs

Agricultural production uses the equivalent of about ten percent of all of the fossil fuel imported into the United States.[25]

Besides the cost of the grain used to produce meat, we can also measure the cost of the fossil fuel energy used compared with the food value we receive. Each calorie of protein we get from feedlot-produced beef costs us seventy-eight calories of fossil fuel, as we learn from the graph below, prepared from the work of Drs. Marcia and David Pimentel at Cornell. Grains and beans are from twenty-two to almost forty times less fossil-fuel costly.

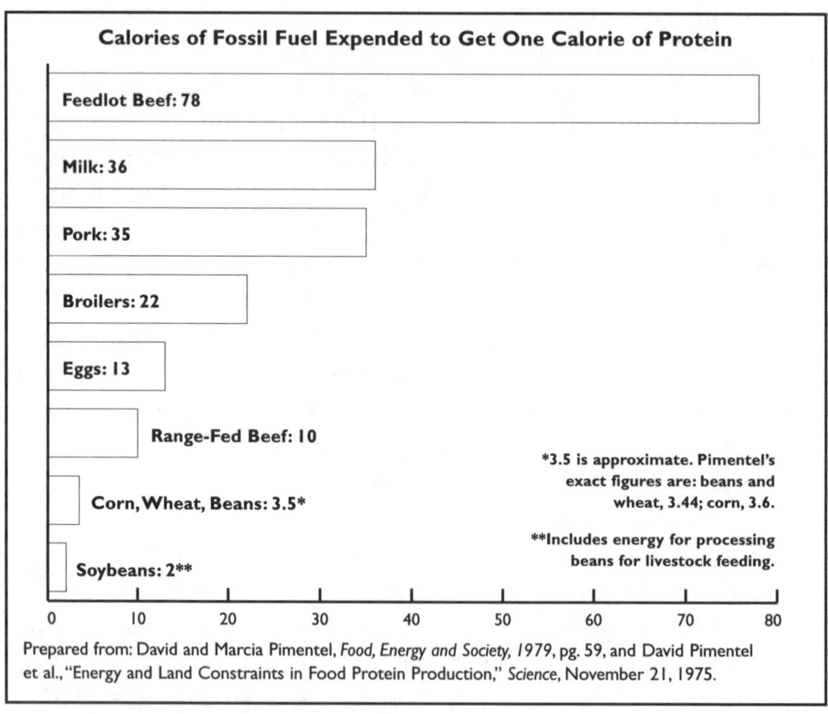

Prepared from: David and Marcia Pimentel, *Food, Energy and Society,* 1979, pg. 59, and David Pimentel et al., "Energy and Land Constraints in Food Protein Production," *Science*, November 21, 1975.

Enough Water To Float A Destroyer

"We are in a crisis over our water that is every bit as important and deep as our energy crisis," says Fred Powledge, who wrote the first in-depth book on our national water crisis, *Water: The Nature, Uses, and Future of Our Most Precious and Abused Resource* (New York: Farrar, Straus & Giroux, 1981).

According to food geographer Georg Borgstrom, to produce a one-pound steak requires 2,500 gallons of water![26] The average U.S. diet requires 4,200 gallons of water a day for each person, and of this he estimates animal products account for over 80 percent.[27]

"The water that goes into a 1,000-pound steer would float a destroyer," *Newsweek* recently reported.[28] When I sat down with my calculator, I realized that the water used to produce just ten pounds of steak equals the household consumption of my family for the entire year.

Based on the estimates of David Pimentel at Cornell, the graph on the following page shows that to produce one pound of beef protein can require as much as fifteen times the amount of water needed to produce the protein in plant food.

Mining Our Water

Irrigation to grow food for livestock, including hay, corn, sorghum, and pasture, uses fifty out of every one hundred gallons of water 'consumed' in the United States.[29] (Some of this production is exported, but not the major share, since close to half of the irrigated land used for livestock is for pasture and hay.) Other farm uses—mainly irrigation for food crops—add another thirty-five gallons, so agriculture's total use of water equals eighty-five out of every hundred gallons consumed. (Water is 'consumed' when it doesn't return to our rivers and streams.)

Over the past fifteen years grain-fed-beef production has been shifting from the rain-fed Corn Belt to newly irrigated acres in the Great Plains. Just four Great Plains states, Nebraska, Kansas, Oklahoma, and Texas, have accounted for over three-fourths of the new irrigation since 1964, and most of that irrigation has been used to grow more feed. Today half of the grain-fed beef in the United States is produced in states that depend for irrigation on an enormous underground lake called the Ogallala Aquifer.[30]

But much of this irrigation just can't last.

Rainwater seeps into this underground lake so slowly in some areas that scientists consider parts of the aquifer a virtually nonrenewable resource, much like oil deposits. With all the new irrigation, farmers now withdraw more water each year from the Ogallala Aquifer than the entire annual flow of the Colorado River. Pumping water at this

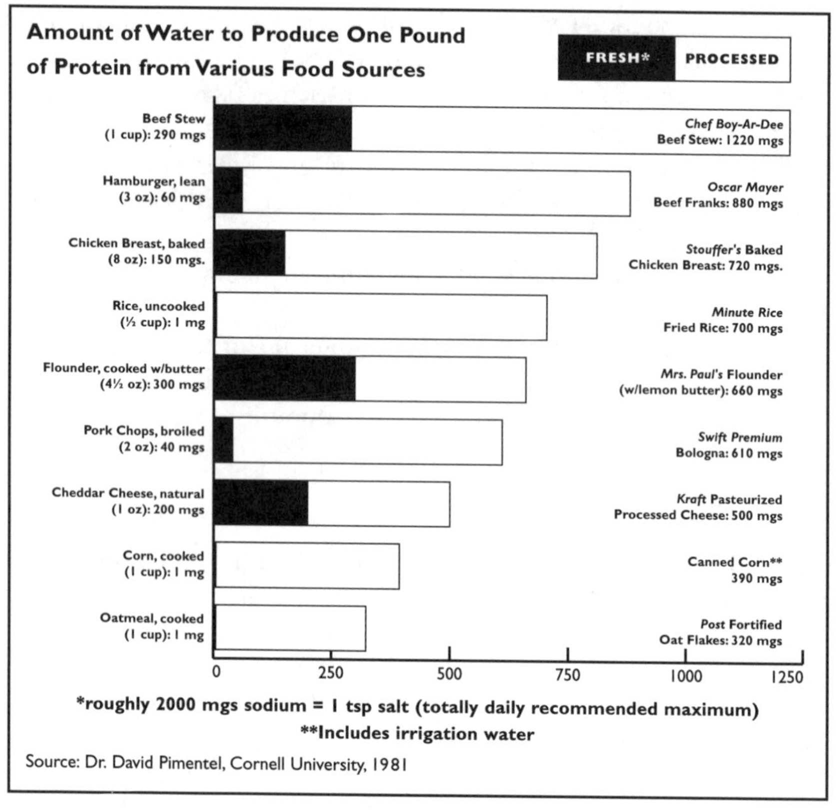

rate is causing water tables to drop six inches a year in some areas, six feet a year in others. And lower water tables mean higher and higher costs to pump the water. The Department of Agriculture predicts that in forty years the number of irrigated acres in the Great Plains will have shrunk by thirty percent.[31]

In only two decades Texans have used up one-quarter of their groundwater.[32] Already some wells in northern Texas are running dry, and with rising fuel costs, farmers are unable to afford pumping from deeper wells. Why is this water being mined in Texas? Mostly to grow sorghum for the feedlots which have sprung up in the last decade.

When most of us think of California's irrigated acres, we visualize lush fields growing tomatoes, artichokes, strawberries, and grapes. But in California, the biggest user of underground water, more irrigation water is used for feed crops and pasture than for all these specialty crops combined. In fact, forty-two percent of California's irrigation goes to produce livestock.[33] Not only are the water tables dropping, but in some parts of California the earth itself is sinking as groundwater is drawn out. According to a 1980 government survey, 5,000 square miles of the rich San Joaquin Valley have already sunk, in some areas as much as twenty-nine feet.[34]

The fact that water is free encourages this mammoth waste. Whoever has the $450 an acre needed to level the land and install pumping equipment can take groundwater for nothing. The replacement cost—the cost of an equal amount of water when present wells have run dry—is not taken into consideration. This no-price, no-plan policy leads to the rapid depletion of our resources, bringing the day closer when alternatives must be found—but at the same time postponing any search for alternatives.

Ironically, our tax laws actually entice farmers to mine groundwater. In Texas, Kansas, and New Mexico, land-owners get a depletion allowance on the groundwater to compensate for the fact that their pumping costs rise as their groundwater mining lowers the water table. Moreover, the costs of buying the equipment and sinking the well are tax-deductible. Irrigation increases the value of the land enormously, but when the land is sold the profits from the sale are taxed according to the capital gains provisions; that is, only forty percent of the difference between the original cost of the farm and its sale price is taxed as ordinary income. The rest is not taxed at all.

Few of us—and certainly not those whose wealth depends on the mining of nonrenewable resources—can face the fact that soon we will suffer for this waste of water. Donald Worster, author *of Dust Bowl: The Southern Plains in the* 1930s (New York: Oxford University Press, 1979), interviewed a landowner in Haskell County, Kansas, where $27.4 million in corn for feed is produced on about 100,000 acres of land irrigated with groundwater. He asked one of the groundwater-made millionaires, "What happens when the irrigation water runs out?"

"I don't think that in our time it can," the woman replied. "And if it does, we'll get more from someplace else. The Lord never intended us to do without water."[35]

The Soil In Our Steaks

Most of us think of soil as a renewable resource. After all, in parts of Europe and Asia, haven't crops been grown on the same land for thousands of years? It's true, soil should be a renewable resource; but in the United States. we have not allowed it to be.

We are losing two bushels of topsoil for every bushel of corn harvested on Iowa's sloping soils, warned Iowa state conservation official William Brune in 1976.[36] "It can take one hundred to five hundred years to create an inch of topsoil," but under current farming practices in Iowa, an inch of topsoil "can wash away in a single heavy rainstorm." Brune said after the spring rains in 1980. On many slopes in Iowa we have only six inches of topsoil left.[37]

Few would argue with Brune. Few would dispute that our topsoil loss is a national catastrophe, or that in the last two decades we have backpedaled on protecting our topsoil or that in some places erosion is as bad as or worse than during the Dust Bowl era. Few dispute that excessive erosion is reducing the soil's productive capacity, making chemical fertilizers ever more necessary while their cost soars. The only dispute is how many billions of dollars topsoil erosion is costing Americans and how soon the impact will be felt in higher food prices and the end of farming on land that could have been abundant for years to come.

Since we began tilling the fields in our prime farming states we have lost one-third of our topsoil.[38] Each year we lose nearly four billion tons of topsoil from cropland, range, pasture, and forest land just because of rain-related water erosion.[39] That four billion tons could put two inches of topsoil on all of the cropland in Pennsylvania, New York, and New Jersey.[40] Adding wind erosion, estimated at three billion tons, we hit a total erosion figure of nearly seven billion tons a year.[41]

Robin Hur is a mathematician and Harvard Business School graduate who has spent the last year documenting the resource cost of livestock production for his forthcoming book. "How much of our topsoil erosion is associated with crops destined for livestock and overgrazing of rangeland?" I asked him. "Most of it—about 5.9 billion tons," he calculates, including erosion associated with exported feed grains. This is true not only because feed crops cover half of our harvested acres, but because these crops, especially corn and soybeans, are among the worst offenders when it comes to soil erosion. According to the Department of Agriculture, one-quarter of all soil erosion in the United States can be attributed to corn alone.[42]

Mining the Soil

The loss of billions of tons of topsoil threatens our food security only if we are losing topsoil faster than nature is building it. The difficulty is knowing how fast nature works. The most widely accepted rule of thumb is that we can lose up to five tons of topsoil per acre per year without outpacing nature's rebuilding rate—yet one-third of the nation's cropland already exceeds this limit, the Department of Agriculture estimates, and one out of eight acres exceeds the limit almost three times over.[43] This is bad enough, but many soil scientists challenge the standard itself, suggesting it applies only to the top layer of the soil. Soil formation from the underlying bedrock may proceed *ten times* more slowly."[44] If these scientists are correct, we are mining the soil on most of our cropland.

Lost Soil, Lower Yields

In some areas we are already experiencing lower yields due to erosion and the reduction in fertility it causes. The U.S. Department of Agriculture estimates the annual dollar value of the loss just from water erosion at $540 million to $810 million.[45] Adding wind erosion may increase that estimate by thirty percent.

"In our area of Nebraska you see hilltops eroded— completely naked," says Marty Strange of the Center for Rural Affairs. "Yet farmers are still getting ninety to ninety-five bushels of corn an acre. Farmers don't believe they are losing productivity." They use chemicals to make up for the soil's lost natural fertility, but the cost of fertilizer has risen two hundred percent since 1967 and is likely to keep rising. Higher production costs must ultimately mean higher food prices.

We also pay in our taxes, for billions of dollars have gone toward conservation measures (although this spending is shrinking, while the need increases). Moreover, the soil washed from farmlands ends up in rivers, streams, and reservoirs. Dredging sediment from rivers and harbors, the reduction in the useful life of reservoirs, and water purification—these costs amount to $500 million to $1 billion a year.[46]

The direct and indirect costs of soil erosion already approach *$2 billion a year.*

But Why?

Why is soil erosion accelerating, despite thirty-four Department of Agriculture programs related to soil and water conservation? There are several reasons:

- The increased tillage of soil so fragile it probably should have remained uncultivated. The government estimates that forty-three percent of the land used for row crops in the Corn Belt is composed of highly erodible soils.[47]
- The increased planting of row crops, especially the feed crops corn and soybeans, which make the land particularly susceptible to the problems of erosion.
- The growing neglect of conservation practices, including the removal of shelterbelts planted during the Dust Bowl era to protect the soil. By 1975 the total real value of soil conservation improvements had deteriorated over twenty percent from its peak in 1955.[48]

These are the reasons, but what are the causes? Unfortunately, they lie in the economic givens that most Americans take as normal and proper. Squeezed between ever higher costs of production and falling prices, farmers must increase their production. They plant more acres, including marginal land susceptible to erosion, and they plant what brings the highest return, even if this means continuous planting of the most erosion-inducing crops, corn and soybeans. "The most erosive production system—continuous corn—produces the highest net income," according to researchers at the University of Minnesota.[49]

Fertilizers: Becoming Import-Dependent

To determine a price for grain which reflects all its costs would also mean looking at the fertilizers required to mask our lost fertility and continually increase production. Higher yields and continuous cropping deplete soil nutrients. so that ever greater quantities of fertilizer must be used. This vicious circle caused our nation's use of chemical fertilizer to increase fivefold between the 1940s and the 1970s. Just in the last ten years, the use of ammonia (for nitrogen fertilizer) has increased by almost two hundred percent and that of potash by almost three hundred percent.[50] Corn, the major national feed grain, which occupies about twenty-three percent of all our cropland, uses more fertilizer than any other crop—about forty percent of the total.[51]

Because fertilizer has been relatively cheap, farmers have been encouraged to apply ever greater quantities in their desperate struggle to produce. As with topsoil and groundwater, we squander fertilizer resources today without considering the consequences tomorrow. One of the consequences of our heavy consumption of fertilizer is increasing dependence on imports. Americans might be alarmed at how our dependence on imported strategic metals can be used to justify U.S. political or even military intervention abroad. Americans would probably be even more alarmed about becoming dependent on imported food. But is being dependent on the fertilizer needed to produce food really much different?

Let's look at the three major types of fertilizer:

Nitrogen fertilizer. We won't run out of nitrogen, since it makes up about seventy-eight percent of our air, but the price of natural gas, used to make ammonia, the most common nitrogen fertilizer, has risen so rapidly that we have begun to import ammonia from countries with cheap supplies of natural gas. We now import about twenty percent of our supplies.[52]

Potash. Today we import about eighty-five percent of our potash (from Canada), and by the year 2000 we are expected to import ninety percent.[53]

Phosphate fertilizer. The U.S. is the world's leading producer, but our high-grade reserves will probably be exhausted over the next thirty to forty years at the current rate of use, according to a 1979 government report. "We will probably move from assured self-sufficiency and a dominant exporter position to one of increasing dependency on possibly unreliable foreign sources of supply," says the ominous report. "Since phosphates are a fundamental necessity to agriculture ... *the situation... is somewhat analogous to that now being experienced with oil*"[54] (my emphasis).

Livestock Pollution

Some people believe that although we feed enormous quantities of high-grade plant food to livestock with relatively little return to us as food, there is really no loss. After all, we live in a closed system, don't we? Animal waste returns to the soil, providing nutrients for the crops that the animals themselves will eventually eat, completing a natural ecological cycle.

Unfortunately, it doesn't work that way anymore. Most manure is not returned to the land. Animal waste in the United States amounts to two billion tons annually, equivalent to the waste of almost half of the world's human population."[55] Much of the nitrogen-containing waste from livestock is converted into ammonia and into nitrates, which leach into the groundwater beneath the soil or run directly into surface water, contributing to high nitrate levels in the rural wells which tap the groundwater. In streams and lakes, high levels of waste runoff contribute to oxygen depletion and algae overgrowth.[56] American livestock contribute five times more harmful organic waste to water pollution than do people, and twice that of industry, estimates food geographer Georg Borgstrom.[57]

Cheap Water for Cheap Grain

In a true accounting, the two bushels of topsoil washed away with every bushel of corn grown on Iowa's sloping land would be seen as a subsidy to our cheap grain. If we were to use all of the conservation measures we know of to prevent this erosion, the cost of producing our grain would go up, as it would it we were to add in all of the costs of dredging the soil from our waterways or charge for feedlot pollution. Failing to account for these costs amounts to hidden subsidies. But in addition, you and I as taxpayers are paying *direct* subsidies right now.

Our tax dollars have paid for more than one-half of the net value of all irrigation facilities in the United States as of 1975.[58] Since the turn of the century the federal government has sponsored thirty-two irrigation projects in seventeen western states where twenty percent of the acreage is now irrigated with the help of government subsidies. A recent General Accounting Office study concluded that even though farmers are legally required to repay irrigation construction costs, in the cases studied the repayments amounted to less than eight percent of the cost to the federal government.[59]

In some of the projects, the irrigators pay even less. Take the Fryingpan-Arkansas Project near Pueblo, Colorado. This $500 million project helps farmers grow corn, sorghum, and alfalfa for feed. The GAO calculated the full cost of water delivered to be $54 per acre-foot, but the farmers are being charged only seven cents per acre-foot.[60] (And the GAO's 'full cost' is based on an interest rate of 7.5 percent.) According to *Fortune* magazine, the huge California Central Valley irrigation project is being subsidized at a rate of $79,000 a day.[61]

Cheap water encourages farmers to grow livestock feed. "Because water is so cheap, its use is based on its price and not its supposed scarcity," observes *Fortune.* "Many farmers . . . use inferior land to grow low value crops that require large amounts of water, like alfalfa and sorghum" for feed.[62]

Government subsidies are so large that "the market value of the crops to be grown with federal water is less than the cost of the water and the other farming supplies used to grow those crops," the government study concluded.[63] But Robin Hur has an even dimmer view of the economics of federal irrigation. After studying federal irrigation in the Pacific Northwest, he calculated that in six major projects the value of the crops produced doesn't cover even the cost of the water alone!

Federally subsidized irrigation water helps keep grain cheap. It also helps make people rich. To a farmer with 2,000 irrigated acres in California's Westlands district, the federal water subsidy is worth $3.4 million—that's how much more the land is worth simply because of what the government contributes to irrigation.[64]

A key 1902 federal law stipulated that beneficiaries of the subsidized irrigation were to be small farmers *only,* those owning no more than 160 acres. But the law has never been enforced, despite suits filed by National Land for People and others seeking the irrigated land they are legally entitled to. Today one-quarter of the federally subsidized irrigated land is owned by a mere two percent of the landholders, who own far more land than the legal limit. In California, over a million and a half acres of federally subsidized water are controlled illegally—that is, by farms over the legal acreage limit.[65] Southern Pacific alone controls land almost seven hundred times the legal limit for an individual. (All told, Southern Pacific owns almost four million acres in three states.[66])

Tax Benefits at the Feedlot, Too

Besides directly and indirectly subsidizing the feedlot system by keeping the price of grain low, we taxpayers also subsidize the feedlot operations themselves. Tax laws favoring feedlot owners and investors in feedlot cattle shift the tax burden onto the rest of us. While these tax advantages were cut back in 1976, there are still 'income tax management strategies' that can benefit cattle owners who contract with feedlots to fatten their cattle, putting on the last two hundred to six hundred pounds of each head.[67] According to a Department of Agriculture report, the law that allows farmers to use cash accounting for tax purposes can also profit investors in feedlot cattle, especially those with high nonfarm incomes seeking to reduce their taxable income. More than a quarter of 'custom feeding' clients in the Southern Plains are such outside investors, including doctors, lawyers, and bankers.[68]

Agricultural economists V. James Rhodes and the late Joseph C. Meisner of the University of Missouri offer this observation of tax favors to feedlot operations:

> Subsidies to large-size feedlot firms, indirect though they be, would tend to lead to survival and growth of those firms on a basis of other than economic efficiency. If the nation seeks to subsidize beef production, direct grants to feedlot firms is an alternative. Then, true economic costs of the subsidies would be more apparent. However, in a world of growing concern for energy supplies, the beef industry would seem to be a most unlikely recipient of national subsidy.[69]

A Fatal Blindness

After reading this account of the resource costs of our current production system, you probably are amazed that more people are not aware and alarmed. I am continually amazed. Again and again I have to learn this lesson: often those with the most information concerning our society's basic problems are those so schooled in defending the status quo that they are blind to the implications of what they know.

As I was preparing this chapter I came across a book that read as if designed to be the definitive rebuttal to *Diet for a Small Planet.* Three noted livestock economists conclude that "total resource use in this [livestock] production has decreased dramatically."[70] To arrive here, they had to ignore such hidden costs as I've just outlined—the fossil fuel used, the water consumed (including groundwater that is irreplaceable), the topsoil eroded, and the domestic fertilizer depleted as we attempt to make up for our soil's declining fertility. They also ignore feedlot pollution and hidden tax subsidies. All this I would have expected. What really shocked me was their attempt to prove that we are producing more meat using *less* resources. Their evidence? A decline in labor used and a dramatic drop in acres devoted to feed grains between 1944 and about 1960, while meat production rose. What they fail to tell us is that about one-third of our total cropland was released from feed-grain production between 1930 and 1955 by the rapid replacement of grain-consuming draft animals by fuel-consuming tractors. *Much of the decline in feed-grain acres had nothing to do with increased efficiency of meal production.* Just as appalling, these economists ignore the fact that livestock eat more than feed grains. Since 1960 there has been a spectacular rise in soybean use as animal feed. Tripling since 1960, acres in soybeans now exceed two-thirds of total acres in feed grains.[71] (Almost half of those acres are used to feed domestic livestock [The protein concentrate made from soybeans is an excellent livestock ration, and the oil extracted is used to make margarine, salad oil, etc.], the rest for export.) Soybeans are not even mentioned by these economists as a resource in livestock production.

While it is useful to keep these gross oversights in mind for the next time we feel cowed by an 'authority' questioning *our* facts, they sidetrack us a bit from the basic argument used by such defenders of the status quo. Most economists defend our current meat production system by arguing that feeding grain to livestock is the cheapest way to pro duce meat. The fatal blindness in this argument is attention only to price. As we have seen, the price of our grain is an illusion. It results from the powerlessness of farmers to pass on their costs of production and the fact that so many of the costs of production—topsoil and groundwater, for example—carry no price at all.

In writing this I came to realize more clearly than ever that our production system is ultimately self-destructive because it is self-deceptive; it can't incorporate the many costs I've outlined here. It can't look to the future. And it blinds those closest to it from even seeing what is happening. The task of opening our eyes lies more heavily with the rest of us—those less committed to protecting the status quo. As awakening stewards of this small planet, we have a lot to learn—and fast.

1 *Raw Materials in the United States Economy 1900-1977*; Technical Paper 47, prepared under contract by Vivian Eberle Spencer, U.S. Department of Commerce, U.S. Department of Interior Bureau of Mines, pg. 3.

2 Ibid. Table 2, pg. 86.

3 U.S. Department of Agriculture. *Livestock Production Units, 1910-1961*, Statistical Bulletin 325, pg, 18, and *Agriculture Statistics 1980*, pg. 56. Current world imports from FAO At Work, newsletter of the liaison office for North America of the Food and Agriculture Organization of the United Nations, May 1981.

4 Pimentel, David, et al. "The Potential for Grass-Fed Livestock: Resource Constraints," *Science*, February 22, 1980, vol. 207, pp. 843 ff.

5 Pimentel, David. "Energy and Land Constraints in Food Protein Production," *Science*, November 21, 1975, pp. 754 ff.

6 Oltjen, Robert R. "Tomorrow's Diets for Beef Cattle," *The Science Teacher*, March 1970, vol. 38, no. 3.

7 The amount varies depending on the price of grain, but 2,200 to 2,500 pounds is typical. See note 13 for more detailed explanation of grain feeding.

8 U.S. Department of Agriculture, Economic Research Service. *Cattle Feeding in the United States*, Agricultural Economics, Report No. 186, pg. 5.

9 Ibid., pg. iv.

10 *Agricultural Statistics* 1979 and 1980, Tables 76 and 77.

11 Norman Borlaug in conversation with Frances Moore Lappé, April 1974.

12 *Agricultural Statistics 1980*, Table 76.

13 How many pounds of grain and soy are consumed by the American steer to get one pound of edible meat?

(a) The total forage (hay, silage, grass) consumed: 12,000 pounds (10,000 pre-feedlot and 2,000 feedlot). The total grain- and soy-type concentrate consumed: about 2,850 pounds (300 pounds grain and 50 pounds soy before feedlot, plus 2,200 pounds grain and 300 pounds soy in feedlot). The actual percent of total feed units from grain and soy is about 25 percent.

(b) But experts estimate that the grain and soy contribute more to weight gain (and to ultimate weight produced) than their actual proportion in the diet. They estimate that grain and soy contribute (instead of 25 percent) about 40 percent of weight put on over the life of the steer.

(c) To estimate what percent of edible meat is due to the grain and soy consumed, multiply that 40 percent (weight gain due to grain and soy) times the edible weight produced at slaughter, or 432 pounds: .4 x 432 = 172.8 pounds of edible portion contributed by grain and soy. (Those who state a 7:1 ratio use the entire 432 pounds edible meat in their computation.)

(d) To determine how many pounds of grain and soy it took to get this 172.8 pounds of edible meat, divide total grain and soy consumed, 2850 pounds by 172.8 of edible meat: 2,850 divided by 172.8 = 16-17 pounds. (I have taken the lower figure since the amount of grain being fed may be going down a small amount.) These estimates are based on several consultations with the USDA Economic Research Service and the USDA Agricultural Research Service, Northeastern Division, plus current newspaper reports of actual grain and soy currently being fed.

14 U.S. Department of Agriculture, Economic Research Service and Agricultural Research Service, Northeastern Division, consultations with staff.

15 In 1975 I calculated this average ratio and the return to us in meat from *Livestock-Feed Relationships*, National and State Statistical Bulletin no. 530, June 1974, pp. 175-177. In 1980 I approached it differently and came out with the same answer. I took the total grain and soy fed to livestock (excluding dairy) from *Agricultural Statistics 1980*. The total was about 145 million tons in 1979. I then took the meat and poultry and eggs consumed that year from *Food Consumption, Prices, and Expenditures*, USDA-ESS, Statistical Bulletin No. 656. (I included only the portion of total beef consumed that was put on by grain feeding, about 40 percent, and reduced the total poultry consumed to its edible portion, i.e. minus bones.) The total consumption was about 183.5 pounds per person, or 20 million tons for the whole country. I then divided the 145 million tons of grain and soy fed by the 20 million tons of meat, poultry, and eggs produced by this feeding and came up with the ratio of 7:1. (Imports of meat are not large enough to affect this calculation appreciably.)

16 Calculated as follows: 124 million tons of grain 'lost' annually in the United States x 2,000 pounds of grain in a ton = 248 billion pounds 'lost' divided by 4.4 billion people = 56 pounds per capita divided by 365 days equals .153 pounds per capita per day x 16 ounces in a pound—2.5 ounces per capita per day—1/3 cup of dry grain, or 1 cup cooked volume.

17 Brokken, R.F., and James K. Whittaker and Ludwig M. Eisgruber. "Past, Present, and Future Resource Allocation to Livestock Production," in *Animals, Feed, Food, and People: An Analysis of the Role of Animals in Food Production*, R.L. Baldwin, ed., An American Association for the Advancement of Science Selected Symposium (Boulder, CO: Westview Press, 1980), pp. 99-100.

18 Martin, J. Rod. "Beef," in *Another Revolution in U.S. Farming?* by Lyle Schertz, et al. (Washington, DC: U.S. Department of Agriculture, 1979), pg. 3.

19 Brady. D.E. "Consumer Preference," *Journal of Animal Science*, vol. 16, pg. 233, cited in H.A. Turner and R.J. Raleigh, "Production of Slaughter Steers from Forages in the Arid West," *Journal of Animal Science*, vol. 44, no.5, 1977, pp. 901 ff.

20 *Des Moines Register*, December 8, 1974.

21 *Cattle Feeding in the United States*, op. cit., pp. 78-79.

22 "Past, Present, and Future Resource Allocation to Livestock Production," op. cit., pg. 97.

23 Ibid., pg. 91.

24 U.S. Department of Agriculture, Economics, and Statistics Service. *Status of the Family Farm*, Second Annual Report to the Congress, Agricultural Economic Report no. 434, pg. 48

25 Quantities of each fuel used from *Energy and U.S. Agriculture: 1974 and 1978* (Washington, DC: U.S. Department of Agriculture, Economic and Statistics Service, 1980). Conversion to BTUs used Cervinka, "Fuel and Energy Efficiency," in David Pimentel, ed., *Handbook of Energy Utilization in Agriculture* (Boston, MA: CRC Press, 1980). Fossil fuel imports from *Monthly Energy Review*, March 1981, U.S. Department of Energy, Energy Information Administration, pg. 8.

26 Georg Borgstrom, Michigan State University, presentation to the Annual Meeting of the American Association for the Advancement of Science (AAAS), 1981.

27 Ibid.

28 "The Browning of America," *Newsweek*, February 22, 1981, pp. 26 ff.

29 To arrive at the estimate of 50 percent, I used *Degradation: Effects on Agricultural Productivity*, Interim Report No. 4, National Agricultural Lands Study, 1980, which estimates that 81 percent of all water consumed in the Unites States is for irrigation. And I used the *Fact Book of U.S. Agriculture*, Misc. Publication no. 1065 (Washington, DC: U.S. Department of Agriculture, 1979), table 3, showing that about 64 percent of irrigated land is used for feed crops, hay, and pasture. Sixty-four percent of 81 percent is 52 percent.

30 Raup, Philip M. "Competition for Land and the Future of American Agriculture," in *The Future of American Agriculture as a Strategic Resource*, Sandra S. Batie and Robert G. Healy, eds. (Washington, DC: A Conservative Foundation Conference, 1980), pp. 36-43. Also see William Franklin Lagrone, "The Great Plains," *Another Revolution in U.S. Farming?* by Lyle Schertz and others, pp. 335-361. The estimate of grain-fed beef's dependence on the Ogallala is from a telephone interview with resource economist Joe Harris of the consulting firm Camp, Dresser, McKee (Austin, Texas), part of a four-year government-sponsored study: "The Six State High Plains Ogallala Aquifer Agricultural Regional Resource Study," May 1980.

31 "The Great Plains," op. cit., pp. 356 ff.

32 "Report: Nebraska's Water Wealth is Deceptive," *Omaha World-Herald*, May 28, 1981.

33 Giannini Foundation of Agricultural Economics. *Trends in California Livestock and Poultry Production, Consumption, and Feed Use: 1961-1978*, Information Series 80-85, Division of Agricultural Sciences, University of California Bulletin 1899, November 1980, pp. 30-33.

34 General Accounting Office. *Groundwater Overdrafting Must Be Controlled*, Report of the Congress of the United States by the Comptroller General, CED 80-96, September 12, 1980, pg. 3.

35 Worster, Donald. *Dust Bowl: The Southern Plains in the 1930s* (New York: Oxford University Press, 1979), p. 236.

36 William Brune, State Conservationist, Soil Conservation Service, Des Moines, Iowa, testimony before Senate Committee on Agriculture and Forestry, July 6, 1976. See

also Seth King, "Iowa Rain and Wind Deplete Farmlands," *The New York Times*, December 5, 1976, pg. 61.

37 Harnack, Curtis. "In Plymouth County, Iowa, the Rich Topsoil's Going Fast. Alas," *The New York Times*, July 11, 1980.

38 Pimentel et al. "Land Degradation: Effects on Food and Energy Resources," in *Science*, vol. 194, October 1976, pg. 150.

39 National Association of Conservation Districts, Washington, DC. *Soil Degradation: Effects on Agricultural Productivity*, Interim Report Number Four, National Agricultural Lands Study, 1980, pg. 20, citing the 1977 National Resources Inventory.

40 Calculated from estimates by Medard Gabel for *The Cornucopia Project* (Emmaus, PA: Rodale Press, 1980).

41 King, Seth. "Farms Go Down the River," *The New York Times*, December 10, 1978, citing the Soil Conservation Service.

42 Ned D. Bayley, Acting Assistant Secretary for Natural Resources and Environment, "Soil and Water Resource Conservation Outlook for the 1980's," 1981 Agricultural Outlook Conference, Washington, DC.

43 *Soil Degradation: Effects on Agricultural Productivity*, pg. 21.

44 Larson, W.E. "Protecting the Soil Resource Base," *Journal of Soil and Water Conservation*, vol. 36, no. 1, January-February 1981, pp. 13 ff.

45 Soil and Water Resources Conservation Act—Summary of Appraisal, USDA Review Draft, 1980, pg. 18.

46 "Land Degradation: Effects on Food and Energy Resources," op. cit., pg. 150, estimates $500 million costs of sediment damage. Philip Le Veen, in "Some Considerations for Soil Conservation Policy," unpublished manuscript, Public Interest Economics, 1981, pg. 29, estimates $1 billion.

47 *Soil Degradation: Effects on Agricultural Productivity*, pg. 28.

48 U.S. Department of Agriculture, Economics, and Statistics Service. *Natural Resource Capital in U.S. Agriculture: Irrigation, Drainage, and Conservation Investments Since* 1900, ESCS Staff Paper, March 1979.

49 *Ag World*, April 1978, citing work of Clifton Halsey, University of Minnesota conservationist.

50 U.S. Department of Agriculture. *Handbook of Agricultural Charts* (Washington, DC: U.S. Department of Agriculture, 1979), pg. 19.

51 U.S. Department of Agriculture. *Fertilizer Situation* (Washington, DC: U.S. Department of Agriculture, 1980), pg. 14.

52 Gabel, Medard. Preliminary Report for *The Cornucopia Project* (Emmaus, PA: Rodale Press, 1980).

53 Walfbauer, C.A. "Mineral Resources for Agricultural Use," *Agriculture and Energy*, William Lockeretz, ed. (New York: Academic Press, 1977), pp. 301-314. See also *Facts and Problems*, U.S. Bureau of Mines, 1975, pp. 758-868.

54 General Accounting Office. *Phosphates: A Case Study of a Valuable Depleting Mineral in America*, Report to the Congress by the Comptroller General of the United States, EMD-80-21, November 30, 1979, pg. 1.

55 *Environmental Science and Technology*, vol. 4, no. 12, 1970, pg. 1098.

56 Commoner, Barry. *The Closing Circle* (New York: Alfred A. Knopf, 1971), pg. 148.

57 Borgstrom, Georg. *The Food and People Dilemma* (Boston, MA: Duxbury Press, 1973), pg. 103.

58 U.S. Department of Agriculture, Economics, and Statistics Service. *Natural Resource Capital in U.S. Agriculture: Irrigation, Drainage, and Conservation Investments Since* 1900, ESCS Staff Paper, March 1979.

59 General Accounting Office. *Federal Charges for Irrigation Projects Reviewed Do Not Cover Costs*, Report to the Congress of the United States from the Comptroller General, PAD-81-07, March 3, 1981, pg. 43.

60 Ibid., pg. 26.

61 Vitullo-Martin, Julia. "Ending the Southwest's Water Binge," *Fortune*, February 23, 1981, pp. 93 ff.

62 Ibid.

63 *Federal Charges for Irrigation Projects Reviewed Do Not Cover Costs*, pp. 3-4.

64 "Ending the Southwest's Water Binge," op. cit.

65 U.S. Department of Agriculture, *Farmline*, September 1980.

66 Moskowitz, Milton and Michael Katz and Robert Levering, Eds. *Everybody's Business: An Almanac* (New York: Harper and Row, 1980), pg. 643.

67 U.S. Department of Agriculture, Economics, and Statistics Service. *Status of the Family Farm*, Second Annual Report to the Congress, Agricultural Economic Report no. 434, pg. 47.

68 Ibid., pg. 45.

69 Meisner, Joseph C. and V. James Rhodes. *The Changing Structure of U.S. Cattle Feeding*, Special Report 167, Agricultural Economics, University of Missouri-Columbia, November 1974, pg. 13.

70 "Past, Present, and Future Resource Allocation to Livestock Production," op. cit.

71 U.S. Department of Agriculture. *Agricultural Statistics*, 1979, pp. 435-438.

72 Winrock International Livestock Research and Training Center. *The World Livestock Product, Feedstuff, and Food Grain System: An Analysis and Evaluation of System Interactions Throughout the World*, with projections to 1985 (Winrock, AR: Winrock International, 1981).

From *Circle of Poison: Pesticides and People in a Hungry World* by David Weir and Mark Shapiro

The Circle of Poison

The export of banned pesticides from the industrial countries to the Third World is a scandal of global proportions. Massive advertising campaigns by multinational pesticide corporations—Dow, Shell, Chevron—have turned the Third World into not only a booming growth market for pesticides, but also a dumping ground. Dozens of pesticides too dangerous for unrestricted use in the United States are shipped to underdeveloped countries. There, lack of regulation, illiteracy, and repressive working conditions can turn even a 'safe' pesticide into a deadly weapon.

According to the World Health Organization, someone in the underdeveloped countries is poisoned by pesticides *every minute*.[1]

But we are victims too. Pesticide exports create a circle of poison, disabling workers in American chemical plants and later returning to us in the food we import. Drinking a morning coffee or enjoying a luncheon salad, the American consumer is eating pesticides banned or restricted in the United States, but legally shipped to the Third World. The United States is among the world's top food importers and ten percent of our imported food is officially rated as contaminated.[2] Although the Food and Drug Administration (FDA) is supposed to protect us from such hazards, during one fifteen month period, the General Accounting Office (GAO) discovered that *half* of all the imported food identified by the FDA as pesticide-contaminated was marketed without any warning to consumers or penalty to importers.[3]

At least twenty percent of U.S. pesticide exports are products that are banned, heavily restricted or have never been registered for use here.[4] Many have not been independently evaluated for their impacts on human health or the environment. Other pesticides are familiar poisons, widely known to cause cancer, birth defects and genetic mutations. Yet, the Federal Insecticide, Fungicide, and Rodenticide Act explicitly states that banned or unregistered pesticides are legal for export.[5]

In this article we concentrate on hazardous pesticides which are either banned, heavily restricted in their use, or under regulatory review in the United States. Some, such as DDT, are banned for any use in the United States; others, such as 2,4-D or toxaphene, are still widely used here but only for certain usages. As we will discuss, even 'safe' pesticides which are unrestricted in the United States may have much more damaging effects on people and the environment when used under more brutal conditions in the Third World.

In the United States, a mere dozen multinational corporations dominate the $7-billion-a-year pesticide market. Many are conglomerates with major sales in oil, petrochemicals, plastics, drugs, and mining.

The list of companies selling hazardous pesticides to the Third World reads like a Who's Who of the $350-billion-per-year[6] chemical industry: Dow, Shell, Stauffer, Chevron, Ciba-Geigy, Rohm & Haas, Hoechst, Bayer, Monsanto, ICI, Dupont, Hercules, Hooker, Velsicol, Allied Union Carbide, and many others.

Tens of thousands of pounds of DBCP, heptachlor, chlordane, BHC, lindane, 2,4,5-T, and DDT are allowed to be exported each year from the United States, even though they are considered too dangerous for unrestricted domestic use.[7]

"You need to point out to the world," Dr. Harold Hubbard of the U.N.'s Pan American Health Organization told us, "that there is absolutely no control over the manufacture, the transportation, the storage, the record-keeping—the entire distribution of this stuff. These very toxic pesticides are being thrown all over the world and there's no control over any of it!"[8]

Not only do the chemical corporations manufacture hazardous pesticides, but their subsidiaries in the Third World import and distribute them.

- Ortho: In Costa Rica, Ortho is the main importer of seven banned or heavily restricted U.S. pesticides—DDT, aldrin, dieldrin, hetachlor, chlordane, endrin, and BHC. Ortho is a division of Chevron Chemical Company, an arm of Standard Oil of California.[9]

- Shell, Velsicol, Bayer, American Cyanamid, Hercules, and Monsanto: In Ecuador these corporations are the main importers of pesticides banned or restricted in the United States—aldrin, dieldrin, endrin, heptachlor, kepone, and mirex.[10]

- Bayer and Pfizer: In the Philippines, these multinationals import methyl parathion and malathion respectively;[11] neither is banned but both are extremely hazardous.

The Ministry of Agriculture of Colombia registers fourteen multinationals which import practically all the pesticides banned by he United States since 1970.[12] And in the Philippines, the giant food conglomerate Castle & Cooke (Dole brand) imports banned DBCP for banana and pineapple operations there.[13]

Pesticides: A Pound Per Person

Worldwide pesticide sales are exploding. The amount of pesticides exported from the U.S. has almost doubled over the last fifteen years.[14] The industry now produces four billion pounds of pesticides each year—more than one pound for every person on earth.[15] Almost all are produced in the industrial countries, but twenty percent are exported to the Third World.[16]

And the percentage exported is likely to increase rapidly: The GAO predicts that during the decade ending in 1984, the use of pesticides in Africa, for example, will more than quintuple.[17] As the U.S. pesticide market is "approaching saturation... U.S. pesticide producers have been directing their attention toward the export potential ...exports have almost doubled since 1965 and currently account for thirty percent of total domestic pesticide production," the trade publication *Chemical Economics Newsletter* noted.[18]

Corporate executives justify the pesticide explosion with what sounds like a reasonable explanation: the hungry world needs our pesticides in its fight against famine. But their words ring hollow: in Third World fields most pesticides are applied to luxury, export crops, not to food staples the local people will eat. Instead of helping the poor to eat better, technology is overexposing them to chemicals that cause cancer, birth defects, sterility, and nerve damage.

'Blind' Schedules, Not 'As Needed'

But the crisis is not just the export of banned pesticides. A key problem in both the industrial countries and the Third World is the massive overuse of pesticides resulting from their indiscriminate application. Pesticides are routinely applied according to schedules preset by the corporate sellers, not measured in precise response to actual pest threats in a specific field. By conservative estimate, U.S. farmers could cut insecticide use by thirty-five to fifty percent with no effect on crop production, simply by treating only when necessary rather than by schedule.[19] In Central America, researchers calculate that pesticide use, especially parathion, is forty percent higher than necessary to achieve optimal profits.[20]

In the United States the result of pesticide overuse is the unnecessary poisoning of farmworkers and farmers—about 14,000 a year according to the Environmental Protection Agency (EPA).[21] But if pesticides are not used safely here—where most people can read warning labels, where a huge government agency (the EPA) oversees pesticide regulation, and where farmworker unions are fighting to protect the health of their members—can we expect these poisons to be used safely in the Third World?

An Inappropriate Technology

In Third World countries one or two officials often carry responsibility equivalent to that of the entire U.S. EPA. Workers are seldom told how the pesticides could hurt them. Most cannot read. And even if they could, labels on banned pesticides often do not carry the warnings required in the United States. Frequently repacked or simply scooped out into old cans, deadly pesticides are often handled like harmless white powder by peasants who have little experience with manmade poisons.

But perhaps even more critical is this question: can pesticides—poisons, by definition—be used safely in societies where workers have no right to organize, no right to strike, no right to refuse to carry the pesticides into the fields? In the Philippines at least one plantation owner has reportedly sprayed pesticides on workers trying to organize a strike.[22] And, in Central America, says entomologist Lou Falcon, who has worked there for many years, "The people who work in the fields are treated like half-humans, slaves really. When an airplane flies over to spray, they can leave if they want to. But they won't be paid their seven cents a day or whatever. They often live in huts in the middle of the field, so their homes, their children, and their food all get contaminated."[23]

Yet the President's Hazardous Substances Export Policy Task Force predicts that the export of banned pesticides is likely to increase as manufacturers unload these products on countries hooked on the agchemical habit. "Continued new discoveries of carcinogenic and other damaging effects of many substances are probable over the next few years," predicts the task force. "In some cases, certain firms may be left with stocks of materials which can no longer be sold in the United States, and the incentive to recover some of their investment by selling the product abroad may be considerable."[24]

The Genetic Boomerang

The pesticide explosion also has a second built-in boomerang. Besides the widespread contamination of imported food, the overuse of hazardous pesticides has created a global race of insect pests that are resistant to pesticides. The number of pesticide-resistant insect species doubled in just twelve years—from 182 in 1965 to 364 in 1977, according to the U.N. Food and Agriculture Organization. So more and more pesticides—including new, more potent ones—are needed every year just to maintain present yields.

A Circle of Victims

Enormous damage is done even before the pesticides leave American shores. At Occidental's DBCP plant in Lathrop, California, workers discovered too late that they were handling a product which made them sterile. Elsewhere in the United States, worker exposure to the pesticides Kepone and Phosvel resulted in terrible physical and mental damage.

As part of our investigation into the 'circle of poison,' we looked at these examples of how the manufacture of hazardous pesticides affects American workers. Since companies are allowed to produce pesticides for export without providing health or safety data, there is no way to be sure they are not poisoning their own workers in the process. In fact there is abundant evidence that workers in the industrial countries are indeed suffering from their employers' booming export sales.

We talked with West Coast pesticide workers who complained of inadequate protection—and information—on the job. Even after two hot showers, one group explained, their hands still carried enough toxic residue of an unregistered pesticide that, when they stuck a finger in a fish bowl, the goldfish died.

These workers in pesticide manufacturing plants are the very first victims in the circle of poison. Add to them all the people who load and unload the chemicals into and out of trucks, trains, ships, and airplanes; and those who have to clean up the toxic spills which inevitably occur. Then the total number of potential American victims of hazardous pesticide exports becomes very large. In addition there are the victims—Third World peasants, workers, and consumers—as well as everyone else in the world who eats food contaminated with pesticide residues. *We* complete the wide circle of victims.

To uncover the story we have had to overcome powerful obstacles. The pesticide industry is a secretive one. The Environmental Protection Agency guards industry production data from the public, press, and even other government agencies. The information made available often seems to defy meaningful interpretation.

We filed over fifty requests under the Freedom of Information Act in order to penetrate the industry's 'trade secrets' sanctuary inside the EPA and other agencies. In assembling hundreds of tiny pieces of the puzzle, we studied trade publications and overseas magazines and newspapers for evidence of hazardous pesticide sales. In addition, we obtained import figures from a number of Third World countries. Finally, we interviewed hundreds of people in industry, government, unions, environmental groups and international organizations. We corresponded with farmers, consumers, and environmental groups in the Third World.

The story told here is intended not merely to shock and to outrage. Its purpose is to mobilize concerned people everywhere to halt the needless suffering caused by pesticides' circle of poison.

A Victim Every Minute

For chemical company executives, exporting hazardous pesticides is not 'dumping.' If one country bans your product, move to where sales are still legal. It's just good business. But 'good business' practice seldom takes account of the human toll inflicted by the massive use of pesticides.

Every minute of the day, on the average, someone is poisoned by pesticides in the Third World. This World Health Organization statistic amounts to 500,000 poisoned people every year. A pesticide-caused death occurs about every hour and forty-five minutes, totaling at least 5,000 each year.[25] Yet these estimates tell us nothing about the number of cancers, miscarriages, deformed babies, and still-births resulting from the use of pesticides.

The rate of pesticide poisoning in underdeveloped countries is more than thirteen times that in the United States, despite vastly greater use here, according to Virgil Freed, a consultant to the U.S. Agency for International Development (USAID).[26] But *why* are there so many more victims in the Third World ? The following accounts from around the world tell why.

Culiac n, Mexico

In Culiacán in Northern Mexico, where large plantations grow tomatoes for American supermarkets. government doctors report seeing two or three pesticide poisonings every week. Sometimes workers are brought in with convulsions. Since they get no paid sick leave, often they return immediately to the fields, where their condition deteriorates. Every two or three weeks a federal hospital in Culiacán treats a farmworker for aplastic anemia, the blood disease linked to organochlorine pesticides used in the area. Approximately half of these victims die.[27]

But *Los Angeles Times* reporters Laurie Becklund and Ron Taylor were told by one group of workers that "someone in their camp dies every two or three days."[28] The farmworkers are routinely poisoned by drifting pesticide sprays and leaking pesticide applicators, according to the reporters.

The workers live along the small patches of earth between the crops and the irrigation canals that receive all of the pesticide run-off. "They wash their babies, their dishes, and their clothes in the canals and then turn back to the canals to fill discarded insecticide tubs with canal water to drink," reports the *Times*. While the workers become ill from contaminated water, modern greenhouses with purified water systems have been erected to nurture the tomato seedlings. "The seedlings are more important than the people," one U.S.-born grower explained.[29]

Central America

More than 14,000 poisonings and forty deaths from pesticides were tabulated between 1972 and 1975 in the cotton-growing Pacific coastal plains of Central America, according to a three hundred page report by the Central American Institute of Investigation and Industrial Technology (AICAITI).[30] The actual total is undoubtedly much higher, but impossible to determine. According to the report, "some of the large cotton producers maintain their own clinics (partly) to hinder public health officials from detecting the seriousness of human insecticide poisonings."[31]

Although the pesticides are applied mainly to cotton grown for export, food crops—mainly corn and beans—are often contaminated simply because they are near the cotton fields. The report says that seventy-five percent of the sprayed pesticide frequently misses the cotton fields completely.[32] And toxic residues contaminate the soil.

Some farmworkers try to wash the pesticide from their skin, the ICAITI study revealed. But they use the irrigation drainage ditches, laced with the toxic runoff of insecticides, thereby compounding their contamination. Washing could not remove much of the parathion anyway, due to its pernicious tendency to concentrate in the oil on the skin, which transmits it directly into the bloodstream.[33]

Parathion, which causes eighty percent of Central America's poisonings,[34] was originally developed for chemical warfare by Nazi scientists during World War II. Slight chemical alterations converted it into a profitable insecticide after the war. The lethal dose of parathion to human beings is about *one-sixtieth* that of DDT: that is, it is sixty times more toxic. Parathion, explains Dr. H. L. Falk, of the National Institute of Environmental Sciences, "breaks down the substance which your body produces to stop the movement of your finger or your eye, for example. So those movements won't stop. You exhaust the muscles until they stop functioning altogether. You go into convulsions and die."[35]

The legacy of heavy pesticide use in Central America is ominous. Average DDT levels in cow's milk in Guatemala are ninety times as high as allowed in the United States. People in Nicaragua and Guatemala carry thirty-one times more DDT in their blood than people in the United States, where the substance has been banned since 1970.[36]

In Guatemala, reports *New York Times* correspondent Alan Riding, "the worst conditions, though the best pay, are on the cotton plantations. Here, pesticide spraying levels are so high that shipments of meat from cattle ranches in the area are frequently rejected by the

United States Department of Agriculture because of their high DDT content. Studies also show that DDT levels in human blood in the cotton districts are eight times higher than in Guatemala City. Yields, though, are among the highest in the world. 'It's very simple,' explained Eduardo Ruiz, a young cotton planter. 'More insecticide means more cotton, fewer insects mean higher profits.'[37]

"But little concern is shown for those living and working in the region," reports Riding. "At the height of spraying (in the Tiquisate area), thirty or forty people are treated daily in the nearby government clinic for the toxic effects on the liver and other organs.

"'The farmers often tell the peasants to give another reason for their sickness, but you can smell the pesticide in their clothing,' a nurse said. 'And we know the symptoms—dizziness, vomiting, and weakness. Only people who die in the clinic are reported. Otherwise bodies are buried on the farms.' "[38]

Pakistan

A world away from Central America, pesticides also kill. In Pakistan, at least five persons died and 2,900 others became ill in 1976 from malathion supplied in part by New York-based American Cyanamid for an American government program to eradicate malaria.[39] Monte-Edison, an Italian chemical company, also supplied the malathion.

Government Silence

Few Third World countries have either adequate pesticide regulations or the capacity to enforce them. As a result, the multinational pesticide producers have a free hand. Central America, for instance, has been turned into "a sort of experimental grounds for pesticide manufacturing companies," concludes the detailed study cited earlier.[40]

Most Third World governments are reluctant to disclose their poisoning statistics, incomplete as they might be. Robert Chambers, who supervised the GAO's investigation of pesticides, cites three reasons the pesticide poisonings are often hushed up.

"One is tourism," he explains. "It doesn't look good to have press reports about contaminated food. Two, no government wants to admit it was poisoning its own people. Would you admit you were allowing dangerous conditions in your country with President Carter's emphasis on human rights? Three, the countries are worried that if they report poisonings, the FDA will start to check their food exports to the United States and find illegal residues. This could have a severe adverse impact on their export earnings."[41]

Poisons in a Coke Bottle

Pesticide poisonings are much more common in the Third World than in the industrial countries not only because of the more brutal working conditions there, but also because of hazards of distributing any poison in societies where most people cannot read and have never had to learn the dangers of manmade chemicals.

"Small shops in Indonesia sell pesticides right alongside the potatoes and rice and other foods," says Lucas Brader of the U.N. Food and Agriculture Organization (FAO). "The people just collect it in sugar sacks, milk cartons, Coke bottles—whatever is at hand."[42]

"The laws in less developed countries typically say no repackaging of pesticides," Fred Whittemore of AID explains. "But in the villages it is done routinely. Parathion in Coke bottles stuffed with newspapers with no label is typical."[43] Gramoxone, which contains the deadly weed-killer paraquat, is not only sometimes sold in Coke bottles—it's the same color as Coke.

In Pakistan and Middle-Eastern countries, peasants sometimes wrap pesticides in their turbans, then place the turbans back on their heads to carry the pesticides to the fields.[44]

"In the rainy season in many tropical countries, the plastic liners used in pesticide bags are used as raincoats," says Whittemore. "That is an acute problem causing poisonings."[45]

Gramoxone killed at least eighteen people during a four-year period in the Western Highlands of Papua New Guinea, where it is used on coffee plantations and home gardens. "On June 16 a pastor conducted a religious service at Tega village near Mt. Hagen. He accidentally gave gramoxone instead of wine for communion to four people. They all died over the next week," Dr. D. J. Wohlfahrt, assistant secretary for Health in the Mt. Hagen district, wrote in the Papua New Guinea *Post Courier* of July 25, 1980. "In mid 1979 a young father bought gramoxone and stored it in a bottle. He asked his young son to go and get him a drink. He accidentally brought back the gramoxone and gave it to his father. After a gulp, the father realized it was not water he had drunk. But it was too late—he died," Wohlfahrt says.

"Gramoxone is legally marketed by the manufacturer in plastic bottles with built-in carrying handles that are just perfect for villagers to store drinking water in after they have used up the weed-killer.

"How many people are we prepared to kill for the convenience of also easily killing weeds?" asks the doctor.

Inadequate labeling or deliberate mislabeling of pesticides also causes poisoning in Third World countries. During 1979 the government of Colombia fined Hoechst and Shell for mislabeling pesticides, and fined Dow, Velsicol, Ciba-Geigy, American Cyanamid, and Hoechst for selling substandard products.[46] A recent check in Mexico disclosed that more than fifty percent of the pesticides sold there were labeled incorrectly.[47]

"One aid post orderly came to collect his medicines at Mt. Hagen Hospital and brought an empty gramoxone bottle to put the cough mixture in. The label read 'Poison' and had all the instructions written in English, but how many plantation laborers or village people can read English?" asks Dr. Wohlfahrt.

"Disposal of pesticides is a major problem, too," says Virgil Freed. "One horrible example is dieldrin in the Cameroon. A couple of years ago too much dieldrin was ordered, and the extra drums were simply placed outside in a jungle area. Now the containers have deteriorated and the dieldrin is spilling all over. I was there and saw the chemical sitting in puddles on the ground. There were people living in huts nearby. There could very well be subtle effects on them."[48]

Indiscriminate, widespread promotion of pesticides is especially disastrous in the Third World. In countries where most people cannot read, what use are warning labels on pesticide packages? In countries which outlaw unions that could protect farmworkers, what chance do peasants have against the crop duster's rain of poison? In countries with neither enough scientists to investigate pesticide dangers, nor enough trained government officials to enforce regulations, should foreign pesticide makers be given a free hand to push products so dangerous they are banned at home?

The Pesticide Boomerang

Pesticide pollution does not respect national borders. As one of the world's largest food importers, we in the United States are not escaping hazardous chemicals simply by banning them at home.

According to the U.S. Food and Drug Administration (FDA), approximately ten percent of our imported food contains illegal levels of pesticides.[49] But that ten percent is deceptive. The FDA's most commonly used analytical method does not even check for seventy percent of almost nine hundred food tolerances for cancer-causing pesticides.[50] (A tolerance is the amount of a pesticide allowed in any particular food product.)

In addition, the FDA frequently finds mysterious, unknown chemicals in imported foods. Government investigators believe that some of these fugitive chemicals come from the millions of pounds of 'unregistered' pesticides the EPA allows U.S. manufacturers to export without divulging any information about their chemical makeup or their effects on people or the environment.

With how little we know, we suspect these statistics from the General Accounting Office (GAO) represents only the tip of the iceberg:

- Over fifteen percent of the beans and thirteen percent of the peppers imported from Mexico, during one recent period, were found to violate FDA pesticide residue standards.[51]
- Nearly half the imported green coffee beans contain levels (from traces to illegal residues) of pesticides that are banned in the United States.[52]
- Freshly cut flowers flown in from Colombia caused a rash of organophosphate poisonings among American florists.[53]
- Imported beef from Central America often contains pesticide contamination. The GAO has estimated that fourteen percent of all U.S. meat is now contaminated with illegal residues,[54] and imports make a significant contribution to that total.

The pesticide residue problem has escalated to such a level that all beef imports from Mexico,[55] El Salvador,[56] and Guatemala[57] have been halted by the USDA. Agricultural practices in those countries, including heavy pesticide use on crops next to cattle-grazing land, have backfired on ranchers raising beef for the U.S. market.

Despite the widespread contamination of imported food, FDA inspectors rarely seize shipments or refuse them entry. Instead, a small sample is removed for analysis while the rest of the shipment proceeds to the marketplace...and the consumer. The rationale is that perishable food would spoil if held until the test results were known. But by the time the test results *are* available—showing dieldrin or parathion or DDT residues—the food has already found its way into our stomachs. Recalls are difficult.

During one recent fifteen-month period, government investigators found that *half* of all the imported food identified by the FDA as pesticide-contaminated was marketed without any penalty to the importers or warnings to consumers! Even products from importers with repeated violations were routinely allowed to pass. Here are some examples:

USDA officials in Dallas noticed a strong 'insecticide-like smell' in a batch of imported cabbage from an importer with a record of shipping contaminated products. Despite USDA's complaint, the FDA allowed the cabbage to go to market. A sample that had been removed for testing later revealed illegal levels of BHC, the dangerously carcinogenic pesticide whose registration was cancelled in 1976 at Hooker Chemical's request. But it was too late to recall the cabbage.[58]

Peppers from a shipment that was sent on to supermarkets turned out to have *twenty-nine times* more pesticide residue than allowed by U.S. law.[59]

In a world of growing food interdependence, we cannot export our hazards and then forget them. There is no refuge. The mushrooming use of pesticides in the Third World is a daily threat to millions there—and a growing threat to all consumers here. Therefore we and Third World people are allies in a common effort to halt the production of hazardous pesticides and contain *all* pesticide use to safe levels.

Pesticides to Feed the Hungry?

"We see nothing wrong with helping the hungry world eat," says an executive of the Velsicol Chemical Company, defending his company's overseas sales of Phosvel after it was banned in the United States.[60] And many would agree with his logic: since we need pesticides to produce more food for the hungry, pesticide dangers are a necessary evil—part of the price of averting famine. "Men will not starve because there are hazards in killing pests," is the way a Rohm and Haas official makes the same point.[61]

But in the course of our investigation, we came to a startling conclusion: over half, and in some countries up to seventy percent, of the pesticides used in underdeveloped countries are applied to crops destined for export to consumers in Europe, Japan and the United States.[62] The poor and hungry may labor in the fields, exposed daily to pesticide poisoning, but they do not get to eat many of the crops protected by pesticides.

In Central America a staggering seventy percent of the total value of agricultural production—mainly coffee, cocoa and cotton—is exported, despite widespread hunger and malnutrition there.[63] Cotton is one of the biggest pesticide users. In tiny El Salvador, cotton production absorbs one-fifth of all the deadly parathion used *in the world*.[64] Twenty-four hundred pounds of insecticides are used each year on every square mile of cotton fields in the country.[65] Yet cotton contributes to the global food supply only in processed cattle feed for Latin America's burgeoning beef production, almost half of which is exported to the United States and Europe.[66] The meat remaining for local consumption is eaten by the rich and the middle classes, not by the hungry.

Herbicides like 2,4,5-T and D (the basic ingredients of the infamous Agent Orange) are also used to help clear huge amounts of forest for grazing land in Latin America. The herbicide 2,4,5-T leaves residues of its contaminant, dioxin, in soil and water. Dioxin, one of the deadliest poisons ever developed, shows up later in birth defects, skin rashes, and miscarriages.

In Indonesia, estate-style farms growing export crops—coconuts, coffee, sugar cane, and rubber—consume twenty times the quantity of pesticides used by the small holders growing food for local markets. This, despite the fact that small holders cultivate seven times more acreage than the estates.[67]

Some might argue that although export crops do not directly feed hungry people, at least the foreign exchange earned benefits them indirectly: it is used to import economic necessities for development. But even the most superficial look at development in most Third World countries belies this assumption. Foreign exchange earned by agricultural exports does not return to improve the lives of the workers through better wages, housing, medical care, or schools. Instead the foreign exchange is most often plowed into luxury consumer goods, urban industrialization, tourist facilities, and showy office buildings—all geared to the budgets and tastes of the top ten or twenty percent living in the cities.

The Perfect Banana

One reason pesticide use is so much more intense on export crops than on subsistence food crops is that the multinational corporations which control the production and marketing of exports demand a blemish-free product. Nothing less, they say, will meet the discriminating standards of the consumers in Europe, North America or Japan.

"The Japs eat with their eyes" is how the manager of a Philippine banana plantation explained why they went to such lengths to produce a blemish-free fruit to ship to Japan.[68] In the United States, too, it is estimated that ten to twenty percent of insecticides used on fruits and vegetables serve only to improve their appearance.[69]

Most people think of multinational food corporations in the Third World as big plantation owners. But over the last twenty years, corporations have become leery of owning land directly. As the U.S. Overseas Private Investment Corporation warns, the possibility of "expropriation, revolution or insurrection {makes} plantations a poor risk."[70] Multinational food producers and marketers such as Del Monte, United Brands (formerly United Fruit), and Castle & Cooke (Dole brand) have hit upon a safer strategy—contract farming. Rather than own land directly, these companies now often contract with large local landowners to produce crops for export to consumers in the industrial countries.

A contract farming boom hit southern Mindanao, the Philippines, in the late sixties. Before that time there were no bananas growing on its rich coastal plains. Small farmers and tenant farmers grew rice and abaca. Then came the multinational corporations, seeking contracts with local entrepreneurs to produce bananas for the lucrative Japanese market. Within ten years the entire area was transformed: now twenty-one giant plantations cover 57,000 acres, and bananas have become one of the country's top agricultural exports.[71] In order to fulfill their banana contracts, the local entrepreneurs had to push small holders, tenants, and 'squatters' off the land. (Some of the so-called 'squatters' had worked the land for more than a generation.)

Although the multinational corporation may not own the land, it still calls the shots. When the corporation signs a local entrepreneur under contract, it specifies not only the amount of fruit or other commodity to be produced but also the amount of fertilizers and pesticides to assure high yields and blemish-free products.[72]

Lifetime Debt to Pesticide Companies

Once locked into the banana export contract, the plantation owner is totally dependent on the multinational firm. "Money is deducted from the banana grower's earnings to pay for things like pesticides and irrigation," explains Father Jerome McKenna, a U.S. missionary who worked in the area. "It's part of the contract. Those banana growers will be in debt to the pesticide companies for the rest of their lives."[73]

Typically pesticides are applied at three stages in the banana production process. Workers with heavy tanks strapped to their backs (and no masks or protective covering) routinely spray every tree. Twice a month a pesticide plane passes over the plantation, blanketing everything, including the drinking water supply. A group of banana workers recently petitioned Castle & Cooke to stop heavy pesticide spraying after local studies showed that the workers have dangerously low oxygen levels in their blood, making them more susceptible to disease.[74]

In the packing sheds, the bananas are dumped in long water-filled troughs to remove some of the pesticides. "What bothers me most," says McKenna, "is that these people have very little protection from the chemicals they come in contact with. The women have their hands in the water up to their elbows all day long. They don't wear any gloves. Their only protection is plastic-type aprons they fashion for themselves."[75] Finally, to protect the fruit during its long ocean voyage, women workers in the packing sheds spray every bunch of bananas with a fungus-killing agent.[76]

McKenna checked at two nearby hospitals for reports of pesticide poisonings. One, run by Castle & Cooke, "didn't have any cases." But the other hospital run independently of the company, had "reports all around of people poisoned by pesticides."[77]

The contract farming system also gives the multinationals an easy way to avoid responsibility for pesticide poisoning. They can simply blame the local plantation owner for being careless.

The examples of cotton in El Salvador or bananas in the Philippines tell us that, in large measure, pesticides in the Third

World actually feed the well-fed, but endanger the poor and the hungry. Since the mid-fifties, the growth rate of export crops—which receive the overwhelming bulk of pesticides—has exceeded that of food crops.[78] Between 1952 and 1967, cotton acreage in Nicaragua increased fourfold while the acreage in basic grains was cut in half.[79] It is hardly surprising that the demand for pesticides in the Third World has soared. What is surprising is how many believe that their principal use is to save crops to feed the hungry.

More Food and Yet More Hunger

While it is true that most pesticides in the Third World are used on luxury export crops, in the last twenty years Third World farmers growing basic food crops—especially rice and wheat—have also been encouraged to use ever greater quantities of pesticides. As part of the 'Green Revolution,' hybrid seeds were developed which produce higher yields, given the correct amount of fertilizer and water; but the hybrids are much more susceptible to pests. Bred in the laboratory and in test fields in a foreign setting over only a few years, these 'miracle seeds' do not have the pest resistance characteristic of traditional seeds, bred over thousands of years in the same locality in which they are used.[80] To make up for this vulnerability, the new seeds must be protected with more pesticides.

Throughout much of the Third World, international lending agencies and government development programs have encouraged the use of these new seeds, often making their use a condition for receiving farm credit.[81] Once Third World farmers begin using the new, more vulnerable seeds, they have no choice but to vastly increase their use of pesticides.

Few dispute that the new seeds and their accompanying inputs—fertilizers and pesticides—have increased grain production, notably in Asia. But growing more food doesn't necessarily mean alleviating hunger. What we have learned is that food production can increase while the poor majority gets even more hungry.

Take the Philippines. It is the home of the prestigious International Rice Research Institute which helped instigate the Green Revolution in Asia. During the 1970s, use of the new seeds spread throughout the country. Accompanying their proliferation, pesticide imports leapt fourfold between 1972 and 1978.[82] As a result of the new seeds and new inputs, rice production almost doubled in the Philippines in little more than a decade.[83] Indeed, in the late 1970s, the Philippines became a rice exporter. But has this production success reduced the hunger of the Philippine poor? No. According to studies by the Asian Development Bank and the World Health Organization, Filipinos are now the worst fed people in all of Asia, with the exception only of war-torn Kampuchea.[84]

How can there be more food produced and yet greater hunger? The answer is that the Green Revolution strategy for producing more food forces more and more people off the land. Mechanization robs them of work. Dependency on irrigation, pesticides, and fertilizers—all required by the new seeds—favors the wealthier, literate farmers who have access to credit and political pull. Without land to produce food or money to buy it, people go hungry no matter how much their country produces.

This dramatic transformation is documented in the International Labor Organization's study of rural poverty. After studying seven Asian countries, comprising seventy percent of the rural population in nonsocialist underdeveloped countries, the ILO reported that the rural poor have become measurably poorer than they were ten or twenty years ago. The study concludes: "The increase in poverty has been associated not with a fall but with a rise in cereal production per head, the main component of the diet of the poor."[85]

Another ILO study of the Green Revolution points to vast increases in wheat yields in the Punjab district of India in the 1960s. Yet simultaneously, the portion of the rural population living below the poverty line increased from eighteen to twenty-three percent.[86] "Economic prosperity has not simply missed these people," the study concludes. "Their ability to supply their own basic needs has been gradually but unrelentingly reduced..."[87]

The Poor: Not a Lucrative Market

The narrow production push embodied in the Green Revolution strategy, helping to enrich well-placed farmers and further impoverish the rural poor, has itself encouraged the shift toward export crop production that we discussed above. This is true in part because impoverished people simply do not make up a lucrative market. So, as in the Philippines, a staple food like rice is exported while

Filipinos—without money enough to buy the rice—go hungry or production shifts from staple foods needed by the poor and toward luxury items demanded by the rich. Corn and bean production in Mexico has declined while production of luxury fruits and vegetables for the U.S. market and feedgrains such as sorghum have greatly increased. Almost thirty-two percent of basic grain staples are now fed to livestock in Mexico.[88] In Brazil the figure is forty-four percent.[89]

The rationale of using more pesticides to protect crops to feed the hungry simply does not hold up. First we discover that most pesticides are not used to protect food crops anyway! Second, pesticides to protect the more vulnerable grain seeds of the Green Revolution are part of a production strategy benefiting the better off. While increasing production, this strategy cannot eliminate hunger because it fails to address the question of *who controls* that production Under these conditions, the extra food which pesticides help to grow is frequently either eaten by the better off, exported or fed to livestock. The whole equation bypasses the fundamental problem: the hungry have neither money to buy food nor land to grow it on.

The Global Pesticide Supermarket

From the billboards of rural Nebraska to shantytown walls in Kenya, pesticide company advertising is part of the scenery. The language may be English or Spanish or Swahili, but the message is the same: you need our brand of pesticide if you want a good crop.

"Whenever a new pesticide hits the area, every farmer knows about it right away," says Dr. Lou Falcon, a University of California entomologist who has studied Central America. "There is heavy publicity by the companies—big billboards, radio, and newspaper ads."[90]

Using sophisticated marketing techniques and their worldwide network of subsidiaries and affiliates, the giant multinational pesticide manufacturers—such household names as Dow, Shell, Chevron, Bayer, Dupont—have created a global supermarket, its shelves stocked with products so dangerous they have been banned in the countries where they have been investigated.

As we have said, the multinationals claim they sell pesticides overseas merely to supply a demand, a demand for their products to help feed a hungry world. But the fact is that multinational companies use sophisticated mass marketing techniques to *create* a demand in the Third World.

"Those pesticide boys are all over the place down there," says Michel Moran of the Interamerican Institute for Agricultural Sciences in Costa Rica. "It's amazing how they get down to the grass roots. Very few places are left in Latin America which are in isolation from the new technologies, including pesticides."[91]

"We have overseas offices in almost every country in Asia," explains Rene Montmeyor, an agricultural product supervisor for Stauffer Chemical Company. "We have exclusive distributorships in most of those countries, too. We have our technical people who instruct farmers how to use our pesticides."[92]

Ads for pesticides appear prominently in Third World agricultural journals. Away from the eyes of U.S. regulators, pesticide companies often extol the virtues of pesticides banned in the U.S.

At a supply center for the Kenyan Farmer's Association in Nairobi, a reporter spotted aldrin, BHC, and chlordane—all banned from most uses in the United States—for sale on shelves and listed in the association's inventory. They were being sold by local subsidiaries of European pesticide companies—ICI, Bayer, and Shell.[93]

Formulating Their Way Around Regulation

To escape regulation in their home countries, the multinationals have discovered a clever strategy: they simply ship the separate chemical ingredients of a banned pesticide to a Third World country, then manufacture it there in 'formulation plants.' From the Third World country, the prepared pesticide can often be re-exported to any third country, free of regulation.

"It's a real Mafia-type operation," says Dr. Harold Hubbard of the U.N.'s Pan American Health Organization. "Global companies are setting up formulation plants all over the world. (They) simply go into less developed countries, give a banned pesticide a local name, and turn around and sell it all over the world under that new name."[94]

"Formulators buy basic ingredients from importers and then put them together and call the product a name like 'Macho' and say it will kill anything," explains Frank Penna, a consultant to the Policy Sciences Center. "Usually it ends up killing the farmer."[95]

('Macho' competes with other chemical weapons with such names as Ambush and Fumazone to battle an army of enemies led by kernel smut, the stinkbug, the whorl maggot, and the black whip, and tip smut.)

The pesticides are dangerous before they ever reach the fields. A plant in Kenya which formulates BHC provides no protection for the workers mixing the chemicals. "The workers' eyes were all sunken, and they looked like they had TB," says a University of Nairobi professor who visited the plant. "There are regulations against this sort of thing, but there is no manpower for enforcing the regulations. And no one complains. The workers are perfectly happy until one of them gets sick, and then he's just fired."[96]

In Latin America, "you can see the dust rising from those formulation facilities for miles," says USAID's Whittemore. "I wouldn't dare walk into some of them. There are no decent health or environmental standards for most of them—it's a terrible problem."[97]

The worst formulators, Penna says, are the "pirate operators—little whiskey-still-like operations." An estimated 8,000 of them have opened in Brazil alone.[98] But the large-scale formulation plants are foreign-owned.

Like many other Third World countries, Brazil offers special incentives to bring foreign chemical plants into the country: deferral of taxes, exemption from import duties, government-sponsored clearing of land for the plants.[99] Shell has put $20 million to $30 million into new plants under these incentives over the past few years. Dow has a 2,4-D plant there.[100] The Swiss firms Sandoz and Ciba-Geigy set up a joint operation.[101] And the largest pesticide company in the world—Bayer—has formulation plants in Brazil as well as in virtually every other country with a market large enough to warrant one.

Formulation plants are also spreading throughout Asia:

India. Many pesticides that have been banned or heavily restricted in the United States are produced in India, including BHC and DDT.[102] Union Carbide, ICI, Bayer, and Hoechst have plants there.[103]

Malaysia. Dow and Shell alone formulate one-quarter of all liquid pesticides here. Three organochlorines banned in the United States—aldrin, DDT, and BHC—constituted 730 of the 960 tons of pesticides manufactured in Malaysia in 1976.[104]

Indonesia. Bayer, ICI, Dow, and Chevron dominate the local pesticide manufacturing industry, accounting for over seventy percent of the total production in 1978.[105]

This trend toward formulation plants is paralleled in many heavily regulated industries which are also moving their production facilities overseas.

Seeds: The Final Round?

The multinational pesticide producers already control the manufacturing, distribution and promotion of pesticides at the global supermarket. Now they are working on a strategy to control an even more basic agricultural 'input,' the seeds themselves.

"Where might a chemical company interested in agricultural chemicals go?" rhetorically asks a high official of the Chemical Manufacturers Association. "Obviously, into seeds," he answers. "Some members of the chemical industry are getting into seed development."[106]

The FAO estimates that by the year 2000, sixty-seven percent of the seeds used in underdeveloped countries will be the 'improved' varieties, which in most cases are more vulnerable to pests.[107] Since virtually all pesticides are produced in the industrial countries, that means more pesticide exports to the Third World.

For the agri-chemical multinationals, plant patenting provides greater inducement to add seeds to their conglomerate families. Championed by the American Seed Trade Association and the USDA controversial legislation to allow the patenting of all U.S. crop varieties has been debated in Congress since early 1980.[108] The bill would extend the patent umbrella to six crop varieties that were excluded from the original 1970 plant protection act.[109]

Already a few multinational corporations, many of them pesticide producers, control the seed patents for several important crops. Of the seventy-three patents granted for beans, for example, over three-quarters are held by just four corporations: Union Carbide, Sandoz, Purex, and Upjohn.[110] Two Swiss-based companies, Sandoz and Ciba-Geigy, alone control most of the U.S. alfalfa and sorghum seed supply.[111]

Chemical companies are buying traditional seed supply firms, and their patentable 'commodities,' at an alarming rate. After the first wave of acquisitions, the international pesticide giants Monsanto, Ciba-Geigy, Union Carbide, and FMC are ranked among the largest seed companies in the United States.[112] In the ten years between 1968 and 1978, multinationals—mainly chemical and pharmaceutical companies—bought thirty major seed companies. Today, the largest seed enterprise in the world is Shell, the oil and petrochemical giant which controls thirty seed outfits in Europe and North America.[113]

Entering the $10-billion-a-year seed industry is a natural for the multinational pesticide producers. They already have the marketing and distribution structures for reaching the small farmer throughout the world, explains *The Global Seed Study*, a $25,000-a-copy investment guide sold to potential seed investors.[114] The study points out how seeds and chemicals can work together, as in the possibility of "seed coatings and pelleting, utilizing the seed as a delivery system for chemicals and biologicals to the field."[115]

By concerning the local seed market, the companies apparently plan to insure the farmers the world over are dependent on their seeds, as well as their fertilizers and pesticides.

"Obviously they're being damn quiet about it," says an industry official. "But some of those high yield seeds require particular applications of fertilizers and pesticides to produce their high yield."[116]

Now that the chemical companies have entered the seed business, they hold the enviable economic position of helping to aggravate the (pest) problem for which they also offer their (chemical) cure. If the chemical industry's monopolization of the world's seed stock is successful, we will be one critical step closer to the ultimate corporate vision of the global supermarket, where every grower in the world is hooked on patented seeds and the pesticides they require.

Genetic Uniformity

Common non-patented varieties often become extinct and disappear as seed varieties are patented. By 1991, the FAO's Erna Bennett estimates, three-quarters of all vegetable varieties now grown in Europe will be extinct due to patenting, which is more advanced in Europe than in the United States.[117]

As fewer seed varieties are used to grower larger crops, the earth's genetic base is narrowing.[118] At the same time, the uniform high-response variety seeds of the green revolution are displacing centuries-old varieties and accelerating their disappearance from the earth's seed stocks.

The implications of this genetic uniformity may be devastating to our food supply, The hybrid, high-yielding seeds do not have an inbred resistance to pests and are usually planted in huge fields that can satisfy swarms of the same type of pest. "If the crop is a monoculture, you no longer have the buffers of different varieties of crops," adds a congressional aide working on the plant patenting issue. "What you've got instead is a superhighway for these insects."[119]

Scientists now suggest that genetic uniformity was the underlying cause of the Irish potato famine in the late 1840s. Then, a single potato variety imported from the Caribbean was struck by blight and over one million people starved to death.[120] More recently, the United States had a glimpse of what this genetic uniformity means, when fifteen percent of the nation's corn crop was destroyed by a pest epidemic in 1970. (Only six seed types make up seventy-one percent of the domestic corn crop.[121])

The world's farmers will become even more dependent on pesticides as they find their seed varieties are less able to resist the diseases and pest epidemics that sweep through local areas periodically.

1 Proceedings of the U.S. Strategy Conference on Pesticide Management, U.S. State Report, June 7-8, 1979, pg. 33.

2 "Report on Export of Products Banned by U.S. Regulatory Agencies," House Report no. 95-1686, October 4, 1978, pg. 28.

3 Ibid.

4 U.S. General Accounting Office. "Better Regulation of Pesticide Exports and Pesticide residues in Imported Foods is Essential," Report no. CED-79-43, June 22, 1979, pp. iii, 39.

5 FIFRA, 7 U.S.C. 1360

6 "New Pesticides Must Now Be Economic Winners," *Chemical Age*, February 17, 1978; Dr. Jay Young, Chemical Manufacturers Association, telephone interview with authors, October 1979.

7 President's Hazardous Substances Export Policy Working Group, Fourth Draft Report, January 7, 1980, pg. 6.

8 Dr. Hal Hubbard, telephone interview with authors, June 1, 1977.

9 "Importacion de Pesticides," Ministerio de Agricultura y Gamaderia, Costa Rica, 1978.

10 "Listudo de Pesticides Registrados en el Departamento de Sanidad Vegetal," Ministry of Agriculture, Ecuador.

11 "Report on Plant Protection in Major Food Crops in Malaysia," U.N. Food and Agricultural Organization, December 1977, pg. 59.

12 "Plaguicidas de Uso Agricola, Défoliantes y Reguladores Fisiologicos de las Plantas Registrados en Colombia," Ministerio de Agricultura, Columbia, June 30, 1979.

13 "Erosion and Soil Depletion-By-Products of Castle & Cooke Operation," Communications, M.S.P.C., April 1978.

14 O'Toole, Thomas. "Over Forty Percent of World's Food is Lost to Pests," *Washington Post*, March 6, 1977.

15 Starr, Douglas. " 'Pesticide Poisoning Alarming,' Says FAO," *Christian Science Monitor*, February 1, 1978.

16 Ibid.; and Francis Moore Lappé and Joseph Collins, *Food First: Beyond the Myth of Scarcity* (New York: Ballantine Books, 1979), op. cit., pg. 64.

17 U.S. General Accounting Office. "Better Regulation of Pesticide Exports and Pesticide Residues in Imported Food is Essential," Report no. CED-79-43, June 22, 1979, pg. 1.

18 Ayres, Jeanie. "Pesticide Industry Overview," *Chemical Economics Newsletter*, January-February 1978, pg. 1.

19 *Food First: Beyond the Myth of Scarcity*, op. cit., pg. 41.

20 "An Environmental and Economic Study of the Consequences of Pesticide Use in Central America Cotton Production," Final Report, Instituto Centro-Americano de Investigation y Technologia Industrial (ICAITI), January 1977, pp. 149, 155, 161.

21 *Food First: Beyond the Myth of Scarcity*, op. cit., pg. 67.

22 Yasuo, Osawa. "Banana Plantation Workers Strike in the Philipines," New Asia News, May 1980, pg. 7.

23 Dr. Lou Falcon, telephone interview with authors, May 21, 1979.

24 U.N. Food and Agricultural Organization. *Agriculture: Toward 2000*. (Rome: U.N. Food and Agricultural Organization, 1979), pg. 82.

25 " 'Pesticide Poisoning Alarming,' Says FAO," *Christian Science Monitor*.

26 Dr. V.H. Freed, personal interview with authors, January 4, 1980.

27 Becklund, Laurie and Ronald Taylor. "Pesticide Use in Mexico—A Grim Harvest," *Los Angeles Times*, April 27, 1980.

28 Ibid.

29 Ibid.

30 ICAITI Report, pp. 97-98.

31 Ibid., pg. 195.

32 Ibid., pg. 2.

33 Riding, Alan. "Guatamala: State of Siege," *New York Times Magazine*, August 24, 1980, pg. 20.

34 ICAITI Report, pg. 164.

35 Dr. H.L. Falk, telephone interview with authors, May 25, 1979.

36 ICAITI Report, pg. 128-132.

37 "Guatamala: State of Siege," pg. 20.

38 Ibid.

39 Baker, Dr. Edward, et al. "Malathion Intoxication in Spray Workers in the Pakistan Malaria Control Program," U.S. Agency for International Development, pp. 3, 7.

40 ICAITI Report, pg. 29.

41 Robert Chambers, GAO, personal interview with authors, March 17, 1980.

42 Lucas Brader, personal interview with the authors, June 7, 1980.

43 Dr. Fred Whittemore, personal interview with the authors, March 17, 1980.

44 U.S. Strategy Conference on Pesticide Management, U.S. State Department, workshop discussions, June 7-8, 1979.

45 Whittemore interview, March 17, 1980.

46 Mensual, Carta. "Boletin Informativo de la Division de Supervision de Insumos Agricolas," Ministerio de Agricultura, Government of Columbia, October 1979, pp. 2-3.

47 Whittemore interview, March 17, 1980.

48 Freed interview, January 4, 1980.

49 "Report on Export of Products Banned by U.S. Regulatory Agencies," U.S. House Report no. 95-1686, October 4, 1978, pg. 28.

50 "Federal Efforts to Regulate Pesticide Residues in Food," Statement of Henry Eschaege, GAO, before Subcommittee on Oversight and Investigations, House Committee on Interstate and Foreign Commerce, February 14, 1978, pg. 12.

51 "Pesticides in Mexican Produce (FY '79)," Chapter 5, FDA Compliance Program Guidance Manual, TN-78-194, December 15, 1978, Attachment H, pg. 1.

52 "Pesticides in Imported Coffee Beans (July 1974-May 1977 and August 1977-October 1978)," FDA Compliance Program Evaluation, report in "U.S. Export of Banned Products," hearings, July 11-13, 1978, pp. 210-11.

53 "Pesticide Contamination of Imported Flowers," Morbidity and Mortality Weekly Report, HEW publication no. CDC-77-8017, April 29, 1977, pg. 143.

54 "Foreign Meat Inspection 1978," U.S. Department of Agriculture, March 1979.

55 Mitchell, William. "Study Finds Meat Can Be Dangerous to Your Health," Knight News Service report in *San Francisco Examiner*, May 1, 1979, pg. 8.

56 "Illegal Pesticides Found in Imported El Salvador Meat," Associated Press report in unknown newspaper, April 1980.

57 Richard Mikita, U.S. Department of Agriculture, personal interview with authors, March 18, 1980.

58 "Better Regulation of Pesticide Exports and Pesticide Residues in Imported Foods is Essential," U.S. General Accounting Office Report no. CED-79-43, June 22, 1979, pp. iii, 39.

59 Ibid.

60 Richard Blewitt, vice-president of public relations, telephone interview with Terry Jacobs, Center for Investigative Reporting, July 31, 1979.

61 "Proceedings of the U.S. Strategy Conference on Pesticide Management," U.S. State Department, June 7, 1979, pg. 30.

62 There are no statistics that could provide a precise estimate. This rough estimate is drawn from impressions we have received from reports and conversations with

government and corporate officials and others with knowledge of Third World agriculture. In their book *Food First*, Lappé and Collins refer to an estimate by the chief of the Plant Protection Service of the U.N.'s Food and Agriculture Service, W.R. Furtick: the "vast majority" of pesticide use in the Third World is on export crops.

63 "An environmental and Economic Study of the Consequences of Pesticide Use in Central American Cotton Production," Final Report, Instituto Centro-Americano de Investigacion y Technologia Industrial (ICAITI), January 1977, pg. 26.

64 "Over Forty Percent of the World's Food Is Lost to Pests," op. cit.

65 Ibid.

66 *Food First: Beyond the Myth of Scarcity*, pg. 289.

67 "Basic Supply and Marketing Data for Agro-Pesticides in Indonesia," ARSAP/Pesticides, FAO (Bangkok), January 1980, pp. 15-16.

68 McCallie, Eleanor and Francis Moore Lappé. *The Banana Industry in the Philippines: An Informal Report,* (San Francisco: Institute for Food and Development Policy, 1977), pg. 8.

69 Pimintel, David, et al. "Pesticides, Insects in Foods, and Cosmetic Standards," *BioScience*, March 1977.

70 Overseas Private Investment Corporation, Annual Report, 1973.

71 *The Banana Industry in the Philippines: An Informal Report*, op. cit., pg. 1.

72 Fr. Jerome Mc Kenna, telephone interview with authors, April 7, 1980.

73 Ibid.

74 Larry Rich, "Castle & Cooke Inc.," report prepared for Big Business Day campaign, March, 1980, pg. 15.

75 McKenna interview, April 7, 1980.

76 Ibid.

77 Ibid.

78 *Food First: Beyond the Myth of Scarcity*, pg. 1.

79 Dorner, Peter. "Export Agriculture and Economic Development," Land Tenure Center, University of Wisconsin, Madison, statement before the Interfaith Center on Corporate Responsibility, New York, September 14, 1976, pg. 6.

80 National Academy of Sciences, Committee on Genetic Vulnerability of Major Crops, *Genetic Vulnerability of Major Crops*, report, Washington, DC, 1972.

81 Benton Rhoades, personal interview with authors, March 21, 1980.

82 "Basic Supply and Marketing Data for Agro-Pesticides in the Philippines," ARSAP/Pesticides, FAO (Bangkok), February 1980, pg. 30.

83 Between 1961-65 and 1976, rice production in the Philippines increased by two-thirds. Andrew Pearse, "A Case for Peasant-Based Strategies," United Nations Research Institute for Social Development, Report No. 79.1, Geneva, May 1979, pg. 36. And rice production has continued to increase, according to the FAO *Production Yearbook*, 1978.

84 Ping, Ho Kwon. "The Mortgaged New Society," *Far Eastern Economic Review*, June 29, 1979, citing Asian Development Bank and World Health Organization reports.

85 *Food First: Beyond the Myth of Scarcity*, pg. 146.

86 "Third World Seen Losing War on Rural Poverty," ILO (International Labor Office) *Information*, February 1980, pg. 8.

87 Ibid.

88 Bachman, Kenneth and Leonardo Paulino. "Rapid Food Production Growth in Selected Developing Countries: A Comparative Analysis of Underlying Trends 1961-76," Research Report 11, (Washington, DC: International Food Policy Research Institute, 1979), pg. 29.

89 Ibid.

90 Dr. Lou Falcon, telephone interview with authors, May 21, 1979.

91 Michael Moran, personal interview with authors, March 12, 1980.

92 René Montmeyor, telephone interview with authors, February 20, 1980.

93 Julie Kosterlitz, letter to authors, May 11, 1979.

94 Dr, Harold Hubbard, telephone interview with authors, June 1, 1979.

95 Frank Penna, telephone interview with authors, March 25, 1980.

96 Julie Kosterlitz, May 11, 1980.

97 Dr. Fred Whittemore, personal interviews with authors, March 17, 1980.

98 Frank Penna interview, March 25, 1980.

99 Ibid.

100 Gary Jones, Dow Chemical Corp., telephone interview with authors, June 11, 1979.

101 Mooney, P.R. *Seeds of the Earth* (London, UK: International Coalition for Development Action, 1979), pg. 117.

102 "List of Standardized Pesticides as Revised by the Pesticide Review Committee on June 16, 1979," provided by G.A. Patel.

103 G.A. Patel, India Institute of Management (Ahmedabad), personal interview with David Kinley, Institute for Food and Development Policy (San Francisco), March 25, 1980.

104 U.N. Food and Agriculture Organization. "Report on Plant Protection in Major Food Crops in Malaysia," December 1977, pg. 17.

105 U.N. Food and Agriculture Organization (Bangkok). "Basic Supply and Marketing Data for Agro-Pesticides in Indonesia," ARSAP/Pesticides, January 1980, pg. 25.

106 Confidential interview with authors.

107 *Agriculture: Toward 2000*, pg. 82.

108 Randolph, Eleanor. "Seed Patents: Fear Sprouts at Grass Sprouts," *Los Angeles Times*, June 2, 1980, pg. 1.

109 H.R. 999, introduced January 18, 1979.

110 *Seeds of the Earth*, pg. 57.

111 Ibid., pg. 62.

112 Ibid., pg. 55, 58.

113 Ibid., pg. 56.

114 Ibid.

115 Confidential interview with authors.

116 "Seed Patents: Fear Sprouts at Grass Sprouts," op. cit.

117 Ibid. See also *Conservation of Germplasm Resources: An Imperative* (Washington, DC: Committee on Germ Plasm Resources, National Academy of Sciences, 1978)

118 Confidential interview with authors.

119 *Food First: Beyond the Myth of Scarcity*, pg. 157.

120 Ibid.

121 *Seeds of the Earth*, op. cit., pg. 14.

From *Food First Backgrounder*, Spring 1997, by Mark Cooper, Peter Rosset, and Julia Bryson

---

WARNING! Corporate Meat and Poultry May Be Hazardous to Workers, Farmers, the Environment and Your Health

> *The people had come in hordes.* (The meatpacker was) *speeding them up and grinding them to pieces, and sending for new ones.*
> —Upton Sinclair in *The Jungle*, 1906

> *With many of the over 150,000 workers in the poultry processing plants thwarted in their efforts to organize, their workplaces remain ever so dangerous. Assembly lines that are constantly accelerated, abnormal temperatures and rapid, repetitive hand motions all contribute greatly to worker skin diseases, crippling hand and arm illnesses, called cumulative trauma disorders, ammonia exposure, infections from toxins in the air, stress and back problems.*
> —A.V. Krebs in *The Corporate Reapers*, 1991[1]

STORM LAKE, IOWA. On his one day off from work this week, forty-five year old Heriberto Solis sits in his uninsulated trailer in a dilapidated mobile home park that the residents of this town of 10,000 in northwestern Iowa call "Little Mexico." While staring out at the piles of snow outside, he laments where destiny has dropped him.

Six days a week, eight hours a day, Heriberto stands on his feet in the refrigerated pork slaughterhouse run by IBP Corporation, the world's fourth largest food manufacturing firm with an estimated $10 billion in sales per year.[2] For take home pay of less than $300 a week, he tediously slices meat off of hog backbones. The work is dangerous and arduous, but he feels he has few other opportunities.[3]

"The company loves to work with illegals," Heriberto says. "When you are illegal you can't talk back. You keep your head down and follow orders. We say you can't do nothing." The workers in the plant are not alone in their concerns about the large corporate packers. Most independent farmers who raise animals are now close to ruin, in part because a wave of mergers and take-overs in the meat industry have driven livestock prices down.[4]

The USDA recently held hearings and asked for testimony from the farmers who produce animals for slaughter. "The powerlessness of the producers in their dealings with concentrated buyers was heard about again and again," the Minority Report by six commissioners stated.[5] "As the number of buyers steadily declines, with only one bidder for cattle in most circumstances, those who incur the dislike of a buyer face economic ruin," it continued. "The fear of the overwhelming power of the packers was raised by beef producers and echoed by in testimony from poultry producers. Retaliation for organizing activities can quickly lead to a producer's bankruptcy."

Corporate Concentration and Falling Wages

Over the past few decades the meat-packing industry has come to be more concentrated than any other in the U.S., as large integrated conglomerates have bought out and squeezed out the independents, and increasingly extended their control over farmers through draconian contract arrangements. In 1973 the top four beef packing companies slaughtered twenty-nine percent of steers and heifers. Today that figure has risen to over eighty percent, and the industry is dominated by three transnational corporations (ConAgra, Cargill, and IBP).[6] Four companies now control forty-five percent of pork production (IBP,

ConAgra, Cargill, and Sara Lee). And four control forty-four percent of boiler production in the poultry industry (Tyson, ConAgra, Gold Kist, and Perdue Farms).[7]

The IBP factory in Storm Lake both dominates and depends upon the work of the six hundred or more Mexican and Central American workers and their families who have come to live here in Storm Lake. Alongside 1,500 Laotians, these immigrant workers are now the majority of the workforce at IBP's massive pork-processing plant.

And not only here in Storm Lake. In a sweeping regional arc that slashes through America's heartland north from the Dakotas, through Minnesota, Nebraska, and Iowa, and then down through Kansas into Northern Texas and Missouri, and now south and east even into the Carolinas and Virginia, scores of meatpacking communities have become the new homes to tens of thousands of impoverished Third World workers. "The entire debate over whether or not immigrants are of economic benefit is disingenuous," says University of Northern Iowa anthropologist Mark Grey, an expert on the restructured packing industry. "No one wants to state the truth—that food processing in America today would collapse were it not for immigrant labor."

Beef, pork, and poultry packers have been aggressively recruiting the most vulnerable of foreign workers to relocate to the American plains in exchange for $6 an hour jobs in one of the country's most dangerous industries. On the Midwestern prairie, these workers now occupy jobs that once went to unionized meatpackers earning three to four times the current wage.

Staggering injury rates—twenty-seven percent in poultry and forty-two percent in meat—and on-the-job stress caused by difficult repetitive work, often means employment for just a few months before a worker quits or the company forces him or her off the job. It's a human chain greased by the grueling work regimen which generates an astonishing worker turn-over rate of eighty to one hundred percent a year — a rate common to the entire industry. "Perfect for the company," says Heriberto. "Most workers just leave before six months is up and the company health insurance begins." As if that were not enough, these new immigrants are being left even more vulnerable by the Clinton administration's welfare and immigration reforms which have a direct and devastating impact on their already fragile existence.

Corporate Megaprofits and Contract Farming

Meanwhile, IBP made a juicy $257 million in profits in 1995, with chairman Robert Patterson receiving a $5.2 million bonus to go with his $1 million annual salary.[8] It's that sort of attractive bottom-line that has over the last fifteen years fueled a revolution—nay a counter-revolution—that has convulsed and redrawn the face of American meat-packing. And if you could boil that counter-revolution down into one slogan it would be: Death to Independent Farmers and Meatpackers! In 1973 there were 795 federally-inspected packing plants, and only two were classified as very large, and they only handled 7.5 percent of the market. Twenty-two years later eighteen megaplants accounted for eighty percent of the market; the independents having largely been driven out.[9]

The immense corporate control of the industry forces farmers into a peonage system over which they have little control. The intensifying vertical integration of the poultry industry, for example, means that broiler and egg production are controlled by a handful of processing companies that contract with individual farmers. Under the contract, the companies supply the birds, feed, and the management scheme (some companies, like Cargill, even hold monopolies on the seeds used to grow the feed.)[10] Farmers are allowed to own their own buildings but are paid by piece rate at levels that often don't cover their costs, let alone a return on investment.[11] The integrity of a farmer as an independent decision-making individual has been lost along with his or her economic viability. As journalist Susan Meeker-Lowry put it, "whole communities' way of life is forever changed as their farms are transformed into meat factories and farmers into production workers."[12]

Dangerous to Your Health?

Concentration and integration are part and parcel of the industrialization of the meat and poultry industries, with significant downsides for consumer health and the environment. Perhaps the most striking has been the much publicized outbreak of 'Mad Cow Disease' in Great Britain, probably associated with Creutzfeldt-Jakob degenerative brain disease in humans. Cattle are thought to acquire the disease by eating commercial feed containing ground up dead animal parts.

It is feared that the disease can be transmitted to humans who consume beef contaminated with Mad Cow Disease.[13] It seems quite a perversion of industrial agriculture to feed ground up dead animals to cattle who are herbivores. Fifteen human deaths have so far been linked to this syndrome, and a recent report in the highly respected scientific journal Nature predicts that future deaths in the U.K. could range anywhere from the low hundreds to the tens of thousands.[14]

Last March the U.S. livestock industry announced a voluntary, partial ban on using certain dead animal parts in feed.[15] The FDA recently proposed making it a mandatory ban, and is now awaiting feedback on the proposed regulation.[16] As if confirming our concern, the industry was just rocked by a scandal in which the USDA found banned spinal cord fragments (the tissue most implicated in Mad Cow Disease) in ground beef.[17]

According to Richard F. Marsh, a University of Wisconsin at Madison Veterinarian "there are reasons to believe that Mad Cow Disease has already risen spontaneously in American cattle, but it apparently has not jumped into the animal feed supply at this point." Yet proposed FDA regulations will still allow cow protein to be fed to fish, chicken, and pigs in hope that if Mad Cow Disease were to appear, a species barrier would stop it from spreading.[18]

On the level of more common sicknesses, data suggests a link between industry concentration and outbreaks of illnesses linked to beef. The median size of E. coli and other disease outbreaks has nearly doubled, from sixteen to thirty-one people per outbreak between 1973 and 1987, the same period during which intense packing concentration occurred. While major meat packing companies make billions each year, five hundred people die and at least 20,000 people become ill from E. coli contamination. Another two million cases of salmonella poisoning occur with up to 2,000 deaths each year. The meat industry disclaims responsibility, calling these slaughterhouse contaminants 'natural' occurrences that can best be cleaned up during food preparation by the consumer. New federal meat regulations simply keep contamination from exceeding current average levels. For chickens that's one in five and turkeys one in two contaminated with salmonella.[19]

As we move increasingly toward factory farming, the chemicals, hormones, and antibiotics given to animals to speed up growth and prevent diseases may pose health risks to humans, as does pesticide contamination. Since the introduction of the use of hormones in livestock production after World War II, there have been a number of claims for an increase of premature sexual development in children, enlarged breasts in men, weight gain, impotence, and infertility.[20] The European Union has banned imports of U.S. beef, fearing human health effects of the hormones used in America. However, the U.S. government disputes the risks and has asked the World Trade Organization in Geneva to overturn the import ban as a barrier to free trade as guaranteed under GATT.[21]

The consolidation of meat packing plants has been made possible in part by the increased use of large feedlots. When animals are concentrated in an area, manure becomes a waste disposal problem and an environmental hazard instead of the source of nutrients for the soil that it once was on diversified family farms. For example, the excessive manure produced on hog farms in North Carolina recently caused numerous spills and runaway algal growth in rivers, killing fish and affecting the health of people who came in contact with remaining live fish. These large hog operations have also caused the contamination of groundwater with nitrates.[22]

Our Tax Dollars at Work

The radical restructuring of American food processing could only be carried out with the acquiescence of local and state governments that have showered the meat-packing giants with millions in tax rebates and subsidies. The space now occupied by IBP was the old Hy-Grade plant. Prior to the Reagan revolution the local unionized work force was averaging $30,000 a year or more—some $51,000 in today's dollars. Refusing to reach agreement with its unions, Hy-Grade closed down in 1981.

After $10 million in local tax subsidies as enticements, IBP reopened the plant a year later, offering six dollars an hour. This pattern of de-unionization and ruralization has been regionwide. One after another, meatpacking plants have moved from the big cities where they were close to labor, into the countryside where they were close to the animals and could save on costly transport. As supermarkets took on more specialty butchers, the processing plants needed more workers with fewer skills. Unions became anathema. The industry's hourly wage rate peaked at $19 in 1980. By 1992 it was down below 1960 levels at $12 an hour and has continued to fall. By 1988 unionization was half of what it was in 1963.

Where the new plants opened labor was in relatively short supply. And even in Storm Lake where hundreds of former Hy-Grade workers re-applied for the new jobs, IBP hired back only thirty. "The company wanted to bar union-experienced workers from the shop floor," says Mark Grey. With just three companies (IBP, Cargill, ConAgra) dominating the field, competition among them was, no pun intended, cut-throat. Production lines were sped up. Injury rates climbed. What was once a stable work force became frenetically mobile.

Over the last five to eight years American meatpacking companies have aggressively sought out and brought in the cheapest and most docile labor force they could find. Along the way they smashed one union after another, sparked wholesale migration and exploited differences in ethnicity, gender, regional and legal status to company advantage, employing methods of labor control and farmer contracts that one group of researchers say "recall systems of peonage."

There is Another Way

Food First co-founder Frances Moore Lappé argued in *Diet for a Small Planet* that a food system in which quality farmland is devoted to raising cattle feed instead of crops is a good way to waste resources, impoverish farmers, and maintain hunger.[23] Yet that doesn't mean there is no place for farm animals in ecologically-managed farming. The rotation of animals with crops, practiced by the world's small farmers since time immemorial, is a highly productive sustainable practice in which the waste products of each component are key inputs for the other: manure fertilizes crops and animals eat crop residues. Leading agroecologists believe that ecologically optimal farming systems should indeed incorporate animals.[24] Furthermore, there are grassland regions which are suitable for grazing though not for crops, where it makes ecological sense to raise range-fed livestock.

The U.S. Humane Society has created guidelines for humane sustainable agriculture, in which meat and poultry can be produced in ways which are humane to the animals, ecologically sound, free of hormones and antibiotics, and certainly don't contain ground up dead animal parts.[25] There is an incipient movement of farmers who offer range-fed or agroecological meat and poultry products, and for those consumers who are not vegetarians, it makes sense to support them.

While these hormone-free products cost more, that can be offset by reducing meat and poultry products in the diet proportionally, which in any event is a healthy choice. Unfortunately we do not as yet have any independent information on labor practices in alternative livestock production, but we can hope they are superior to those described in this piece.

1 Krebs, A.V. *The Corporate Reapers: The Book of Agribusiness* (Washington, DC: Essential Books, 1991).

2 Welsh, Rick. *The Industrial Reorganization of U.S. Agriculture: An Overview and Background Report*, Policy Studies Report no. 6 (Green Belt, MD: Henry A. Wallace Institute for Alternative Agriculture, 1996).

3 Unattributed quotes and background information were obtained by Marc Cooper while researching his story, "The Heartland's Raw Deal: How Meatpacking is Creating a New Immigrant Underclass," which appeared in *The Nation*, February 3, 1997.

4 Strange, Marty and Annette Higby. *From the Carcass to the Kitchen: Competition and the Wholesale Meat Market* (Walthill, NE: Center for Rural Affairs, 1995), and *The Industrial Reorganization of U.S. Agriculture.*

5 USDA Agricultural Marketing Service. "Concentration in Agriculture: A Report of the USDA Advisory Committee on Agricultural Concentration," Minority Report 1, 1996.

6 *From the Carcass to the Kitchen: Competition and the Wholesale Meat Market*, op. cit.

7 Meeker-Lowry, Susan. "Challenging the Meat Monopoly: Massive Corporate Control of Land and Farmers," *Z Magazine*, March 1995, pp. 28-35. For a succinct history check out www.eh.net:80/Archives/h-business/apr-96/0010.html.

8 Hedges, Stephen and Dana Hawkins and Penny Loeb. "The New Jungle," *U.S. News and World Report*, September 23, 1996.

9 *From the Carcass to the Kitchen: Competition and the Wholesale Meat Market*, op. cit.

10 "Challenging the Meat Monopoly: Massive Corporate Control of Land and Farmers," op. cit.

11 Strange, Marty. *Family Farming: A New Economic Vision* (Lincoln, NE: University of Nebraska Press, 1988).

12 "Challenging the Meat Monopoly: Massive Corporate Control of Land and Farmers," op. cit.

13 Greger, Michael. "The Public Health Implications of Mad Cow Disease," www. envirolink.org/arrs/AnimaLife/spring96/madcow2.html

14 Highfield, Roger. "CJD Experts Fear 100,000 New Cases," *The London Telegraph* (UK), January 16, 1997.

15 "USDA, U.S. Public Health Service Announce Additional Steps, Support for Industry Efforts to Keep U.S. Free of BSE," USDA Release no. 0159.96, March 29, 1996.

16 U.S. Department of Health and Human Services. "FDA Proposes Precautionary Ban Against Ruminant-to Ruminant Feeding," FDA Release P97-1, January 2, 1997.

17 "Ground Beef Sometimes Contains Banned Parts, But USDA Denies There's a Health Threat," *San Francisco Chronicle*, February 22, 1997.

18 Blakeslee, Sandra. "Fear of Disease Prompts New Look at Rendering," *The New York Times*, March 11, 1997.

19 Meeker-Lowry, Susan and Jennifer Ferrara. "Meat Monopolies: Dirty Meat and the False Promises of Irradiation" *Food and Water* report, 1997, pg. 18.

20 "Challenging the Meat Monopoly: Massive Corporate Control of Land and Farmers," op. cit.

21 U.S. Department of Agriculture. "Glickman and Barchefsky Announce WTO Panel to Review EU Hormone Ban," USDA Press Release no. 0265.96, May 20, 1996.

22 *From the Carcass to the Kitchen: Competition and the Wholesale Meat Market*, op. cit.

23 Lappé, Frances Moore. *Diet for a Small Planet* (New York: Ballantine Books, 1971)

24 Altieri, Miguel A. *Agroecology: The Science of Sustainable Agriculture*, second edition (Boulder, CO: Westview Press, 1995); Roberto Garcia Trujillo, *Los Animales en los Sistemas Agroecológico.* (Havana, Cuba: Pan para el Mundo, 1995)

25 Fox, Michael W. *The Place of Farm Animals in Humane Sustainable Agriculture* (Washington, DC: Humane Society U.S.A., 1991)

From *Family Farming: A New Economic Vision* by Mark Strange

The Faustian Bargain: Technology and the Price Issue

In a competitive economy, the market price of commodities should reflect the relationship between supply and demand, and supply is a function of (among other things) the technology available to the producer. To restate the competitive paradigm: If technology is available to make two bushels of corn grow where only one has grown before (to make the cost of producing each bushel of corn lower), then those who use the latest technology will produce larger volumes of corn and the price of corn will fall until those who refuse to use the technology (or can't muster the capital to buy it) will be forced out of business, giving up their land to those who, in the vernacular of rational economics, are more 'progressive.' Low commodity prices fuel the competitive process, but technology is the engine that drives it.

If the purpose is to save the family farm, why not strip the competitive process of its principal fuel, low farm prices? There probably isn't a demand more universally appealing to farmers than that of higher commodity prices.

And why not? Intuitively, higher commodity prices mean more income for farmers. As a general rule, farm income and commodity prices do vary directly. But even when commodity prices and farm incomes were relatively high, including the golden years of the 1940s and 1970s, farmers continued to leave farming. But why? Why, instead, don't more people enter farming when farm income is high?

The answer is that because agriculture operates as a market economy in which the rewards of adopting new technologies are transformed into the accumulation of ever-more-valuable farmland, the relationship between commodity prices and the well-being of farmers is not as simple as it seems. Are farmers well off if they farm successfully, live frugally, and plow every available dollar into more machinery and more farmland? How can such farmers be made better off? Would such farmers be better off with higher commodity prices that make it possible both to buy more land and to live better? Would their neighbors be better off?

There was a time when higher prices meant higher incomes for all farmers, pure and simple. But that was when most people were farmers, when farmers were more or less similar in size and character, when they produced most of the inputs they used right on their own farms, when land was plentiful and relatively cheap, when debt was the exception, not the rule, and when tools were more important than machines and the primary factor of production was still human and animal labor. Because farmers purchased and borrowed little, there were few leaks in the farm financial boat through which to lose the benefits of an increase in commodity prices. If the price of corn went up, the additional money went into the farmer's pocket.

But labor is no longer the chief input farmers provide. They now provide more capital, technology, and management sophistication. And increasingly, they buy, rent, or borrow these inputs from others who, by supplying them, gain a claim on some of the income from farming. Even labor, which farmers increasingly hire, is a cash cost of production. As the industrialization of American agriculture continues, farmers' incomes become increasingly dependent on more than the price of the commodities they receive. Their income also depends on the price they pay for the inputs they use, and on the amount they choose to invest in land, buildings, and machinery. The problem of commodity prices is intertwined with other issues: landownership, land prices, and the proper uses of technology.

While some farm organizations have from time to time argued strenuously for public attention to the commodity-price issue, most have not been so eager to address these related issues, perhaps because they require a critical inward look at the behavior of farmers themselves.

To understand these issues better, consider in more detail how the 'cost' of farming has changed over time. Time was, farmers produced most of their own inputs. They raised horses for draft power and the oats with which to feed them. They recycled animal manure as fertilizer, saved seed from one crop to plant the next, and controlled pests by planting in late spring after one crop of weeds had been destroyed, cultivated later weeds, and rotated crops from year to year. They let the crop dry naturally in the field. Land, labor, tools, and some machines were their main inputs. Farming was a matter of hard work, good timing, and knowing how to cooperate with nature.

Technology has changed that. Farmers buy a host of intermediate products—purchased inputs like hybrid seeds that grow bigger crops, chemical fertilizer and pest controls, bigger and more specialized machines. There has been an endless stream of new technologies, many of them, in recent years, by-products of military and space-exploration research. These technologies allow farmers to intercede in biological and chemical processes, to short-circuit nature's constraints on production. By regulating when natural processes occur, they standardize the production process, allowing labor to be routinized. They alter the environment, making it possible to conduct economic activities in climates and under conditions in which they would not naturally occur. They eliminate the balance between predators, making it possible to intensify crop production over vast acreages without diversification or rotation. They influence the pace, scale, location, and character of modern agriculture.

To acquire these powerful technologies, farmers borrow money in large sums and pay interest on the loan, which itself is another input. To use this technology, they rent more land or borrow to buy more land. More interest follows.

The full impact of the post-World War II technological revolution on how farmers spend the money they receive for their crops is evident in table 1. First, note that their gross income—the amount of money they have to spend—has climbed fifteenfold, from $10.8 billion in 1940 to $162.3 billion in 1984 (column 2). Adjusted for inflation, this still represents a doubling of gross income. This doubling reflects a dramatic increase in the *output* of all the farm commodities, and it has occurred despite the fact that prices of those commodities have fallen relative to inflation. Between 1950 and 1982, yields per acre for wheat, rice, and cotton have roughly doubled; corn yields have tripled. These yield boosts, coupled with the increased acreages of all these crops (except cotton), mean total production has increased even more—171 percent for wheat, 198 percent for corn, and 296 percent for rice. While prices of these major commodities have fallen in real terms by forty percent to sixty percent during the same period, gross farm income has doubled because they have had so much more to sell.

How does the picture change when we subtract production expenses from this increase in gross income? The rest of table 1 reveals that although farmers have more to spend, they spend much more to earn it, and as a result, have relatively less for themselves.

The portion of gross income kept as net return to all operators (column 11) has declined from thirty-eight percent in 1940 to as low as ten percent in 1980. In 1984, the net was $28,354 million, which is equivalent to $3,758 million in 1940 dollars. That is ten percent less than the $4,092 million net that farmers actually received in 1940, despite a real doubling of gross farm income. Where does the money go?

Most of it goes to purchase off-farm inputs. In 1940, these cash operating expenses consumed forty-four percent of gross income,

TABLE 1. Where the Money in Farming Goes ($ millions)

		Cash Operating Expenses						Land Costs		
Year (1)	Gross Farm Income (2)	Intermediate Products (3)	Capital Consumption (4)	Business Taxes (5)	Non-Real Estate Interest (6)	TOTAL (7)	Hired Labor (8)	Rent to Landlords (9)	Real Estate Interest (10)	Returns to Operators (11)
1940	10,756	3,491	661	382	186	4,720	1,029	672	243	4,092
1950	31,923	10,984	2,301	810	354	14,429	2,811	1,822	225	12,636
1960	36,747	15,940	3,773	1,373	719	21,805	2,754	1,491	549	10,148
1970	55,769	25,188	5,890	2,383	1,618	35,079	3,906	2,360	1,586	12,838
1980	138,675	73,420	17,847	3,607	8,717	103,591	8,270	5,760	6,920	14,134
1984	162,293	75,925	19,233	4,088	10,396	109,642	8,976	5,442	9,879	28,354

Source: U.S. Department of Agriculture *Economic Indicators of the Farm Sector: National Financial Summary*, 1986, table 4, pp. 13, 14.

but by 1984 took sixty-seven percent (column 7). Intermediate products—seed, fertilizer, and fuel, for example—increased their share from thirty-two to forty-seven percent of gross income receipts (column 3). Capital consumption (depreciation and damage to buildings, vehicles, and equipment—column 4) doubled its share (from six to twelve percent), and interest on loans to pay for these items (column 6) tripled its share (from two to six percent). Business taxes (column 5), so often the rhetorical villain, did not significantly change their share of the take of income.

These figures underscore the growing dependence of farming on expensive technology and capital. But they also reveal something more. A Faustian bargain has been made between the user of the technology and the impersonal market. The bargain is this: In return for the splendid increase in production made possible by this technology, the producer must both accept less money for each unit sold, and share the rewards of increased output with those who produce the technologies. The suppliers of the technology that make cheap abundance possible must be paid, and they take an ever-growing share of the rewards while farmers take less. In return for a chance to profit, farmers have condemned themselves to an endless race against their neighbors.

Although the profits among most farm input suppliers cannot be documented easily, it is safe to say they are substantial. If these profits were added to net farm-income figures, the overall rate of profit from food production would not appear nearly as dismal as it does. The fact that those profits are not farmers' profits reflects the extent to which farmers have paid dearly for the benefits of no longer producing their own horsepower, fertilizer, and weed control.

With less net income to spread among farmers, many cannot continue to farm. They go out of business while those who make use of the new technologies expand. In 1940 there were 6.35 million farms. Today, there are about 2.33 million. Over four million farms simply failed to outrun others, and were caught by the bear. The remaining 2.33 million are still running.

Because today there are so many fewer of them to share the profits, the *average* return to a farm operator has actually more than doubled, after adjustments for inflation, even though the total real net income to the farm sector as a whole decreased by ten percent. Of course, we have seen in what averages hide: this bounty is hardly distributed evenly among farmers. Three-fourths of the net income to farmers is concentrated among the largest five percent, the ones who have run fastest to escape the bear.

In effect, the profit in agriculture has not been diminished as much as the presence of thousands of broke farmers suggests. It has merely been redistributed in two ways: from farmers to the companies that sell them inputs, and from smaller farmers to larger farmers.

Will mechanically raising commodity prices change that? Not alone. For while farmers compete with each other vigorously, the firms that supply them with technology do not compete in the same way. As a result, when commodity prices increase, suppliers can raise their prices, too. The benefits of higher commodity prices end up in the pockets of the suppliers, not the farmers.

Input suppliers are better able to reap the gain because there are fewer of them and they are less competitive than the farmer. If there were many input suppliers in competition with each other, a rise in commodity prices would not affect the prices they charge farmers for inputs. Farmers would be able to keep the additional income from a commodity price increase. But there are only three major manufacturers of farm machinery, a handful of highly specialized chemical suppliers, and a rapidly dwindling number of seed companies. While the pricing practices of these and other farm input industries are poorly documented, the evidence is sufficient to support skepticism that competition among them would prevent them from raising their product prices enough to capture at least part of any increase in farm income.

A 1981 analysis of the major farm-input industries by the staff of the Federal Trade Commission (FTC)[1] revealed that none of the four major input industries—fertilizer, chemicals, seed, and machinery—were clearly competitive. The FTC staff analysis suggested that the most competitive was the fertilizer industry, despite the tact that, historically, it has been the subject of a large number of antitrust actions. The nitrogen fertilizer subsector—easily the largest of the fertilizer subsectors—is dominated by natural gas and petrochemical companies. In the other input sectors the level of concentration certainly invites monopoly pricing practices. This is especially so within particular product lines within those sectors.

For example, two companies dominate the market for hybrid seed corn, with the leader holding a third of the market and earning a return on equity as high as twenty-five percent in 1980. Four companies account for 87.1 percent of the sales of herbicides used on corn, two of them alone producing 92.4 percent of the sales of herbicides used to control grass weeds and the other two producing 76.1 percent of the broadleaf weed herbicides.[2] From seventy to eighty-one percent of the sales of tractors, plows, and harvesting and haying machinery are controlled by four companies.

Likely, a rise in farm prices would simply encourage these input suppliers to raise the price of their products. That would be particularly true if commodity prices were actually pegged to input prices, so that any increase in input costs to the farmer would automatically be reflected in higher commodity prices. In that case, the farmer would be insulated from the consequences of increases in input costs, but the rest of society would not. There would be little incentive for farmers to conserve on these inputs and the farmer would become officially what he is now unofficially—a pass-through account for these suppliers to charge consumers for their services. Farmers would probably be a little better off, but only because they would be on the same gravy train as the suppliers.

Failing to come to grips with the economic power of these input suppliers has made farmers vulnerable to the unjust uses of their market power. But tying commodity prices to input prices, as many family farm advocates have suggested, would do nothing but let this power loose on society. As it is now, these concentrated industries must moderate their prices enough to allow some profit in farming, lest they kill the goose that lays the golden egg. It was revealing hat, after the bottom fell out of the farm economy in the 1980s, suppliers lowered prices for fertilizer, chemicals, and tractors, causing some thoughtful farmers to wonder how much they had been overpriced before. If farmers were guaranteed sufficient income to pay these companies whatever they charged, there could be no limit to their profit taking. A concentrated market place is likely to be an unfair market place, but protecting only farmers from the unfairness is not the right solution.

At an even more fundamental level, the issue is whether even a genuinely competitive market, through the mechanism of commodity prices, should be the sole determinant of the technology used in modern agriculture. It is the technology, brilliantly deployed, that has altered the market equation between supply and demand. No amount of quibbling about the greed and inequities at work in the market can overpower this central fact. If we are going to use the full range of technologies available in agriculture without caution or restraint, the price of commodities will continue to fall. The low price that most farmers painfully experience is but a symptom—a revelation—of the power of technology. Ultimately, the issue is not whether to raise farm prices to relieve the stress in farming, but whether the technology that causes the stress should be restrained. If it should not be restrained, then the bargain has been made and farm policy can do little to resolve the difficulties of those farmers who fail to outrun the bear. But if for any reason the technology does not increase the well-being of society as a whole, then it should not be used (or allowed to be used). The price of commodities will then rise to reflect the restraints placed on the unbridled use of technology.

Will it work the other way? Would simply mandating higher commodity prices dissuade farmers from adopting new technologies that are harmful to society? Would higher prices slow down the treadmill? Not likely for long. It is true that those who are not inclined to make new investments in technologies—the old, the satisfied, or the ethical—would be sheltered from lower prices and would survive despite the impact of the technologies. The overall rate of adoption would therefore be retarded. But for the willing, the rewards of adopting new technologies would only be that much greater. They would then produce more and sell at the higher, protected prices. The benefits of technological innovation would be concentrated among them. The rest of society would gain nothing. It would, of course, be stuck with whatever social and environmental costs are associated with the new technology.

An alternative is to protect prices by limiting production, reducing incentives to produce more even among the willing adopters of new technologies. Of course, limiting production diminishes incentives to adopt any new technology, harmful or not. As much as some would like to, you can't have it both ways. Wanting both higher commodity prices and the right to unbridled use of technology is like wanting the coin to come up both heads and tails.

1 Leibenluft, Robert F. *Competition in Farm Inputs: An Examination of Four Industries*, Policy Planning Issues Paper (Washington, DC: Federal Trade Commission, 1981).

2 A USDA study from about the same period found that the top four corn herbicide companies had ninety-four percent of the market, the top two seventy-four percent, and that similar levels of concentration in crop herbicide subsectors existed in most other commodities. (Eichers, Theodore. *The Farm Pesticide Industry*, Agricultural Economic Report no. 461 [Washington, DC: Economics, Statistics, and Cooperative Service, USDA, 1980].)

Part Three

Global Policies and Hungry People

Trade, Aid, Debt, and Dying

The movements of grains, pesticides, farm workers, and energy around the world in the global food system are guided, to paraphrase Adam Smith's famous phrase, by invisible hands. These currents follow channels which are created for them by another kind of flow: that of money. And the paths of money in the world market are not preordained by God; they depend on decisions made by human beings. This section examines those decisions which are most critical to the world food system, on policy questions dealing with world trade, foreign aid, and the global structure of debt.

Governments worldwide are aggressively promoting agriculture exports, and the free trade agreements which facilitate them. NAFTA, the Uruguay round of GATT, and fast track are the 1990s manifestations of this trend, by which the survival of peasant farmers, environmental protection, and food security all become secondary to the goal of increasing international trade. Inevitably, resistance grows, as people decide that the right to survive must precedence over the right to trade without restraint. And sometimes, as the overthrow of the three-decades-old Suharto dictatorship in Indonesia in 1998 showed, the political consequences can be dramatic.

The sustainability of agriculture, too, is threatened by free trade policies, as Mark Ritchie points out in his analysis of NAFTA in *Trading Freedom: How Free Trade Affects Our Lives, Work, and Environment.* The free trade vision of the economy demands wide-open borders; restrictions on pesticide use, rain forest destruction, or worker exploitation are all seen as restraints on trade. Markets are won by whoever can produce at the lowest cost, irrespective of the damage to the environment or the long-term sustainability of agriculture. Any policy considered to be a subsidy, whether for hungry consumers, family farms or worker protection, is overridden by the international treaties guaranteeing agribusiness the right to trade freely.

Foreign aid, as Lappé, Collins, and Kinley show in *Aid as Obstacle*, has been guided by the same vision. Despite the humanitarian rhetoric, the underlying goals of U.S. aid have long been to develop foreign markets for agribusiness and to dispose of agriculture surpluses. Aid is almost entirely in the form of loans which must be repaid, and the chief recipients are not the most needy, but the most able to pay. The result is that surpluses are dumped into foreign markets, undercutting peasant farmers and driving them out of business.

Susan George, in *A Fate Worse Than Debt* shows how the debt crisis has allowed international lending institutions such as the World Bank and the International Monetary Fund to pressure Latin American nations into increasing exports to and imports from the U.S., even when the cost is more poverty and inequality at home. In nation after nation, the result has been hunger and misery for the majority, while small ruling elites (such as the generals linked to the cocaine trade in Bolivia) become even richer. Inspired by studies such as George's, a broad movement against IMF policies has developed in the late 90s; its strength was shown by the Congressional reluctance to replenish IMF funds used to 'help' economies in Asia, Russia, and Latin America in 1998.

The ways that international lenders pressure governments were revealed in detail in the early 1980s, when internal World Bank documents leaked to Food First and other progressive groups showed how, in countries such as the Phillipines, the Bank had used its power to serve the needs of U.S. business (*Development Debacle: The World Bank in the Philippines*). Failing in its strategy of pacification through development programs, the World Bank turned to growth policies which required cutting costs by repressing labor. Implemented by the authoritarian regime of Ferdinand Marcos, this technocratic approach led to economic collapse and eventually Marcos' overthrow by a democratic revolution.

Nonetheless, the neo-liberal policies which have caused so much misery continue to guide the World Bank and the IMF, now under the rubric of 'structural adjustment.' Bello, Cunningham, and Rau in "Creating a Wasteland: The Impact of Structural Adjustment on the

South, 1980–1994" tell how such programs serve to force cutbacks in social programs, currency devaluations, reductions in real wages and the elimination of restraints on foreign capital. The environment may be plundered and the majority of the people made even poorer, but debt payments are guaranteed and the world is made safe for foreign capital.

The trade, aid, and debt policies established by the governments of rich nations and the international lending institutions which they control, are a powerful cause of world hunger. They make food self-sufficiency impossible, instead forcing poor countries to sacrifice their people and their natural resources on the altar of the free market. They are one of the principal pillars of the food system which leads to the poverty and starvation of so many of our sisters and brothers around the world.

From *Trading Freedom: How Free Trade Affects Our Lives, Work, and Environment*, by Mark Ritchie

Free Trade versus Sustainable Agriculture

Two contrasting visions have emerged concerning the future of agriculture in North America. One approach, often referred to as sustainable agriculture, calls for social and economic regulations to protect the environment and family farms. This approach seeks to protect our soil, ground water, and fossil-fuel resources, and to promote economically viable rural communities.

Sustainable agriculture emphasizes the use of farming practices which are less chemical and energy-intensive, and it places priority on reducing the time and distance between production and consumption. This maximizes the freshness, quality, and nutrition, while minimizing processing, packaging, transportation, and preservatives.

A competing vision, often referred to as the 'free market' or 'free trade' approach, pursues 'economic efficiency' aimed at delivering crops and livestock to agri-processing and industrial buyers at the lowest possible price, ignoring almost all social, environmental, health, and taxpayer costs. Based on traditional economic theories dating from the seventeenth and eighteenth centuries, this approach argues that any government intervention in the day to day activities of business diminishes 'economic efficiency.' Free-market and free-trade policies are heavily favored by the agribusiness corporations involved in the trading and processing of farm commodities.

Supporters of the free-market approach argue for the de-regulation of food production, under the rallying cry of "getting the government out of agriculture." They seek to scale back, eliminate, or de-couple farm programs such as price supports, supply management, and land-use regulations. In world trade, they support opening borders to unlimited imports and exports.

Debate concerning the relative merits of these two views has now become the central argument over modern agriculture policy. Agricultural trade negotiations under the auspices of the General

Agreement on Tariffs and Trade (GATT) have pushed this controversy onto the front and editorial pages of the world's leading newspapers. President Bush's call for a Free Trade Agreement between the U.S., Mexico, and Canada will intensify this debate even further.

Agriculture and the Canada-U.S. FTA

The U.S.-Canada Free Trade Agreement was a serious setback for sustainable agriculture on both sides of the border. Canada had to weaken its stricter regulations on pesticides and food irradiation, and there was a concerted effort to get rid of supply management in the Canadian poultry, egg, and dairy industries. Both were seen as 'bad examples' by agribusiness, which feared that U.S. consumers and farmers would begin to demand similar programs. The U.S.-Canada deal, negotiated by right-wing governments in each country, shows that international agribusiness wins while family farmers, farmworkers, food-industry workers, and consumers tend to lose. The proposed continental free trade agreement will only heighten these inequitable gains and losses.

Agriculture and NAFTA

The principle threat to farmers in each country is wide-open borders. If governments cannot regulate the flow of goods coming into their countries, farmers, the environment, and national economies suffer devastating effects.

The U.S. imposes strict controls on the level of beef imports, based on legislative authority in the Meat Import Act of 1979. Fast-food hamburger retailers have pushed hard to make sure that any new GATT or regional trade agreement will lift such controls, allowing the companies to import beef. Since beef can be produced more cheaply on cleared rainforest land in Central or South America, a sharp increase in U.S. imports from this region, causing accelerated rainforest destruction, would almost certainly follow.

Unlimited beef imports would also obviously lower family income for U.S. cattle producers. U.S. producers would have to sell at lower prices to compete with rainforest beef, potentially the 'cheapest' (from a short-term perspective) in the world. They also would sell less, since much of the U.S. supply would come from elsewhere. With more beef coming from overseas there would be a smaller market for U.S.-grown hay, corn, and other foodstuffs. Replacing U.S. beef with rainforest-fed beef thus not only devastates beef farmers in the U.S. but also those who produce feedgrains.

FTA proponents claim that U.S. grain farmers will increase shipments to Mexico under free trade. But the real impact is up for question. The Mexican government recently completed a barter deal with Argentina to bring over half a million tons of grain into Mexico, grain which could even be re-sold abroad if Mexico badly needed foreign exchange. Mexico, which is not even a major producer of grains, could actually end up as an exporter.

The implications of deregulated trade for environmental sustainability are staggering. Beef cattle again can serve as an illustration. Currently many U.S. beef cattle are fed in huge, environmentally damaging feedlots. Many others, however, graze on the hillsides and meadows of the Upper Midwest. In Minnesota, there is the generally poor soil north of the Twin Cities, with the exception of the Red River Valley. It is hilly with thin topsoil and quite fragile. The only suitable agriculture for this land, and needed by this land, is cattle grazing, either for beef or dairy. As beef comes across borders and drives down the prices, Minnesota's diversified, small family beef operations will go under. The land will likely be put into row crops, soybeans or corn, to pay taxes and rent, soon depleting the thin soil. The political battle to encourage family farms and regulate factory-type feedlots would be blocked by free trade.

Eliminating Winter Produce in the U.S.

U.S. fruit and vegetable production also will be seriously threatened by free trade. U.S. producers currently operate under substantial regulations concerning chemicals and workers. They pay higher taxes and extend more worker benefits than producers in most countries. Even if U.S. and Mexican produce growers were regulated in the same fashion, the U.S. Food and Drug Administration inspects only two percent of the food coming across the border. There is little chance that violators of food safety regulations will be caught.

The entire U.S. winter-produce industry could be threatened. If farmers in Florida, Texas, and California are to take the often enormous risks inherent in winter production they must be confident of steady markets, profitable enough to allow economic survival during years of complete crop losses. Unlimited imports would push weather and other risks to an unacceptably high level, eventually displacing

U.S. production. The consequent dependence on imported fruits and vegetables could have dire effects on U.S. food safety and food security, and importing fruits and vegetables that can easily be grown in the U.S. unnecessarily worsens the trade deficit.

Pillsbury's Green Giant division moved a frozen food packing factory from Watsonville, California to Mexico in anticipation of a Free Trade Agreement which will allow them to bring products formerly produced in Watsonville across the border without tariffs and with few controls. Low wages and weaker environmental and safety regulations make the advantages obvious. Such agro-maquilas are the wave of the future should free trade advocates realize their dream.

According to Edward Angstead, president of the Growers and Shippers Association of Central California, total production costs of frozen broccoli in Mexico are less than pre-harvest costs in California. The biggest difference is the cost of labor. Angstead estimates the cost of farm labor in Mexico at three dollars per day, compared with five to fifteen dollars per hour in California. The loss of Watsonville's Green Giant factory means that the farmers in the areas who grew crops for the plant lost a market. Farmworkers who picked those crops and the California cannery workers also lost jobs. The struggling community is already suffering.

NAFTA and Organic Farming

Free trade between the U.S. and Mexico will deliver a 'double whammy' to farmers on both sides of the border trying to grow organic produce. The general lowering of prices on much commercial produce will make it harder to charge prices high enough to cover the organic producer's additional costs. Second, expansion of fruit and vegetable production in Mexico will increase the overall use of chemicals, further disrupting and interfering with natural pest control patterns. Organic farmers cannot apply pesticides to control pests driven to their fields by neighbors' spray. They are dependent on proper predators for their own biological pest management.

Conclusion

A *real* trade and development agreement would address wide range of trade-related issues for agriculture throughout North America. Export dumping whereby subsidized U.S.-based grain-trading corporations over-supply the market and drive down market-prices in

Mexico and Canada by selling grain at prices far below the cost of production, would be explicitly banned. Common food-safety regulations would strive for maximum protection of agricultural workers and consumers in all countries. Food security would be enhanced. The key question remains: Can advocates for sustainable agriculture forge their own positive vision for a future set of economic, political, and social relations among these nations?

From *Aid as Obstacle: Twenty Questions about our Foreign Aid and the Hungry* by Frances Moore Lappé, Joseph Collins, and David Kinley

QUESTION: Don't U.S. food aid programs channel American abundance to hungry people around the world?

OUR RESPONSE: Perhaps the most important fact to remember about the U.S. food aid program is that the bulk of U.S. food aid is not given away. Over the years, about two-thirds of U.S. food aid has gone under Title I of Public Law 480 (PL 480).[1] Under the provisions of Title I, foreign governments take out long-term loans from the U.S. government to purchase surplus agricultural commodities. The bulk of what is called "food aid" is actually purchased by foreign governments *that then may do with it as they please.* Generally the food is sold on the local market—meaning that those who can pay for it, get it. (The hungry, of course, cannot pay.)

Before we look at the impact of food aid, it is instructive to examine its purpose. The $30 billion worth of food shipped abroad as aid since 1950 is often presented as a clear expression of the generosity of the American people. Undoubtedly, the humanitarian intentions of ordinary Americans supporting food aid are genuine. But the actual motives behind the program are something else.

During different periods of time, the U.S. food aid program served many purposes for diverse interest groups, but has its primary purpose been to feed the hungry. In fact, the humanitarian intent was not even written into the food aid law until 1966. From its very inception in 1951, the food aid program has been an extension of foreign policy, farm interests, and corporate interests, which in most cases have been mutually supportive. Public records unequivocally show that U.S. policymakers have viewed the food aid program as a means to:

- rid U.S. markets of price-depressing domestic surpluses
- open new markets for commercial sales of U.S. farm products and thereby offset trade deficits
- provide support for U.S. military interventions in the Third World
- extend the reach of U.S. agri-business corporations into food economies abroad
- pressure foreign governments to accommodate U.S. economic and military interests

The Origins of Food Aid

The direct origins of Public Law 480 (PL 480, or simply food aid) go back to 1951. In that year, India made an emergency request to the United States for grain to stave off famine precipitated by monsoon failure. Since the end of World War II, India had embargoed exports of monazite sands, which contain thorium, a material necessary for the production of nuclear weapons. The U.S. government seized the threat of famine as an opportunity to have the embargo lifted. Congressman Charles J. Kersten (R-Wisconsin), put it bluntly, "In return for the wheat we are asked to give to India, the very least we should ask of India is that it permit the United States to buy some of these strategic materials..."[2] The result was the India Emergency Food Act of 1951, the direct predecessor of Public Law 480.

Public Law 480 (later called "Food for Peace") passed in 1954 and addressed a crisis that was much closer to home than the Indian famine: the great American grain surplus. During the 1940s, U.S. grain production had grown by almost fifty percent while domestic consumption lagged well behind, increasing only about thirty percent. Higher crop yields, resulting from more fertilizers, pesticides, and improved seed varieties, as well as from price supports which encouraged farmers to peak production, were creating enormous surpluses costing taxpayers one million dollars per day just for storage.

The farm lobby did not want the surpluses put on the domestic market. If dumped on the world market, grain prices would drop by a dollar a bushel; the giant grain-trading corporations opposed such a disruption of their international commercial market. At its national convention in 1952, the American Farm Bureau, a group representing large and medium-sized farmers, proposed a solution. They suggested the creation of a secondary foreign market by allowing food-deficient countries to pay for American food imports in their own currencies instead of in dollars. While not interfering with the commercial dollar price demanded from higher-income countries, PL 480 permitted Third World governments, which otherwise would not constitute a market at all, to buy surplus American food. In terms of exporting grain, PL 480 meant the United States could have its cake and eat it too.

Food Aid and Local Production

Even before we began our research on food aid, we had long been familiar with the claim that U.S. food aid shipments depress the incentive of foreign farmers to grow their own food. We were tempted to reject this conclusion simply because it sounded like an unfounded critique of welfare ("If you feed 'em, they won't want to work").

We did not understand an important distinction. Many critics were charging that if you give people food they will not *want* to grow food for themselves. The fact is that dumping large quantities of low-priced American grain in underdeveloped countries makes it economically *impossible* for the small domestic producers to compete. Unable to get a fair return for their grain, such producers are frequently asked to sell their land and become landless (and often jobless) laborers. A study in 1969 concluded that for every pound of PL 480 cereals imported, there was a net decline of almost one-half pound in Indian domestic production over the following two years, because of the reduced return to the farmer.[3]

South Korea has been the second largest recipient of U.S. food aid and has purchased more U.S. agricultural goods than any other underdeveloped country.[4] A basic purpose of the U.S. food aid, along with more than $13 billion in direct economic and military assistance to Seoul since the end of the Korean War, has been to maintain a low-paid, 'disciplined' labor force for export-oriented multinational corporations that dominate the South Korean economy.

The U.S. grain imported into South Korea has allowed the government to maintain a 'cheap food' policy, undercutting many Korean farmers. Throughout the 1960s, prices that the government paid to rice producers barely approached the costs of production.[5] Not surprisingly, Korea's rural population fell from one-half to slightly more than one-third of the total population between 1963 and 1976.[6] People lost their livelihoods and were forced to seek jobs in the cities. Pressure from the remaining farmers forced some increases in the government's mandatory rice purchase price in the 1970s, but prices still fall below production costs, according to the Korean Catholic Farmers Association.[7] (According to association members we interviewed in 1979, farmers who have dared to circulate a petition asking the government to pay a fair price for their rice have been harassed, arrested, and beaten.)

Nonetheless, former U.S. Assistant Secretary of Agriculture Clayton Yeutter proclaimed, "South Korea is the greatest success story worldwide of the Food for Peace Program (PL 480) in terms of contribution to the growth of that nation."[8]

Colombia provides another dramatic example of the effects of PL 480 shipments. Between 1955 and 1971, Colombia imported from the U.S. over one million tons of wheat that could have been produced more cheaply locally. The marketing agency of the Colombian government fixed the price of the imported grain so low that it undercut domestically-produced wheat. This dumping resulted in fifty percent lower prices to Colombian farmers. From 1955 (the first year of PL 480 shipments) to 1971, Colombia's wheat production dropped by sixty-nine percent while its imports increased 800 percent. By 1971, imports accounted for ninety percent of domestic consumption.[9]

Moreover, two-thirds of the 407,550 acres that were pushed out of wheat production by subsidized wheat imports have not been replanted in other crops for local consumption. The fertile Sabana de Bogotá Valley which once grew wheat is now used for cattle grazing—primarily for export.[10] Such feeding operations were abetted by PL 480 loans that went to subsidiaries of U.S. multinational corporations such as Ralston Purina, Quaker Oats, Pfizer, and Abbott Laboratories to build plants for processing feed and producing veterinary drugs. Large landowners now making greater profits on beef, flowers, and vegetables for export, expand their operations, evict their tenants and, in general, exclude more and more ordinary farmers from the land. Without land and without jobs, those needing food cannot buy the food aid. In this traditionally corn-eating country, the imported wheat goes to meet the 'demand' of the 'Americanized' minority who can afford processed, brand-name foods.

The impact of American food aid to Bolivia has been similar.[11] But an additional turn of the screw came in 1972, when the United States stopped accepting payments in local currency and started demanding dollars for food aid shipments, albeit on easy terms. Despite its rich agricultural potential and high rural unemployment, Bolivia had come to depend on United States imports, and local wheat production had stagnated. Millers had become primarily flour-importing companies because importing was more profitable than milling. Thus, even after local currency was no longer accepted to repay PL 480 shipments, Bolivia had to continue to import flour. The big difference, however, was that Bolivia was forced to use its scarce foreign exchange to purchase the flour in dollars, foreign exchange that might have gone to purchase what it could not easily produce itself, such as productive industrial goods.

Bolivia is an example of a country which has, as the U.S. Department of Agriculture would phrase it, 'graduated' from PL 480 status to that of a regular commercial buyer. Such a graduation is not always by choice. To receive food aid, the recipient sometimes has to agree to purchase U.S. agricultural commodities on commercial terms in the future. Agreements for PL 480 food aid to the Dominican Republic, Egypt, and South Korea have been tied to such commercial purchases.

Haiti is one of the countries from which we have received first-hand reports that food-for-work programs and other channels of food distribution act as a market disincentive for local production. In 1978, we received the following report from Haiti:

> In theory, these foods are given as a supplement to the local diet and go directly to the poorest people. Thus, the food aid is not supposed to decrease the incentive for local production, since it bypasses the local marketing system. In reality, the PL 480 food

is available in almost every market in Haiti and competes directly with locally-produced food… I met one Haitian swine and chicken-feed processing owner. His source of grain for this operation for several years has been PL 480 food. He would not elaborate on his source of PL 480 food, but indicated that his purchase price has been decreasing each year and that last year he paid one dollar per 100-pound sack.[12]

Although the disincentive effect of food aid on local production continues to be debated, a 1975 U.S. General Accounting Office research survey concluded, "Leading world authorities now indicate that such food assistance by the United States and other countries has hindered the developing countries in expanding their food production and thus has contributed to the critical world food situation."[13]

Market Development: Boon to Corporations

In the first five years after it was passed, PL 480 succeeded in exporting over $5 billion worth of American grain, or twenty-eight percent of all American agricultural exports. Even this was not enough, however, to unload U.S. grain surpluses. By 1959, the United States held its highest grain stocks in history. Merely responding to food aid requests was not enough. Policymakers decided to take an active role in creating markets. The goal spelled out in the preamble to PL 480 included these words: "…to develop and expand export markets for United States agricultural commodities." The goal was clear; the question was how to achieve it.

Part of the answer, policymakers thought, was 'development assistance.' Assistant Secretary of State W.L. Clayton testified that World Bank financing for capital goods from the industrial countries "would certainly be…very good…for U.S. agricultural exports, because as you help develop them (underdeveloped countries) industrially, you will shift their economy to an industrial economy, so that I think in the end you would create more markets for your agricultural products."[14] Added to this, PL 480, by allowing countries to import food *without* using dollars, until 1972, made it more likely that poor governments would have dollars available to import U.S.-manufactured capital goods for light industrialization.

The other part of the answer was direct support to U.S. corporations. Between 1959 and 1971, PL 480 provisions had allowed up to twenty-five percent of the local currency generated by sales of U.S. food aid to be lent at low interest rates to American corporations in order to finance their entry into a foreign country. Eldridge Haynes, chairman of Business International, the service organization for multinational corporations, told the U.S. House Committee on Agriculture of the need to expand the U.S. food processing industry into the underdeveloped world in order to create a commercial demand for American agricultural exports.

"We are not exporting bread," he testified. "We are exporting wheat. Somebody has to turn it into bread. If they do not, if there are not facilities to make bread, it will not be consumed."[15] Haynes said the same is true of American tobacco and cotton. He suggested that U.S. companies get 'Cooley loans' to invest in plants for making cloth, cigarettes, and other products from agricultural commodities. These 'Cooley loans' were named after the U.S. House Committee on Agriculture chairman, Rep. Cooley (D-North Carolina). During the 12-year period from 1959 to 1971, PL 480 directly subsidized 240 private American and foreign businesses overseas through the Cooley loan program. The U.S. government loaned $419 million worth of local currencies collected by U.S. embassies in repayment of previous food aid shipments, making it possible for companies in thirty-one countries to start or expand abroad with little or no capital outlay.[16] In India alone, Cooley loans have gone to Wyeth Labs, Union Carbide, Otis Elevator, Sylvania, Rockwell International, Goodyear, CPC International, Sunshine Farms, First National City Bank, the Bank of America, and American Express, among others.

In 1972, when PL 480 repayment arrangements were changed from local currency sales to dollar sales, Cooley loans were changed into Private Trade Entity loan form. Under this loan program, the U.S. government now grants credits to American companies overseas to buy agricultural commodities in the United States. The overseas subsidiaries then use the proceeds from local resale of the products for their own financing.[17]

Building a Feedgrain Market: The Case of Cargill, Inc.

In 1968, Cargill, Inc.—the multibillion dollar grain conglomerate—decided to build a complete poultry operation in South Korea, breeding chicks, producing chicken feed, and retailing chicken. The

U.S. government—the American taxpayers—provided ninety-five percent of the financing for what looked like a very profitable operation. Almost $500,000 came as a Cooley loan. An additional $1.9 million loan from the U.S. government came to Cargill under the Private Trade Entity provision of PL 480.

This huge government subsidy was not enough for Cargill to succeed. By 1972, the Cargill operation was in trouble. Cargill had used all possible PL 480 credits for importing grain. Cargill approached its friends in the U.S. government to persuade South Korea to relax domestic price controls and import restrictions that interfered with its feedgrain import operations. The State Department instructed the embassy in South Korea to see that the 'poultry and livestock industries' received special consideration from the government. Finally, when all else failed, Cargill sought and received a deferment of payment on its two PL 480 loans from the U.S. government.[18]

PL 480 credits also enabled Ralston Purina and the Peavey Corporation to establish poultry operations in South Korea. The net effect has been to make South Korea heavily dependent on imported feed. Where South Korea had previously imported no feed grain, it imported about one million tons from the United States by 1974.[19] By 1974, when South Korea had become dependent on American feedgrain imports, the United States raised the price to three and one-half times its original (1970) price.

Building a Wheat Market

PL 480 has also succeeded in creating markets for wheat among the world's original rice lovers. PL 480 "was the best thing that ever happened to the wheat industry," observed one market development specialist, pointing to the tremendous increase in wheat consumption in such countries as Japan, Taiwan, and South Korea. Wheat aid credits to the Chiang Kaishek government in Taiwan allowed it to export the people's staple, rice, while it exhorted the population to embrace the new diet by such slogans as "eating wheat is patriotic."[20] South Korea now has 7,000 bakeries, and Koreans eat Italian-style noodles made from wheat flour.[21] "We taught people to eat wheat who did not eat it before," bragged an official from the U.S. Department of Agriculture.[22]

PL 480 has perhaps proved that people like what they eat, rather than eat what they like. At any rate, American corporations have taught people to eat what they, the corporations, have to sell. This achievement was lauded in 1974, before the Senate Foreign Relations Committee, in testimony by former Secretary of Agriculture Orville Freeman (now president of Business International—the same organization whose chairman, seventeen years earlier, had urged the use of PL 480 to create markets for American agricultural exports). Freeman noted, "In the last seven years, our agricultural exports to Taiwan have climbed by 531 percent and those to [South] Korea by 643 percent because we created a market." PL 480 "makes very good sense," he added.

But it does not make such good sense for South Korea. The country as a whole has become heavily dependent on the highly volatile international market for grain—dominated by only five multinational corporations. Nor should Americans forget the political repercussions in our own country of collaborating with such corrupt regimes; e.g., the attempts by Korean grain merchant Tongsun Park to bribe members of the United States Congress to support even more food to the Korean dictatorship.[23] Tongsun Park maneuvered his way into a key middleman position, receiving substantial commissions for arranging sales between U.S. rice exporters (Connel Rice and Sugar, and Rice Growers Association of California) and the Korean government. He then used this fortune to buy influence in Washington and in various state governments. He bribed Rep. Otto Passman (D-Louisiana), a member of the House appropriations subcommittee which oversees the U.S. aid program. He also gave substantial cash gifts to the governor of rice-producing Louisiana, Edwin Edwards. Due to his dealings with Tongsun Park, California Congressman Richard Hanna was convicted of defrauding the United States.

In spite of the Koreagate scandal, South Korea has continued to receive substantial quantities of PL 480 commodities under Title I: \$59.5 million in 1978, and \$40 million in 1979; \$40 million has been proposed for 1980.[24]

Food as a Weapon

While commercial forces worked from one end, broader political and military considerations influenced the PL 480 program from the other end. Thus, while our food aid has been distributed to 130 countries over the course of its history, at any particular period a few strategic countries have been the dominant recipients. Of these select few, U.S. military allies have been most favored: Israel and Turkey during the 1950s; South Korea, Taiwan, and Pakistan throughout; and South Vietnam and Cambodia during the period of U.S. military intervention in Indochina.[25] By 1973, almost half of all U.S. food aid was going to South Vietnam and Cambodia. Between 1968 and 1973, South Vietnam alone received twenty times more food aid than was received during the same period by the five African countries most seriously affected by the Sahelian drought.

In 1980, U.S. food aid remains highly concentrated.

QUESTION: What about emergencies? Aren't there times of crisis—a flood, a crop failure, or war—when food aid is essential to prevent massive starvation?

OUR RESPONSE: Food aid is useful and can alleviate suffering only when it is in response to a short-term emergency, such as a severe drought or a flood that actually destroys food supplies, or to assist refugees.

We have come to think of the above statement as a working hypothesis, given the mounting evidence of the negative impact of food aid around the world. Rather than assuming that food aid *must* be helpful, we now think that the opposite assumption is more appropriate. Only by assuming that food aid will become part of the problem is a food aid program in response to a genuine emergency likely to be carried out with a great deal of care and knowledge of the local situation. In such an emergency, food aid should be given only after all means of garnering food within the afflicted country have been pursued, including rationing and measures to prevent hoarding. A more appropriate form of 'food aid' in many instances would be the donation of transportation to help a local government move the necessary food from one agricultural area within the country to the emergency site. It must be remembered that even in disaster-stricken countries there is seldom an absolute shortage of minimum supplies.

We have learned that even, or especially, in times of emergency, food aid can enhance the security of the rich and powerful to the detriment of the poor majority.

Ethiopia. During the final two years (1973–1975) of the U.S.-supported Haile Selassie regime, some 100,000 Ethiopians died of starvation due to drought. At least half the amount of grain needed to keep those people alive was held in commercial storage facilities within the country.[26] In addition, Emperor Selassie's National Grain Corporation itself held in storage 17,000 tons of Australian wheat which it refused to distribute. While commercial interests thrived by selling hundreds of tons of Ethiopian grain, beans, and even milk to Western Europe and Saudi Arabia, the Ethiopian government received 150,000 tons of free food from aid donors. Government officials at all levels withheld stored food from the market, awaiting higher prices even as "...peasants could be seen starving within a few kilometers of grain storage." At one point, the Ethiopian officials offered to sell 4,000 tons of stored grain to the United States with the idea that the United States could then donate it back for relief inside the country.[27]

Upper Volta. Reports from local observers reveal at least seventy-six percent of the relief aid during the 1976 drought and subsequent political 'unrest' was distributed to the better-off dwellers in the capital city and the largest provincial towns, leaving very little for the hard-hit rural areas.[28] At the same time, moneylenders-*cum*-merchants exported the grain they collected from debt-shackled peasants to the Ivory Coast.

In some incidences, extreme official corruption and callous disregard for life in countries receiving U.S. food aid have been cited by the U.S. government as reasons to stop food aid programs.

Ghana. In 1977, the U.S. government suspended the Title II emergency food distribution program, which was intended to provide relief for drought victims in Ghana's Upper and Northern Regions. According to the USAID examination report of the program, "Large quantities of food provided under the emergency program did not reach the high-priority recipients because of unexplained losses in transit, diversion, and issue to other than the most needy population."[29] In four of the districts of Ghana's Northern Region, the report explains that "...forty percent, forty-nine percent, fourteen percent, and twenty-five percent respectively of the U.S.-provided grain was distributed to Government of Ghana employees."

Food Aid Disaster

While visiting Guatemala in 1977, we learned that even in times of natural disasters, food aid can undermine the livelihood of poor local farmers.

The terrible 1976 earthquake hit many small Guatemalan farmers who, prior to the earthquake, had exceptionally large harvests. Following the earthquake, they needed cash to help rebuild their homes and farms. To get that cash, these farmers needed to sell part of their stored corn and other grain.

Immediately following the earthquake, however, food aid from the United States almost quadrupled. CARE and Catholic Relief Services (CRS) handled most of the distribution. This increased availability of free food from the United States was one factor which helped to lower the prices for locally-grown grain, just when farmers most needed cash for their grain.[30] As a result, food aid *stood in the way* of reconstruction.

Even when the Guatemalan government finally asked voluntary agencies to stop bringing grain into the country, the food aid kept coming. CARE and CRS simply switched to grains pre-blended with other foods such as dry milk, since the ban specified 'grain.' One year after the earthquake, U.S. food aid was still sixty-nine percent above pre-earthquake levels.

Ironically, two other voluntary organizations attempted to counter the ill effects of the indiscriminate distribution of free food. Oxfam gave a special loan to a cooperative in the stricken Chimaltenango area that had been organized earlier with the help of World Neighbors. The loan was used to buy crops from the farmers at a price above the depressed level and to establish a grain bank to help stabilize sinking grain prices. The scheme helped the farmers in the area get the cash they needed to rebuild their lives.

William Rudell and Roland Bunch, who have long experience with rural cooperatives in Guatemala and who were on the scene at the time, told U.S. that even where there was a need for food during the first days following the earthquake, the food could have been bought from areas in Guatemala that were not affected by the earthquake. Such purchases would have been a boost to farmers in those villages. Moreover, supplies from within the country could have been *curtailed* more easily once farmers in the recipient villages had dug out their stored harvests. Bunch commented, "If the Guatemalans were sending wheat into the United States this year as their own version of a PL 480 donation and giving it out to American consumers, American farmers would be screaming bloody murder about it."[31]

Interestingly, it was the local people who most counseled against the distribution of free food and free materials. Instead, they advised the outside agencies to provide building materials and to sell them at subsidized prices. Oxfam, World Neighbors and USAID did just that—with notably positive results.

Where disasters destroy rather than bury food supplies; all the lessons inherent in the Guatemalan food aid disaster might not apply. Yet, even in genuine short-term emergencies, relief food should be purchased, as much as possible, from local and national producers whose families' livelihood depends on selling their grain.

In addition, we must note the insensitivity of CARE and CRS to the harmful effects of the U.S. food aid programs they were administering not only on prices, but also on the basic ability of local communities to cope with disaster using their own resources, knowledge and leadership. Because so many people had to stand in line waiting for free U.S. food, it was hard for local farmers to find the labor to harvest their wheat. Bunch noted how the continuation of food aid handouts undermined the genuine leadership in the community: "Largely because of the giveaways, the villagers began to turn more to leaders who could produce free things, whether they were honest or dishonest, rather than turn to the leaders they'd been putting their trust in for years...Groups that had worked together previously became enemies over the question of the selection of recipients for free food."[32]

QUESTION: Are you proposing that chronic food aid be terminated?

OUR RESPONSE: Yes. In evaluating our response, please keep in mind these facts.

- U.S. food aid is not focused on those countries where there is the greatest hunger and the least local productive potential. It is focused on countries like Bangladesh, South Korea, Indonesia, and Pakistan, because the U.S. government perceives the governments of those countries as allies to U.S. corporate and Cold War interests.

- Most U.S. food aid is *bought* by the recipient governments which then sell the food to their own people. Funds raised through the sale of food aid serve as general budgetary support, including maintenance of the police, military and bureaucracy necessary for unpopular regimes to stay in power.
- The influx of food aid can lower prices, making it harder for local small farmers to earn their livelihoods.
- Food aid can allow elite-controlled governments to continue avoiding the redistributive changes necessary to increase local production of food.
- Food aid distributed through food-for-work programs in fact benefits rural elites to the detriment of the rural poor. Such projects often improve the land of the elites and provide them with additional patronage possibilities, while at best providing low-paid work during the slack season to the poor.

Emergency food aid might be helpful under some conditions. Certainly the 1979–1980 famine in Cambodia was an economically and politically created famine in which Americans have a responsibility to help alleviate the tremendous suffering of virtually an entire population. In such rare instances, however, it is probably the voluntary agencies who have a better chance of effectively mobilizing and distributing food aid.

Other than to meet such famine emergencies, the United States should announce a date for the termination of food aid to all countries where a narrowly based elite controls the economic system. This would most definitely include the largest recipients of U.S. food aid, such as Bangladesh, India, Indonesia, Pakistan, and South Korea. Even a 1976 U.S. Senate study mission to Bangladesh concluded that food aid should be phased out over a five-year period.[33]

Concerned Americans should not think of food aid as *the* way to help the hungry. Dwelling on food aid—how much and what criteria should be used—diverts attention from the *process* of how hunger is created. It allows U.S. to forget that the overriding impact of the United States on the ability of people to become food self-reliant is not through food aid but through the corporate, military, economic, and covert involvement of the United States in their countries. We are advocating not only an end to chronic food aid but an end to all forms of government aid to countries where there is not already under way a fundamental democratization of control over productive resources.

1 U.S. Agency for International Development. U.S. *Overseas Loans and Grants and Assistance from International Organizations, Obligations, and Loan Authorizations, July 1, 1945 September 30, 1978* (Washington, DC: U.S. Agency for International Development, 1979)

2 Reibel, James. "Food Aid to India," unpublished thesis, pg. 1.

3 Mann, J.S. "The Impact of Public Law 480 on Prices and Domestic Supply of Cereals in India," *Journal of Farm Economics*, no. 49, February 1969, pg. 143.

4 Morgan, Dan and Don Oberdorfer. "Impact of U.S. Food Heavy on South Korea," *Washington Post*, March 12, 1975, pg. 1.

5 Ibid., pg. 12.

6 Fessler, Loren. *Population and Food Production in South Korea*, Field Staff Report 22–2, East Asia Series (New York: American University Fieldstaff, 1975).

7 Korean Catholic Farmers' Movement. *Report: Study about the Production Cost of Rice-1977* (Daeheungdon, Daejon: Korean Catholic Farmers' Movement, 1978), pg. 189.

8 "Impact of U.S. Food Heavy on South Korea," pg. 1.

9 Dudley, Leonard and Roger Sandilands. "The Side Effects of Foreign Aid: The Case of PL 480 Wheat in Columbia," *Economic Development Cultural Change*, January 1975, pp. 331–332.

10 Ibid.

11 Burke, Melvin. "Does 'Food for Peace' Assistance Damage the Bolivian Economy?" *Inter-American Economic Affairs*, no. 25, 1971, pp. 9, 17.

12 We have received a number of communications (some anonymous) from Americans working in Haiti. For a more thorough report of the role of U.S. food aid in Haiti see *Food Monitor*, no. 10, May-June 1979, pp. 8–11.

13 U.S. General Accounting Office. *Disincentives to Agricultural Production in Developing Countries*, Report to the Congress (Washington, DC: U.S. General Accounting Office, 1975), pg. 25.

14 W.L. Clayton, assistant Secretary of State, testimony on HR 2211, Breton Woods Agreement Act to House Committee on Banking and Currency, 79th Congress, first session, March 9, 1945, pp. 275, 282.

15 Eldridge Haynes, testimony to Senate Committee on Agriculture and Forestry, *Policies and Operations Under PL 480* (Washington, DC: U.S. Government, 1957).

16 Morgan, Dan. "Opening Markets, Program Pushes U.S. Food," *Washington Post,* March 10, 1975, pg. A-1.

17 Ibid.

18 "U.S. Grain Arsenal," *NACLA's Latin America and Empire Report,* vol. 9, no. 7, October 1975, pg. 27.

19 "Opening Markets, Program Pushes U.S. Food."

20 Ibid.

21 "Impact of U.S. Food Heavy on South Korea."

22 "U.S. Grain Arsenal," pg. 23.

23 Morgan, Dan. *Merchants of Grain* (New York: Viking Press, 1979), pp. 280ff.

24 *USAID Congressional Presentation, Fiscal Year 1980,* pg. 128.

25 "U.S. Grain Arsenal," pg. 14.

26 Ott, Hans. "Ethopia 1973–1975," in *IPRA Food Group Circular Letter,* no. 4, January 1978, pp. 36–40.

27 Ibid.

28 Michael Behr, Oxfam field representative, personal correspondence to Roger Newton, June 3, 1977, Ougadougou, Upper Volta.

29 U.S. Agency for International Development. *Report on Examination of the PL 480 Title II Emergency Food Program,* Audit Report no. 3–641–78–1, Area Auditor General, Africa (Nairobi, Kenya: U.S. Agency for International Development, 1977)

30 Riding, Alan. "U.S. Food Aid Seen as Hurting Guatemala," *The New York Times,* November 6, 1977.

31 "The Relationship Between PL 480 Food Distribution and Agricultural Development in Guatemala," edited interview with Roland Bunch and William Ruddell in *Antigua Guatemala,* August 21, 1977.

32 Ibid.

33 "World Hunger, Health, and Refugee Problems: Summary of Special Study Mission to Asia and the Middle East," report prepared for the Senate subcommittee on health, of the Senate Committee on Labor and Public Welfare, and subcommittee on refugees and escapees, of the Senate Committee on the Judiciary, 94th Congress, second session, January 1976, pg. 104.

From *Food First Backgrounder,* Winter 1995, by Li Kheng Poh, research intern, with Peter Rosset.

New Food Aid: Same as the Old Food Aid

Mogadishu, Somalia, March 17, 1993, 3 PM—Thirteen year-old Younis Abdi Mohamed was relaxing near the Bokhara market. He was sitting in front of a building taking in the sights and minding his own business. A minibus drove by and suddenly, shots rang out. Bullets whizzed past him. Gunmen in the minibus were exchanging bullets with a nearby U.S. Marine patrol. Like a slow motion clip in a movie, a U.S. Marine knelt and took aim at Younis. Terrified, Younis whipped around and started to run for his life. Almost immediately a sharp pain struck the inside of his thigh. To collapse and give in to the pain would mean death. Taking quick, tortured steps, Younis limped all the way home to his mother. She brought him to the hospital in a wheelbarrow.

More than a hundred days after the U.S. invasion, the number of Somalis killed by the U.S.-led armed forces stood at over two hundred. Younis almost became one of the two hundred. What happened to 'Food for Peace,' as U.S. food aid is officially known? The U.S. invasion was supposed to bring humanitarian aid in the name of hope and peace. Yet we delivered death and destruction. How did food aid evolve this way?[1]

The History of Food Aid

The ignominious history of U.S. government food aid for developing countries has been well documented by Food First over the years.[2] Food aid began in 1954 as Public Law 480, as a way to dispose of U.S. agricultural surpluses and open new export markets for United States commodities. The intended beneficiaries were U.S. farmers and companies involved in agribusinesses such as processing, bagging, fortification and transportation. During the Cold War era food aid was further used as a carrot to serve U.S. foreign policy interests. Feeding the hungry was always a secondary objective which lent respectability and justification to the existence of food aid.

When Food First initially published these findings on food and foreign aid, waves of concern swept through the food aid industry. Angry responses came from government; officials at the U.S. Agency for International Development (USAID), policymakers, non-governmental organizations (NGOs) and private voluntary organizations (PVOs). The resulting uproar made public a privately acknowledged fact—that food aid was creating long-term dependency on food donors and actually increasing poverty and hunger. In 1990, in response to years of criticism by Food First and others, food aid legislation finally underwent what was hailed as landmark changes that directed assistance to "combat world hunger and malnutrition and their causes."[3] In this article we ask if this 'new food aid' delivers on the promise that it is different from the 'old' food aid. We begin with a review of the past.

Historically there have been three types of U.S. food aid, as established by PL 480. The first type was called Title I, where low interest loans were given to food-deficit countries to purchase U.S. food stocks. The food was then resold to certain citizens of the recipient countries (usually the owners of mills, large bakeries, and pasta factories). Title II was the humanitarian portion of PL 480, under which food was donated to feed the hungry in emergency and non-emergency situations. The main agents of distribution were private voluntary PVOs (usually U.S.-based), intergovernmental organizations such as the World Food Program, and international and local NGOs. Title III was similar to Title I except that the loans could be forgiven as deemed appropriate under PL 480 provisions.

From 1954 through 1990, food aid was shrouded in the humanitarian myth. It was presented in the media as free aid donated to food deficit countries. Food First shattered this myth. The reality was that two-thirds of all food aid was sold under Title I, rather than donated.[4] Those who were too poor to buy the food when resold still went hungry. U.S. agribusinesses, who were the real beneficiaries, not only enjoyed increased sales, but secured permanent positions in foreign markets through Title I loan repayments: twenty-five percent of loan repayment funds were lent at low interest rates to finance entry into foreign markets.

Food aid furthered U.S. economic interests in other ways. Certain conditions required that recipients implement austerity and free trade programs such as structural adjustment, or sign agreements to purchase U.S. agricultural commodities on commercial terms in the future. Under structural adjustment, food security and self-sufficiency were sacrificed for export-oriented agriculture. Another of the conditions stipulated that any goods and services purchased with U.S. aid funds had to come from the U.S., even if they could be purchased more cheaply elsewhere.

Another food aid objective was to further U.S. foreign policy interests. At the height of the Vietnam War between 1968–1973, South Vietnam received twenty times more food aid than the five most drought affected countries in Africa.[5] The top ten recipient countries not only received three-quarters of all food aid, but were also the top recipients of U.S. military and economic aid.[6] Food aid was not focused on areas with the greatest hunger but on governments who served the geopolitical interests of the powerful U.S. government. Table 1 shows this was clearly the case in 1980.

Food aid, in serving U.S. interests, helped prop up antidemocratic governments around the world. The money generated from the resale of Title I food was used as budgetary and balance-of-payments support by recipient countries. In 1980, twenty percent of Bangladesh's

TABLE 1: Allocation of 'Old' and 'New' Food Aid

'Old' Food Aid, 1980			'New' Food Aid, 1992		
Country	U.S. $ Millions	% of Total	Country	U.S. $ Millions	% of Total
Egypt	219.6	22.5	Ethiopia	144.9	12.8
India	110.5	11.3	India	140.1	12.4
Indonesia	107.5	11.0	Peru	97.2	8.6
Bangladesh	78.0	8.0	Bangladesh	84.5	7.5
Pakistan	41.5	4.3	Mozambique	64.2	5.7
Portugal	40.0	4.1	Liberia	52.0	4.6
South Korea	40.0	4.1	Bolivia	48.4	4.3
Peru	31.3	3.2	Sri Lanka	45.0	4.1
Haiti	27.7	2.8	Zambia	39.8	3.5
Sri Lanka	26.1	2.7	Somalia	38.5	3.4
% of Total Aid to Top Ten		74.0	% of Total Aid to Top Ten		66.7

budget came from the resale of food aid. The amount of money spent on the militarization of the government made up twenty-seven percent of the budget.[7] Without food aid, it would have been impossible for the government to arm itself so heavily and so easily suppress popular movements demanding social, economic, and political justice. Furthermore, with the assured supply of food through PL 480, the government could afford to neglect agricultural development. This neglect meant that the poor, a majority of whom were involved in agriculture, would remain powerless and at the mercy of oppressive and brutal landowners, as well as military and government officials.

The inflow of food aid to recipient countries has proved time and again to be detrimental to local farm economies. Cheap subsidized or free U.S. grains undercut the prices of locally-produced food, driving local farmers out of business and into cities. The medium-term abilities of countries to feed themselves are undercut. As local farm economies disappear, many countries have become dependent on the U.S. to meet a large portion of their food needs. South Korea, for example, became the largest Third World importer of U.S. agricultural goods after years of food aid coupled with intensive marketing of wheat products by USAID.[8] This marketing campaign changed the South Korean diet drastically be creating an insatiable demand for wheat.

Title II free food has been distributed in many countries both during emergencies and in non-emergency situations, when food aid has principally been distributed through so-called "Food-for-Work" programs. Supposedly society as a whole would benefit from this sort of program—the jobless would get to eat through food payments for manual labor, while the rest of society would gain through road improvements, irrigation systems and other infrastructural improvements. Yet this argument is seriously flawed: Food-for-Work actually benefits the rich disproportionally, while the poor received no long-term gains.

A typical example makes this disparity clear. In a Food-for-Work program in a village in Haiti, one family controlled the local government and community offices.[9] When a U.S. PVO came to the village with a Food-for-Work program, this one powerful family was chosen to administer it. Jobless villagers built roads and tended the gardens of one of the community leaders, which took them away from their lands five days a week. The wealthy family gained additional benefits through the improvement of their lands, better access to markets for their produce and increased patronage power. The workers gained temporary work which provided food during the slack agricultural season, at the cost of not attending to their own plots, but they did not gain long-term, fundamental changes involving the alleviation of hunger and a way out of structural poverty. Similar stories dot the landscape of Food-for-Work programs.[10]

While Food-for-Work fails to deliver structural changes that would truly benefit the poor, it functions in the local economy just as do other food donations—as dumping. It replaces food that was supplied by local farmers and creates long-term dependency on imports. With food flowing into the recipient country through PVO 'development projects,' indigenous commercial food distribution networks are deprived of profitable opportunities. Food-for-work represents all the usual ills of food aid under a different name.

What about emergency aid in response to disasters? Does it really alleviate hunger? During emergencies food aid often enhances the power of the rich at the expense of the poor. In the Ethiopian drought of 1973–1975, some 100,000 Ethiopians died,[11] a tragedy that happened while Ethiopian milk, wheat, and beans were being exported to Saudi Arabia and Europe by commercial interests. Meanwhile, because of the drought, the U.S.-backed Haile Selassie government received donated wheat from the U.S. and European countries. Corrupt Ethiopian officials, waiting for the price of wheat to rise so that they could sell it themselves at a handsome profit, stored the wheat and watched the poor die of hunger.

In emergency situations the real aid that is needed are funds and local transportation to purchase and move food from still productive agricultural regions to the areas of food shortage. Additional options such as temporary food rationing and anti-hoarding measures can often be more effective than food donations. Only if these fail should international food aid be considered.

Outside efforts to help the poor—but which are administered by local elites—only strengthen the wealthy by further concentrating power in the hands of a few. To affect real long-term changes, the very economic and political structures that create poverty and hunger need to be addressed. But repressive regimes supported by U.S. government aid have long prevented these long-term changes.

The 1990 Farm Bill and the Reform of Food Aid

Food aid was problematic partly because it was used as a weapon during the Cold War. With the collapse of the socialist bloc it suddenly became acceptable to acknowledge the long-standing Food First critique of food aid. In the 1990 Farm Bill the language of the legislation enabling food aid was changed to "address real needs of the hungry."[12] As a result, most U.S. food aid is now distributed through PVOs and NGOs rather than governments, as they are more cost effective and are better able to reach grass-roots groups. Management also has been changed to remove bureaucratic barriers to effective food delivery.[13]

While these changes seem to bode well, they require critical evaluation, because they only address certain technical problems with the delivery of food aid, and most definitely not the root causes of hunger. As numerous Food First studies have demonstrated, the only real answer to eliminating hunger is to make productive resources accessible to all.[14] The 'new' food aid still fails to offer any more than fleeting options to the poor. The issue of why people are hungry continues to be whitewashed by focusing on how 'humanitarian' food aid now is because it goes more directly to the hungry.

The goal of combating world hunger has ostensibly become the first priority in the new legislation, and a stated official objective is the elimination of surplus dumping. The other multiple objectives are similar to those in the 'old' legislation.[15] The final version of the new legislation affirms, just as had the old legislation, that the purpose of food aid is to "promote the foreign policy of the United States by enhancing the food security of the developing world."[16] What does that mean? Very simply: creating markets for U.S. exports. USAID proudly proclaims in 1995 that food aid "benefited American farmers and strengthened U.S. foreign policy... Nine of today's top ten importers of U.S. agricultural products are former recipients of this programs."[17] Food aid is still a supply-driven dumping of surplus agricultural commodities. Yet the creation of export markets is inherently contradictory to the goal of combating world hunger, which can only be achieved by strengthening local capacity to produce food.

Food aid is part and parcel of official U.S. foreign assistance, and must also be considered in that light. Current U.S. law still requires that ninety percent of all military and food aid and fifty percent of all bilateral assistance must be spent on U.S. goods and services.[18] More than anything else, food aid is corporate welfare just as it was in the past. According to testimony given by the Coalition for Food Aid[19] before the Senate subcommittee on agriculture in March 1995, food aid "funds are spent primarily on U.S. commodities, processing, bagging, fortification and transportation."[20]

Just as with the old food aid, dumping surplus grains destroys the local agriculture of food aid recipients, as in the case of Somalia. Little has really changed, although food aid now functions less as major budgetary support to repressive governments. This is because the proportion made up by Title I sales (which governments resell) is now less than one-third of all food aid, and aid has been shifted toward countries where hunger and poverty are major concerns (see Table 1)—though the inclusion of India, a net food exporter, is certainly questionable. The net impact of food aid is still negative, though the balance of why this is so has shifted from propping up anti-democratic governments more toward undercutting rural economies.

Private Voluntary Organizations and Food Aid in the 1990s

PVOs have become the main agents of food aid distribution because donated food (Title II) now makes up two-thirds of PL 480.[21] Aside from emergency relief, most donations are used in various development activities conducted by PVOs. What are these development activities and their impacts on recipients?

Non-emergency food aid is spent on food-related or non-food-related activities such as Food-for-Work, cash-for-work, and mother and child health, nutrition and education programs.[22] Infrastructure such as roads are built, and agricultural projects such as soil and water conservation are undertaken using food or cash as payments to participants. The cash used is raised by selling food aid in the recipient country. Lunch programs are started as an incentive for children to stay in school, and mothers at health centers receive food when they bring their babies to be treated.

While many of these activities seem at first glance to be laudable, we need to look deeper. First and foremost, food aid is still about the injection of food into the economy of recipient countries, which still results in the distortion of food markets. Just as it did with the old food aid, this distortion weakens commercial food distribution channels, drives local farmers off the land, and ultimately creates long-term dependency on imported U.S. agricultural commodities.[23] These effects remain whether we are talking about Food-for-Work, food-for-cash, or other food aid activities. That is not to say that all programs supported by food aid are misguided. Rather, other ways are needed to carry them out.

Mother and child health and nutrition programs can offer substantial benefits to recipients. Michael Maren, well-known writer and one-time food aid project manager, says that the preferred alternative to using food aid in these programs should be to purchase food from surrounding areas.[24] This system would strengthen farmers, local merchants, and the economy of the country, as local expertise, knowledge, and resources are utilized to produce the food. As income is generated internally, communities become self-sufficient and sustainable. This is not possible when the local communities rely predominantly on expatriates and outside resources.

However, food aid-based development projects continue to depend on expatriate expertise, knowledge and outside resources to generate income. These projects are not self-sufficient nor are they sustainable when the aid ends. Not surprisingly, food aid-based development projects carried out by PVOs have historically been failures. As testified to by Michael Maren, "Africa is littered with the ruins of such projects."[25]

On the whole, the 'new' food aid still does more harm than good. It remains an industry built for dumping surplus commodities and furthering U.S. interests, even as projects are trendily prettified with the phrase "sustainable development." As Alex de Waal of Africa Rights has said, "the central issue of the marginality of relief aid itself was never fully acknowledged… The idea of proposing less relief aid was taboo—not surprising given the institutional commitments of those involved in the debate"[26]

Conclusion

The recent history of food aid can be neatly summed up in the phrase *plus ça change, plus c'est la méme chose.* While the new food aid does increase the efficiency of food aid distribution, it still does not address the real failings of the old food aid or the root causes of hunger. It still furthers U.S. interests to the detriment of recipients. As such, private voluntary organizations, for all their good intentions, become a tool for dumping U.S. agricultural surpluses.

Food aid is about profits for ourselves. U.S. shippers, agribusinesses, and processors all profit from food aid. While U.S. farmers presumably also benefit, past research by Food First has shown that the lion's share of the profits go to large grain companies.[27] PVOs benefit: managing food aid accounts for a huge chunk of the budgets of many of the largest agencies and supports hefty salaries for executives and managers in the food aid industry.

The recipients of food aid face the distortion of their economies caused by the massive dumping of American agricultural commodities. Prices slump, production decreases, and local development is stunted. Contributing to this scenario are the debris of food aid-based development projects that proved not to be self-sustaining when project aid ran out. In short, recipients very often become permanently dependent on U.S. agricultural products and expatriate experts, while gaining little lasting improvement for the poor.

In helping ourselves by keeping others down, we prevent far-reaching changes from taking place. As Food First has continually stressed, to affect real change, alliances should be made with those who are fighting for the restructuring of wealth and power. The official purpose of U.S. foreign assistance is to further U.S. national interests. Yet does food aid really serve our national interests if it keeps people poor by chaining them to external resources? Keeping recipient countries poor means job flight from rich to poor countries; keeping people poor depresses potential markets for other U.S. exports; denying people the opportunity of a rural livelihood growing food for local consumption forces them to crowd into cities where few jobs await them. All of these factors represent real economic and security issues for the U.S. Our own security will only really be assured when the poor achieve economic and political security.

Food First calls for a cessation of all international food aid except that of short-term emergency relief. When food must be delivered to starving people in an emergency situation, it should be bought within the recipient country's borders if at all possible. The next choice is to promote regional food security between nations by buying needed food from neighboring countries. U.S. commodities have no useful role to play in resolving world hunger. The only durable way to alleviate hunger is to support the grassroots efforts of local people to change the way food is grown, distributed and consumed inside their own country.

The Somalia Food Aid Fiasco

By the time the U.S. invaded Somalia in December 1992, the worst of the famine was already over: the death rate stood at seventy per day, down from a peak of three hundred per day in September.[28] Good crops of rice, sorghum, and corn from the agricultural regions of Afgoye and the Shebell River Valley had already been harvested.[29] But as free food aid flooded Somalia, the demand for and prices of these local grains plummeted. A 110 pound bag of sorghum cost 50,000 shillings in February of 1993, a quarter of the price in October 1992.[30] Somali farmers, unable to make a living by selling their produce, were forced to abandon their lands and join the lines for food handouts.

If aid agencies had bought local grains to supplement food aid, the depressed prices could have been avoided. Instead, food aid substituted for local grains. Mrs. Faaduma Abdi Arush, a Somali farmer, tried to sell her corn to six PVOs.[31] None would buy it. The institutional structure of food aid made such a life-saving transaction impossible: the U.S. government pays PVOs only to deliver free American food.[32] Another farmer, Mr. Hassan, felt that the PVOs had done a great disservice to him because food aid had pushed the price of corn so low that he had to sell at a loss or let his corn rot.[33]

The dilemmas of Mr. Hassan and Mrs. Arush illustrate that world hunger cannot be fought with an aid program that has multiple and contradictory objectives. Buying indigenous commodities would strengthen local food producing capacity as the increased demand would ensure farmers a better return for their grains. However, local purchases would violate the other food aid objective of creating new markets, and are largely not permitted under existing regulations. U.S. producers and exporters would be harmed because of lost opportunities to provide U.S. agricultural commodities.

During the Cold War, relief agencies operating in countries whose governments were propped up by superpower interests had very little room for maneuvering. They were strictly mandated to provide food and little else. They were accountable to the host government and risked ejection from the host country if they stepped out of line. After the Cold War ended however, *real politik* ceased to matter as much, especially in sub-Saharan Africa, and some host governments collapsed. Sometimes the result of having propped up repressive unpopular governments for so long was war or chaos.

As Western powers distanced themselves from sub-Saharan Africa, aid was mostly channeled through non-governmental organizations such as PVOs. As such, PVOs grew and became more powerful organizations with a lot of say in the internal affairs of host countries. PVOs in Somalia became the sole providers of health care and social services, essentially taking over the role of the government, but without any political accountability. Unlike governments, PVOs are not obliged to provide such services, thus what they do is considered charity. Recipients of PVO largess cannot hold PVOs accountable for their actions.

Given multiple mandates and lack of accountability, opportunistic motives sometimes lead PVO agendas. The easiest contracts to gain are during emergencies. Fund raising success leaps when there is a highly publicized 'humanitarian' event such as the invasion of Somalia.[34] When PVOs encountered difficulties in delivering donations in Somalia—food shipments were stolen by local functionaries or looted by the hungry, relief workers were threatened—American PVOs led the call for a military intervention to facilitate food delivery, a seriously misguided role for humanitarian agencies to play.

With the presence of foreign troops, Somalia turned into a nightmare: the civil war worsened and crime and violence increased. It became more dangerous—rather than less—for PVOs to operate feeding centers, because it became clear to participants in the local power struggle that the PVOs were not apolitical, that the PVOs were in a sense additional competitors in the war for control over the country. How could PVOs not be players in the civil war, they reasoned, when they took on the role of government and made bold calls for military intervention, and when they had foreign armed forces on their side? Providing food aid in a disaster is only one small element in solving a larger problem. Foreign occupation and the assumption of the role of government by foreign agencies are not solutions to a problem created largely by superpower meddling in local politics in the first place.

Somalis must find their own peaceful settlement. The under-utilized capacity of Somali farmers to feed the hungry in Somalia should be tapped and maximized. Flooding the country with free food at bayonet point is no help. In Galcaio, a central Somalian town that has no foreign assistance, peace had endured for more than a year by July 1994, with no help at all from U.N. forces. Commerce and trade had resumed, and young men who had been fighters were now trading livestock for food and other goods. This was possible because "the community leaders, religious figures, businessmen, students, and representatives of the factions"[35] met to produce a local peace accord in May 1993.

1 Africa Rights. *Somalia—Operation Restore Hope: A Preliminary Assessment,* May 1993, pg. 36.

2 Lappé, Frances Moore and Joseph Collins. *Food First: Beyond the Myth of Scarcity* (New York: Ballantine Books, 1977); Frances Moore Lappé et al., *Aid As Obstacle: Twenty Questions About our Foreign Aid and the Hungry* (San Francisco: Food First Books, 1981); and James Wessel with Mort Hantman, *Trading the Future: Farm Exports and the Concentration of Economic Power in our Food System* (San Francisco: Food First Books, 1983).

3 Smith, Mark E. et al. *Provisions of the Food, Agriculture, Conservation and Trade Action of 1990*, Economic Research Service, Agriculture Information Bulletin no. 624, (Washington, DC: U.S. Department of Agriculture, 1990) pp. 64–73.

4 *Aid As Obstacle,* pg. 93.

5 Ibid., pg. 102.

6 The ten countries, in descending order of the amount of food aid received: Egypt, India, Indonesia, Bangladesh, Pakistan, Portugal, South Korea, Peru, Haiti, and Sri Lanka. United States Agency for International Development Congressional Presentation, Fiscal Year 1980, pg. 128.

7 *Aid As Obstacle,* pg. 107.

8 Ibid., pg. 95.

9 Ibid., pg. 113.

10 Rehman, Sobhan. *Agrarian Reform and Social Transformation Preconditions for Development* (London, UK: Zed Books, 1993).

11 Ibid., pg. 114.

12 PL 480 was reauthorized under the Food, Agriculture, Conservation, and Trade (FACT) Act, Title XV, PL 101–624 (also known as the Farm Bill). It was originally authorized in 1990 under the Agriculture Trade Development and Assistance Act (United States Agency for International Development, 1994, "Food for Peace: America's Bounty Serves the World From 1954 into the 21st Century").

13 Titles II and III were placed under the sole authority of the U.S. Agency for International Development, and Title I under the authority of the USDA.

14 For our most recent statement on this issue, see Peter Rosset, John Gershman, Shea Cunningham, and Marilyn Borchardt, "Myths and Root Causes: Hunger, Population, and Development." *Food First Backgrounder,* Winter 1994.

15 The rest of the multiple objectives are the promotion of sustainable and broad development, the expansion of international trade, the expansion of U.S. export markets, and the fostering of private enterprise and democracy in developing countries (Mark E. Smith and David R. Lee, "Overseas Food Aid Programs," in *Food,*

Agriculture, and Rural Policy into the Twenty-First Century: Issues and Trade-Offs, edited by Milton C. Hallberg, Robert G. F. Spitze, and Daryll E. Ray [Boulder, CO: Westview Press, 1994]).

16 Ibid.

17 "Food For Peace Celebrates 40th Anniversary," *U.S. AID Developments*, Winter 1994.

18 Cohen, Andrew. "Clinton Doctrine: The Help That Hurts," *The Progressive*, January 1994, pp. 27–30.

19 The Coalition for Food Aid is actually a pro-food aid lobby of relief agencies, or what might be called the "food aid industry."

20 Testimony of John Donnelly, deputy director, Catholic Relief Services, on behalf of the Coalition for Food Aid before the House Appropriations Committee, subcommittee on agriculture, March 22, 1995.

21 For fiscal years 1992–1995, total Titles II and III averaged seventy percent of all food aid (in $). Of the seventy percent, Title III is increasingly becoming insignificant because by 1995, Title III only made up about thirteen percent of all food aid. Calculated from *USDA Budget Summary FY 1996*, press release, February 6, 1995, pg. 36.

22 United States General Accounting Office. "Management Improvements Are Needed to Achieve Program Objective," pp. 20, 26.

23 Maren, Michael. "Good Will and Its Limits in Somalia," *The New York Times*, August 27, 1993.

24 Interview with Michael Maren on May 12, 1995.

25 "Good Will and Its Limits in Somalia."

26 de Waal, Alex. *African Encounters*, Index On Censorship, (London, UK: Writers and Scholars International, 1994) pp. 14–31.

27 *Trading the Future*, op. cit.

28 Perlez, Jane. "General is Wary of Sending Force to Somali Interior," *The New York Times*, December 14, 1992.

29 Mitchell, Alison. "A New Question In Somalia: When Does Free Food Hurt?" *The New York Times*, January 13, 1993.

30 Dahlburg, John-Thor. "Aid Workers Worry About Helping Somalis Too Much," *Los Angeles Times*, February 16, 1993.

31 Mitchell, Alison. "A New Question In Somalia: When Does Free Food Hurt?" *The New York Times*, January 13, 1993.

32 A failure to deliver means an end to further lucrative contracts. The survival of many PVOs depends on these contracts not just for the direct payments but also for the publicity gained in the media. It is this publicity which motivates the public to give money to PVOs (Michael Maren, "The Food-Aid Racket," *Harper's Magazine*, August, 1993).

33 Perlez, Jane. "General is Wary of Sending Force to Somali Interior," *The New York Times*, December 14, 1992.

34 Interview with Michael Maren on May 12, 1995.

35 Maren, Michael. "Leave Somalia Alone," *The New York Times*, July 1994.

From *A Fate Worse than Debt: The World Financial Crisis and the Poor* by Susan George

Latin America: Going to Extremes

> *If a man owe a debt and the storm god inundate his field and carry away the produce, or if, through lack of water grain has not grown in the field, in that year he shall not make any return of grain to the creditor, he shall alter his contract-tablet and he shall not pay the interest for that year.*
> —The Code of Hammurabi, King of Babylon, circa 2250 BCE

If the Babylonian farmer who lost his crop had been forced to pay back his creditor anyway, he would soon have sold off his possessions, his animals and his land. Instead of living as a more or less prosperous and independent farmer, he would have joined the ranks of the landless poor, surviving on casual labor from hand to mouth. A wealthy neighbor or merchant would have bought the farmer's property on the cheap and grown richer exploiting it. Without the King's Code, Babylon would have been a more polarized society, like so many in the Third World today, where debt thrusts millions into hunger and a marginal existence.

Because of the debt crisis polarization and marginalization are accelerating, both *within* and *between* countries. Heavily indebted societies quite literally go to extremes: the middle class tends to disappear, while a very few rich people able to escape the pernicious effects of debt dominate millions who can barely survive. Zaire is a particularly severe case of this phenomenon, as we've seen. At the international level rich countries also grow richer at the expense of poor ones that are obliged to sell off their property (commodities, labor) at 'famine prices.' Marginalization begins to strike whole nations because they must cash in their assets and cannot save or invest anything in the future.

Acts of God aggravate the process of going to any extreme, but, the Code of Hammurabi does not apply at the International Monetary Fund (IMF). Where his Code made allowances for the unpredictable storm god, the IMF does not see natural disasters as mitigating circumstances. A few days after the Mexico City earthquake of

September 1985, the Fund's flying squads were back in town, demanding their due. A cartoon in *Le Monde* shows two IMF types knocking on Mexico's door, behind which lies a heap of rubble. "What excuse do you suppose they'll come up with this time?" asks one bureaucrat of the other. About the same time a Mexican cartoon pictured a vulture labeled IMF perched above the ruins from which signs of life are beginning to emerge. "Feeling better, Mexico?" asks the predator. "Just remember we still have some unfinished business to settle."

Mexico has about as much control over the price of oil as it had over the earthquake or the Babylonian farmer had over the ravages of the storm god. Yet when oil slumps precipitously, wiping out a large proportion of Mexico's (or Ecuador's or Venezuela's) export revenues, that is no excuse for non-payment of debt. The Fund and other 'sado-monetarists' (Dennis Healey's phrase) think that it is still possible to adjust. Their prescription? Devalue, in order to export more.

Exportomania and Polarization

Even those analysts who do not share the Fund's eagerness for deep cuts in social services tend to see eye to eye with the IMF on the issue of exports. A World Bank summary of the studies on the "social costs of recession" states unequivocally that conditions in Latin America are bound to grow worse because of the "collapse of investment and the resulting deterioration of the physical condition of hospitals and schools." This document pleads for preserving basic services for the poor but still claims that "it is much harder to maintain, let alone improve, the lot of the poor in the absence of economic growth...the lesson, then, is that a resumption of growth is by far the most certain way to stabilize and eventually improve the condition of the poor... *Resumed growth will depend on export expansion*" (my emphasis).[1]

When "expand exports" is incised on the tablets of our present-day Codes, one wonders how the lot of the poor can ever be 'maintained,' let alone 'improved.' The lot of the poor can be improved only when the poor have something useful and productive to do, like providing basic necessities, including food and shelter, to others within their own societies. Those who argue for expanding exports as a road to growth put their faith in comparative advantage, and while they focus on what exports may earn, they rarely tell U.S. what they cost.

When these exports are agricultural products the first thing they cost is land—land that *could* be devoted to food crops, that *could* be devoted to feeding the millions of hungry people in the indebted countries. But it won't be because that would not be 'economical.' Mexico has devoted enormous efforts to expanding beef cattle production, almost exclusively for export, since so few Mexicans can afford to eat beef themselves. Shipments to the United States more than doubled between 1985 and 1986—from 577,000 to 1.2 million head. Mexico did not, however, double its revenues because each steer was worth $100 *less* in 1986 than in the previous year. Whether or not these animals fetch a decent price on the U.S. market, they take up more and more space in Mexico, as do the forage crops they eat.[2]

One usually thinks of the United States as an agricultural exporter, which it is, but it is also, somewhat surprisingly, the world's largest food importer ($20.8 billion worth in 1986). Mexico is now the premier agricultural supplier to its northern neighbor: over $2 billion worth in 1986, mostly fresh fruits and vegetables, as well as beef. Meanwhile, Mexico imported $1.5 billion of agricultural products from the U.S. in 1985, mostly in basic grains and oil seeds. Surely this is comparative advantage in action? Isn't everyone better off concentrating on his strong suit and importing what the other fellow produces more cheaply?

That would be too blissfully simple. For one thing, Mexico has a comparative advantage in fresh fruits and vegetables grown on the huge, irrigated farms of the northern states of Sinaloa and Sonora only because government spending on the agricultural sector has been so heavily concentrated on these farms. Long before the debt crisis, from 1940 to 1970, sixty percent of all government investment in agriculture went to these northern states, where only nine percent of the small peasant holdings (*ejidos*) are located. Today rich farmers in these rich provinces still get water from government irrigation schemes at less than a third of its real cost. If successive governments had decided to invest in other crops—which is to say, in other social classes—Mexico could perfectly well have a comparative advantage in corn, beans, and oil seeds today.

There is little that is natural in 'natural advantage.' This theory works only if the government can collect on its export revenues, which it needs in order to purchase from foreigners the staple foods that the theoreticians claim it doesn't need to grow. Unfortunately for Mexico, the large farms that benefit from exports also pay very low taxes, and their owners are as sophisticated in exporting capital as cucumbers. So the rich grow richer and the poor hungrier.

A further pernicious consequence of investing in large exporting farms and neglecting small growers is the huge outflux of people from the countryside. If Mexico grew more of its basic foodstuffs, life in rural areas would become more prosperous and attractive to the peasantry and the cities would be less congested. The sprawl of Mexico City has already reached crisis proportions. In the year 2000 the population of this megalopolis will top thirty million at present growth rates. Some of its residents already compare it with Dante's *Inferno*, though there is some argument about exactly which circle they inhabit.

A turnaround would not be impossible. One estimate, cited by the Economist Development Report, says that if "the government redirected the perks which it is giving to farmers with irrigated land, to improving inputs and credit for rain-fed (small) farmers, it could bring five million extra hectares under cultivation, eliminate food imports and create five million jobs."[3] But that would not conform to the iron code of the IMF (nor to the interests of the richest and most powerful Mexicans).

Wealthy Wetbacks

Mexico has registered spectacular success in its exports of one extremely valuable commodity—rich and educated people. They are leaving in droves and settling in the United States.[4] Official statistics on human capital flight are unavailable, but some researchers believe that over 100,000 upper-middle-class, highly skilled Mexican professionals headed north between 1982 and 1985.

Another, smaller, group, which one sociologist calls the "post devaluation exiles," may not be especially smart but, man, are they rich! "We're buying California back from the U.S.," boasts a member of one of the twenty-eight families interviewed by this sociologist. Each one of these families arrived with $4 million or $5 million. They, and dozens like them, are buying businesses and luxury housing as if there were no tomorrow. So many Mexicans have bought condominiums in high-rise buildings near La Jolla, California, that racist Anglos have christened these apartments the "Taco Towers."

The exodus of the elite—and of the elite's money—makes the burden for those who must remain behind even greater. Yet which of us individually, can cast a stone at these Mexicans? I have that sinking feeling I might join them were I in their hand-made crocodile shoes. They describe Mexico as a four-C society—rife with cynicism, corruption, crisis, and crime. Kidnapping and murder aimed at the rich have become commonplace. Mexicans who have a choice say they no longer want to bring up their children in their own country.

All of them watched their fortunes (in pesos) melt with each successive devaluation. One manufacturer who had moved to Texas explained that for him the last straw was the bribes he had to pay simply to maintain garbage collection and other basic services. As one escapee says, "Mexico is my home, but now there is no hope there." Debt, devaluation, and human/material capital outflow feed on each other, and rich Mexicans move to the 'North Pole' extreme on the polarization scale.[5]

BOP-ping in Brazil

Brazil is too far away from the U.S. to export as many people as Mexico but compensates by exporting more food. Vincent Leclercq, an economist specializing in Brazilian agriculture at the French National Institute of Agronomic Research (INRA) notes some alarming trends, especially since January 1983, when the IMF first applied its standard analysis and structural-adjustment package to Brazil.[6]

One of the Fund's chief goals was to reduce domestic consumption in spite of huge excess manufacturing capacity and already low demand for food compared with the size of the population. In Fund language this is called "demand management." "Demand" applies only to those requirements that can be expressed in terms of money; it has nothing to do with real needs. The Fund's corrective measures were aimed at restoring a positive balance of payments (BOP) to the exclusion of any other goal—which in turn meant relying on export agriculture, Brazil's biggest money-earner. The economy was declared 'adjusted' when Brazil's international books returned to the black—thanks also to sharp curtailment of imports.

Brazil's emphasis on export crops to the detriment of food crops cannot be blamed exclusively on the IMF. Successive governments, heavily influenced by the largest landowners, have consistently encouraged exports and showered credit on the farmers who produce them. Brazil now trails only the U.S. in agri-export revenues, with soybeans, orange juice, chicken, and coffee heading the list. Land planted to soybeans increased over ninefold in the 1970s. Brazil has also set up the world's largest fuel-alcohol program to cut down oil imports, with enormous sugar-cane plantations providing the raw material. Cane's 'green plague' has brought misery to thousands of smallholders by forcing them off of their land and into a marginal existence.

These deliberate policies have taken their toll on food production and consumption. The vast majority of poor Brazilians depend on six staple food crops: cassava, corn, beans, rice, potatoes, and wheat. In 1977 total production of these six mainstays was just over sixty million tons, with cassava and corn accounting for twenty-six and nineteen million tons respectively. Any country where people count on cassava for forty-three percent of their staple food supply is already in bad nutritional shape unless they get a good amount of protein from beans or other legumes.

Two disastrous years of unfavorable weather in 1978 and 1979 dragged food crops down by twelve percent. By 1982 production had crept slowly back up to sixty-two million tons. The IMF program was implemented in January 1983. That same year production of the mainstay staples dropped by thirteen percent to fifty-four million tons, the lowest level since the very poor year of 1978. Bean harvests fell forty-five percent and a further eleven percent in 1984. These declines could not be ascribed to bad weather. Meanwhile the population was growing at 2.5 percent yearly. The only food crop that actually did better was wheat, a small factor in Brazil with only about two million tons grown annually. Wheat, mostly imported, has none the less been a godsend to ordinary urban Brazilians and has helped to keep their diet from being even worse than it is.

Against this fragile backdrop the IMF had decided to (1) eliminate wheat-consumption subsidies, (2) squeeze rural credit, (3) push agricultural exports harder than ever.

The Fund argues that subsidies cost the state too much and encourage inflation: that imports of subsidized products (like wheat) cancel out export gains; that subsidies distort the market and prevent other staple food products from competing. Fair enough. Brazilian wheat imports cost over $500 million in 1978 and nearly $900 million in 1980—a startling sixty-four percent jump and a lot of money when one is expected to build up a positive BOP. But wiping out the subsidy when the wheat habit was firmly entrenched (consumption was increasing by ten percent yearly) could only worsen an already precarious national diet.

In the first half of the 1980s Brazil averaged half a billion dollars worth of imported wheat from the United States alone but in 1985–1986 slashed its purchases from 450,000 to 48,000 tons! Can people compensate for the loss of wheat by consuming other staple foods? No, because the IMF stabilization program prevents this too: agricultural credit has been drastically curtailed. Farmers consequently cut back on inputs—better seeds, fertilizer—so their yields suffer. Staple foods seem to be everyone's lowest priority—except for the poor and hungry who, by definition, don't count.

The only real hope for Brazilian consumers and peasants alike is land reform. As of the end of 1986, promised reforms had been stymied. The Sarney government announced sweeping changes, then backtracked under concerted assault from the land barons, who pretend that any reduction of their privileges is part of the 'international communist conspiracy.' Their hired thugs murdered five hundred peasants and their supporters (lawyers, priests, trade unionists) in 1985–1986. In Brazil two percent of the farms occupy fifty-eight percent of all farmland, while eighty-three percent of the farms share only fourteen percent of the land. Four hundred mega-farmers between them own an area equivalent to eighty-five percent of Great Britain.[7]

Bloodshed in Santo Domingo

The final exhibit in the gallery of Latin American extremes is the Dominican Republic, a small country that holds a dubious record. Here, in April 1984, occurred the greatest number of deaths in the hemisphere stemming directly from an IMF riot.

The country stands out in other ways as well, none of them, unfortunately, beneficial to the inhabitants. The Dominican Republic is the eastern half of the island of Hispaniola; Haiti is the western half. Columbus landed there in 1492, making the island the first outpost of European colonization of Latin America. The original inhabitants, the Carib Indians, were slaughtered in short order. Black Africans were then imported to work on the sugar plantations. Life in the countryside has improved only marginally since. Today 0.07 percent (7/1000) of the landowners monopolize forty-five percent of the arable land, and 300,000 rural families have no land at all—a lot of people for a population of only 6.4 million.

Income is about as well distributed as land. Four hundred families have annual incomes of over one million pesos (about $325,000); another 2,000 families make more than $15,000 yearly; the 'middle class' is composed of some 50,000 households with incomes between $5,000 and $15,000. At the bottom are 3.5 million people who try to manage on less than $400–600 a year. A third of the population—over million—lives in the capital, Santo Domingo, at least seventy percent of them in slums, known as *cordones de miseria*, for obvious reasons. The *official* estimate of unemployment is thirty percent, the legal minimum salary about $US 80 a month. The debt is $3.6 billion, and interest payments amounted to ten percent of export earnings in 1984.

This, briefly, is the social and economic background of the government agreement with the IMF, signed in January 1983 and requiring the usual measures: devaluation of the currency, reduction of government expenditure, and others. In April 1984 the Dominican Republic received a loan of $100 million from the IMF. The blood money required in exchange included price increases of up to two hundred percent for basic necessities, including bread—and, a nice touch, the new prices were announced during Easter week.

The protests that began the following Monday, April 23, were spontaneous, certainly, but also the result of consultation and conscious decision among literally hundreds of popular organizations in the *cordones*. Tens of thousands took to the streets; President Jorge Blanco announced he would take all necessary steps to "maintain public order." On the 24th all the official trade unions joined in demanding a change in the government policy and higher salaries. They, along with the slum dwellers, were met with fierce repression.

After three days of the uprising the people were finally put down. Estimates of the number of deaths vary. One Dominican organization counted 186, plus hundreds of wounded and at least 5,000 people arrested. Newspapers published the identities of seventy-one people and listed fifty other unidentified bodies. The dead were mainly young people under twenty, and there were at least eighteen women, including a thirteen-year-old girl and a seventy-year-old woman. President Blanco, in his speech of April 25, declared, "The armed forces and the National Police have given an example of restraint (*ecuanamidad*) displaying their high degree of professionalism, elevated human feeling and respect for life... they have kept their reactions within the limits of reasonable prudence and displayed excellent training."[8]

Words fail me.

1 World Bank. *Poverty in Latin America: The Impact of Depression*, Report No. 6369, 6 August 1986, paragraphs 48 and 49.

2 "Country Life," *Economist Development Report*, September 1984, and FATUS (Foreign Agricultural Trade of the United States), USDA, November-December 1986, various tables. The export figures given are for fiscal years (October 1 to September 30).

3 Ibid., both sources in note 2.

4 See "Nation in Jeopardy: Mexico's Crisis Grows as Money and the Rich Both Seek Safer Places," *Wall Street Journal*, October 11, 1985; and Larry Rohter, "Exit of the Skilled Dims Mexico's Future," *International Herald Tribune*, October 28, 1986.

5 Quoted in "Exit of the Skilled Dims Mexico's Future."

6 The official IMF program came to an end in 1985, but Brazil continues to apply most of the same measures on its own. Vincent Leclercq, "Politique d'ajustement structurel et politique agricole au Bresil 1980–1985," paper presented at the Journees d'Etude, Reseau Strategies Alimentaires, Paris, June 10, 1985.

7 Postal, Patrick. "L'Enlisement de la reforme agraire," *Le Monde Diplomatique*, November 1986.

8 This account is compiled from Christian Rudel, "Bidonvilles en Amérique Latine," CCFD Dossiers no. 86–11, Paris, November 1986; Romon Quinones, "IMF Plunges Dominican Republic into Acute Economic, Social Crisis," IDOC Internazionale 85/3, dossier on Third World indebtedness; Françoise Barthelémy, "République Dominicaine: La porte à droite," *Le Monde Diplomatique*, July 1986; and information compiled by Dra. Josefina Padilla of CIAC (Centro de Investigacion y Apoyo Cultural), Santa Domingo, especially for the event of April 1984. The translation of Blanco's speech is mine. The relevant passages of the original run: "las Fuerzas Armadas y la Policia Nacional han dado un ejemplo de ecuanimidad revelando su grado de professionalizacion con alto sentimiento humano de respeto a la vida... asi como que mantuvieron sus reacciones dentro de una prudencia razonable y con una preparacion excelente."

From *Development Debacle: The World Bank in the Philippines* by Walden Bello, David Kinley, and Elaine Elinson

Development Debacle

On June 30, 1981, Robert McNamara resigned from the World Bank. The atmosphere surrounding his departure was reminiscent of that which prevailed thirteen years earlier when he left the Pentagon at the height of the Vietnam War.

There was however, one difference: this time the fusillade was coming from the right, which had just won the American presidency. The *Wall Street Journal* accused McNamara of having created a highly paid, 6,000-person strong bureaucracy doling out $12 billion worth of welfare checks annually to the Third World. Other critics, such as the Heritage Foundation, went further: the Bank, they said, was encouraging socialism in the Third World because its lending policies were biased toward state enterprises rather than private businesses. The institution even had its resident 'Maoist,' they claimed, in the person of Mahbub Ul-Haq, the director of McNamara's elite corps, the Policy Planning and Program Review Department. An international civil servant in the British mold, Haq had come to be known as the 'brains' behind McNamara's antipoverty, 'basic needs' approach.

But not all criticism came from the outside. Within the Bank's beleaguered headquarters in Washington, DC, bureaucrats joked cynically about McNamara's obsession with statistics and the game of generating meaningless numbers that this had produced. Others grumbled about the overcentralization and authoritarianism which put a premium on loyalty rather than creativity or objectivity. And almost everyone concurred that staff morale had degenerated significantly in the four years leading up to McNamara's departure.

In the midst of this crisis of confidence, the Bank was subjected to fire from snipers on the left. In late 1980 and throughout 1981, the Congress Task Force, *CounterSpy* magazine, the Institute for Food and Development Policy, and the Southeast Asia Resource Center fed the international press with a steady stream of confidential documents on Bank programs in the Philippines, Indonesia, South Korea, and China.[1] The leaked reports dramatically shattered the myth of a pro-socialist Bank advanced by the right. They showed, especially in the case of the Philippines, the instrumental role that the Bank played in refashioning Third World political and economic structures to better serve the needs of U.S. multinational corporations and the U.S. government. The reports revived the memory of McNamara's manipulation of World Bank policy to serve U.S. foreign policy in 1971, when he cut off assistance to the Allende administration in Chile as part of the Nixon-Kissinger strategy to destabilize that government.

The debate between left and right about the role of the Bank was 'settled' in February 1982—in favor of the left. The umpire was an unlikely one, the U.S. Treasury Department, which arrived at its conclusions with the help of some of the documents which we had earlier leaked to the international press.[2] In the landmark report, *U.S. Participation in the Multilateral Development Bank*, the Treasury Department dismissed the right's allegations about the pro-socialist bias of Bank policies and painted a picture of an institution solidly dominated by the United States and faithfully promoting not only strategic U.S. economic goals but short-term political objectives as well.

Failure of an Experiment

The internal World Bank documents on the Philippines, however, did not only illustrate the essential character of the Bank as a valuable instrument of U.S. foreign policy. They also were a testament to the failure of a key model for Third World economic development.

As we have seen, the Bank set in motion in the Philippines a development program with two key objectives: the 'pacification' and 'liberalization.' The pacification component consisted of rural and urban development programs aimed at defusing rural and urban unrest. The liberalization referred to the drastic restructuring of Philippine industry and external trade strategy to open up the country more completely to the flow of U.S. capital and commodities. To implement this strategy of 'technocratic modernization,' the Bank encouraged the formation of an authoritarian government and carefully cultivated a technocratic elite.

The strategy failed, largely because of its internal contradictions. Rural development was undone by the overwhelming focus on productivity combined with the absence of any serious effort to alter the relations of political and economic inequality that were themselves thwarting productivity. Urban development fell apart because of the Bank's insistence on applying the criteria of cost recovery and other principles of capitalist finance to its urban housing and development projects for the poor and its refusal to grant 'beneficiaries' any meaningful role in making decisions on issues that affected their lives. Instead of social peace, these programs spawned popular resistance.

With the failure of the pacification program, the legitimacy of the Bank's overwhelming presence in the Philippines came to rest mainly on its strategy of export-led industrialization (EOI)—a program that would allow the country a modicum of 'industrialization' while at the same time integrate it more fully into a global economic system dominated by multinational corporations. But EOI was a short-lived panacea, undone by its internal contradictions and the disappearance of the two pillars on which the model rested: docile cheap labor and an expanding international market.

With the collapse of rural development, urban development, and export-oriented industrialization, the Bank and its sister agency, the International Monetary Fund (IMF), were left with a purely repressive program of liberalization. The main tenets of this strategy were battering down tariff walls, destroying the national capitalist class, transforming the financial structure to better serve the needs of U.S. corporations and banks, and debauching the currency. This course was suicidal for the Philippines since it was being made more completely dependent on the world market at a time when the market was being savaged by an international recession that threatened to turn into an international depression. The World Bank's and the IMF's institution of direct rule in the form of a technocrat cabinet that could only justify the imposition of the repressive strategy with the metaphor of 'bitter medicine,' was a confession that a debacle of major proportions had overtaken the Philippine development effort.

The Philippine debacle was not just the failure of a program in one country. With its westernized elite and neocolonial ties to the United States, the Philippines had opened itself up in the 1960s and 1970s to the influence of U.S. and U.S. dominated agencies to a greater degree than any other Third World country. By the time the World Bank entered in force in 1972, the country had a reputation as one of the most "exciting places" for experiments in development. Over the decade the Philippines became the testing ground for many World Bank projects in rural development. It was also the site for the Bank's first urban-upgrading project. Together with Kenya, Bolivia, and Turkey, it served as the guinea pig for the new structural adjustment loan. Most important, the World Bank effort in the Philippines was the first coordinated, broad-front experiment in technocratic, authoritarian modernization. What went down the drain in the Philippines was not just a country program but a larger model for Third World development.

Politics, Ideology, and the Bank

This conclusion still leaves many apparent questions. Why did the Bank continue to plug away in the face of evidence of repeated failures? Or, more fundamentally, how could an agency supposedly devoted to development contribute instead to underdevelopment, to the creation of an economic disaster?

Bureaucratic inertia is part of the answer, but only a small part. A more popular explanation resorts to conspiracy theory. McNamara's rhetoric about the Bank's meeting basic needs and alleviating poverty is, in this view, nothing more than a smokescreen for diametrically opposite motives. While it is easy to see why many in progressive circles might subscribe to this Machiavellian image of the Bank, reality in this instance is much more complex.

Many World Bank bureaucrats, especially those at the middle and lower levels who are directly involved in operations in Third World countries, are well-intentioned individuals. Many are likeable Benthamites who believe that what they are doing will bring about the "greatest good for the greatest number." Indeed, some are people with considerable sensitivity—the very sensitivity that turns into cynicism after years of serving the Bank and realizing that their work has reaped more bad than good.

Where then should we locate the problem?

The explanation can be arrived at by examining three areas: the position of the Bank in the international economic and political power structure; the influence of the ideological assumptions that Bank technocrats bring to their work from their academic training; and the role of ideological predispositions that spring from the Bank's character as an authoritarian and technocratic organization. It is only by addressing these theoretical underpinnings which define the efforts of the Bank that we can really understand why the Bank does what it does despite resistance from the people it claims to serve and the subsequent failure of its programs. In the following pages, we become more theoretical than usual, but hopefully, not hopelessly so.

Politics and Planning

The World Bank is a pillar of the current international structure of wealth and power. Development within this system is the mandate implicitly entrusted to technocrats by the Western governments that dominate the institution, in particular the U.S. government. This task is problematic, since maintaining the current structure and providing development that benefits the majority of Third World countries are essentially contradictory goals.

If U.S. multinational corporations—and American conservatives—had their way, they would assign just one role to the Bank: that of directly supporting private financial and investment efforts in the Third World. They would have the World Bank perform nothing more than the "accumulation function," (directly support profit-making enterprises) to use James O'Connor's term.[3] But the World Bank has gone far beyond this basic function to become a highly active political and ideological institution; its role is also to legitimize the international structure of power among those countries that objectively suffer, or draw only marginal benefits, from the current system. Promoting 'development' is a necessary task for the Bank, in the same manner that some agencies of the liberal capitalist state must specialize in co-opting low-income, structurally disadvantaged classes. The conservative view of the Bank as the equivalent of the welfare department within a liberal international order is, in many ways, correct. Where the conservatives are wrong is in their corollary opinion that neither the welfare department nor the World Bank is necessary for the maintenance of the established order.

The tension between the accumulation and legitimation functions of the Bank appears in the form of 'negative' and 'positive' guidelines that frame the work of the World Bank technocrat. The key negative guideline is that development plans must never infringe on the interests and operations of the multinational corporations and banks of the Western industrialized countries. The main positive guideline is that, whenever possible, development planning must incorporate benefits for the corporations and banks. If the interests of the poor can be somehow fitted into the framework, all the better. Indeed, a basic principle which guides the Bank's rural development programs actually applies to its whole approach to planning development within an international system dominated by a few rich countries: "It may frequently be desirable to design a project so that all sections of the community benefit to some degree...avoiding opposition from the powerful and influential sectors of the community is essential if the program is not to be subverted from within."[4]

These very powerful political constraints on the conceptualization and application of policy prescriptions explain the emergence and evanescent popularity of 'export-oriented industrialization' growth as a strategy for development. For this strategy appeared to offer a way of responding to Third World demand for indigenous industrialization without disturbing—and, in fact, reinforcing—the prevailing international structure and distribution of economic power. It offered, it seemed, the best of all possible worlds: Third World countries would be placated with the prospect of industrial development, multinational firms would get their cut-price labor, and the advanced capitalist countries would be assured of a flow of cheap, labor intensive commodities.

As we have seen, export-oriented industrialization was an ill-conceived strategy that papered over the very real contradictions that eventually torpedoed it as a model for Third World development. But it was the best that the Bank could offer the Third World. Its swift collapse revealed how very little, indeed, the Bank—biased and hemmed in by the constraints of the prevailing economic power structure—could offer its clients in the way of meaningful development. That is, the Bank was unable to support strategies of growth that would enable the poor countries to leave the subordinate and passive roles to which the current system assigned them.

Ideology and Development

The work of the Bank technocrat must not be vulgarized as merely a series of negotiated attempts to iron out unstable compromises between opposing interests to the advantage of the dominant economies. What makes the process complex is that the crude realities of power, inequality, and poverty that confront the technocrat are filtered through the lens of ideology—a set of deeply held assumptions and propositions that reorganize reality, as it were, in order to make it manageable.

Most World Bank technocrats, particularly the economists, are recruited from the cream of the American academic establishment. Trained at such prestigious institutions as Yale, Harvard, Berkeley, Stanford, and Princeton, their intellectual formation takes place in the context of the all-pervasive neo-Keynesianism that is sacrosanct doctrine in these places.

In the neo-Keynesian world view, the relations of conflict and exploitation that govern economic life disappear. The actors in the economic drama are depoliticized and transformed into functionally complementary categories of consumers, savers, and investors. Economic development becomes, first and foremost, a technical problem that involves raising the rate of savings in an economy, channeling a substantial portion of savings into productive investment, and filling the gap between domestic and planned investment with external capital. Translated into public policy, this theory involves removing obstacles to entrepreneurial activity, increasing the rate of taxation to provide the government with funds to finance new infrastructure projects that spark economic activity, and facilitating the inflow of foreign investment and loans.

With development viewed as a 'technical question,' the power and control that accompany substantial foreign investment hardly figure as a problem. When technocrat Gerardo Sicat expressed the view that at this stage in Philippine history, it does not matter who controls the economy, he was exposing the bias and the blindspot of established economic ideology.

Instead of power and control, 'efficiency' is the *problematique* of the dominant paradigm. Labor unions, as well as protective barriers to imports and nationalist checks on foreign investment, are viewed primarily as obstacles to efficient production. In the rarefied world of the neo-Keynesian technocrat, the real interest of the Filipino people and that of foreign investors coincide in the commodities which the latter can allegedly produce more efficiently and cheaply than can national producers. The classic formulation is found in the U.S. State Department's justification of the Bank's Structural Adjustment Program: "Lowering of tariffs and easing of import restraints...is ...directed at creating competition for consumption goods that should lower household costs generally..."[5] 'Consumer sovereignty' becomes the ideological buttress for a pervasive foreign economic presence. It is not, however, simply made up out of nowhere. rather, it flows 'logically' from the ideological assumptions of the neo-Keynesian perspective.

Like foreign capital, poverty is abstracted from the context of the unequal relationships of power that create and perpetuate it. The problem of poverty is transformed principally into a problem of scarcity. The solution to scarcity is economic growth. And the key to economic growth is efficient production. Redistribution of wealth is a secondary issue in the neo-Keynesian paradigm—and one which establishment economists expect will become less important with growth since the resulting larger income pie, though still distributed unequally, will provide larger absolute slices for all. Moreover, redistribution is conceptualized in a very superficial way; it is viewed as a modification in the sharing of national income, rather than a manifestation of a deeper problem—control over the means of production by those who have no interest in the needs of the majority.

In spite of the disclaimers of McNamara, Mahbub Ul-Haq, vice president Hollis Chenery, and other theoreticians of the basic needs approach, the World Bank never really abandoned the 'trickle down' theory. The sidestepping of the structural causes of poverty in rural development pro grams was motivated not only by conscious political decision. *It was also a reflection of the secondary and minor status of the distribution question in established economic ideology.*

The 'antipoverty' strategy of McNamara thus came to resemble his friend Lyndon Johnson's Great Society effort to promote Black enterprises in the ghettoes in the 1960s. Rural development aimed to make a limited number of peasants "more efficient producers." This was like throwing a life preserver to a swimmer caught in a typhoon. For the real dynamics of the system were such that 'cost effectiveness' went to those with bigger landholdings, bigger blocks of capital, more political influence, and more access to advanced mechanical and biochemical technology.

The result was growth in production going hand in hand with an increase in absolute poverty. This could not ; however, go on indefinitely since the 'minor' issue of distribution created new limitations, this time in the form of a limited domestic market which restricted further growth. Faced with this contradiction but still unwilling to tackle the redistribution question, Bank bureaucrats saw a solution in 'export-led growth'—that is, in hitching industrial growth to foreign rather than domestic markets. But no sooner had one obstacle emerged than another took its place: once growth became dependent on achieving cost competitiveness in the ruthless international market, *growth became dependent on constricting and depressing wages, that is, on increasing poverty and repression.*

When international recession and the rising protectionist wave in export markets banished the vision of limitless growth offered by export-oriented development, the last emergency exit was slammed shut in the neo-Keynesian ideological trap. There was no more escaping the specter of radical redistribution of wealth and power—the social necessity which the whole elaborate edifice of neoclassical economic ideology had been erected to counter. In the Philippines, this specter took the form of a spreading revolutionary movement. By the late 1970s, reality was breaking through the ideological defenses of many Bank bureaucrats, who began to shed the bankrupt economic ideology. Afraid for their positions, however, they did not oppose the dominant current and the structure of power it represented. Instead they sank into a corrosive cynicism.

The Technocratic Bias

The ideological assumptions that Bank bureaucrats bring to their work are not only those they carry over from their academic training. The Bank itself cultivates certain ideological predispositions. Bank technocrats are encouraged to look upon themselves as an elite corps of experts who have the last word in development planning.

In the bureaucratic mind, 'politics' is counterposed to 'expertise.' 'Politics' is the irrational variable that sabotages the smooth flow of the planning process, of economic development. 'Politics' is associated first and foremost with the give and take of the democratic process. This distaste for democracy is captured in the authoritarian and hierarchical decision-making structure of the Bank—a structure which became even more centralized and more hierarchical under McNamara, the technocrat par excellence. General policy is formulated at the presidential level (with crucial input from the U.S. government), rubber stamped at the Executive Board level, and adapted to each country at the program and project offices.

It is hardly surprising that this elitist institution had no qualms in teaming up with the equally authoritarian Marcos regime to impose a program of development from above. For both administrative elites, control of the development process was a high priority. For both, democracy was something to pay lip service to but not to take seriously. When the slum dwellers of Tondo, the tribespeople of Chico, and the local government of Cagayan de Oro demanded a say in development projects that threatened their very existence, both World Bank and regime technocrats instinctively recoiled. The people demanded democracy—and this was something that challenged the unacknowledged but fundamental assumption of the technocratic approach: that policy making is best left in the hands of experts.

To conclude, the World Bank debacle in the Philippines was not just the failure of a country program. It was not merely a case of an experiment in Third World development that came apart. It was also the 'Tet' of neo-Keynesian technocratic economics—a sobering lesson in how established ideology can be blasted by reality.

1 Newspaper and periodical articles based on exposés headed up by the Congress Task Force and *CounterSpy* include the following:

"A Chiller for Manila: Reduce Borrowing and Cut Growth Targets, IMF Tells the Philippines, " *Far Eastern Economic Review*, April 30, 1982; "President Marcos' Authority is Eroding, According to Study for the World Bank," *Wall Street Journal*, December 3, 1980; "Bank Criticizes War on Poverty in the Philippines," *Asian Wall Street Journal*, January 16, 1981; "Philippines May Need to Find New Creditors as U.S. Banks Near Caps, World Bank Says," *Asian Wall Street Journal*, January 20, 1981; Walden Bell and John Kelly, "Les pays occidentaux protecteurs des Philippines s'inquiètent de la dégradation financière du pays," *Le Monde*, February 3, 1981; "Seoul Encounters Problems Instituting Major Reforms Urged by World Bank," *Asian Wall Street Journal*, January 5, 1981; Apolonia Batalla, "World Bank Lending," *Bulletin Today* (Manila), December 8, 1980; Walden Bello and David O'Connor, "McNamara's Second Vietnam," *Asia Record*, June 1981, Walden Bello, "Rural Debacle: The World Bank in the Philippines," *Food Monitor*, July-August, 1981; Teodor Valencia, WB-IMF Formula Could Sink RP Economy," *Philippines Daily Express*, March 2, 1981; Walden Bello and David Kinley, "McNamara's Second Vietnam," *Pacific News Service*, July 7, 1981; "Study Asserts Philippines Overestimates Energy Potential," *Asian Wall Street Journal*, February 24, 1981; Walden Bello and David Kinley, "La Politique de la Banque mondiale a l'heure de l'orthodoxie libérale," *Le Monde Diplomatique*, September 1981; Walden Bello, "The World Bank in the Philippines: A Decade of Failures," *Southeast Asia Chronicle*, December 1981; John Kelly and Joel Rocamora, "Indonesia: A Show of Resistance," *Southeast Asia Chronicle*, December 1981; "The Poverty Puzzle," *Far Eastern Economic Review*, March 27, 1981; Walden Bello, "Multinationals Under Marcos," *Multinational Monitor*, February 1981; Walden Bello, "Secret World Bank Document on Marcos: An Alliance Coming Apart?" *CounterSpy*, February-April 1981; "Dustere Thesen für Zukunft der Philippinen," *Tager Anzeiger* (Zurich), December 6, 1980; "The World

Bank and Marcos," *Journal of Commerce*, January 21, 1981; Walden Bell and John Kelly, "The World Bank Writes Off Marcos and Co.," *The Nation*, January 31, 1981; "Indonesia," *Far Eastern Economic Review*, May 29, 1981; Walden Bello and John Kelly, "China: World Bank Report Sets Stage for Taiwan-Style Development," *Multinational Monitor*, February 1982; "Philippine Mayor Resists Slum-Improvement Project," *Asian Wall Street Journal*, September 4, 1981; Walden Bello, John Kelly, and Robin Broad, "Comment Washington intervient dans la politique economique de Hondoras," *Le Monde Diplomatique*, May 1982; "IMF Urges that South Korea Devalue its Currency," *Wall Street Journal*, May 10, 1982.

For these exposés, the Congress Task Force and *CounterSpy* magazine were denounced by *Human Events*, President Reagan's favorite newspaper, as being made up of "far left extremists, some of whom...are dedicated enemies of the Unites States." *Human Events*, September 5, 1981. See also " 'CounterSpy' Counterattacks," *Human Events*, October 17, 1981.

2 U.S. Treasury Department. "Assessment of U.S. Participation in the Multilateral Development Banks in the 1980s," consultation draft, Washington, DC, September 21, 1981.

3 O'Connor, James. *The Fiscal Crisis of the State* (New York: St. Martin's Press, 1973), pp. 5–8.

4 World Bank. *Rural Development: Sector Working Paper* (Washington, DC: World Bank, 1975), pg. 40.

5 State Department response to "Questions Asked by the Committee" at House of Representatives Committee on Foreign Affairs, subcommittee on Asia-Pacific affairs, *Hearings*, Washington, DC, November 18, 1981.

From *Food First Action Alert*, Winter 1993 by Walden Bello, Shea Cunningham, and Bill Rau

---

Creating a Wasteland: The Impact of Structural Adjustment on the South, 1980 to 1994

On the eve of the 21st century, most of the South is in a state of economic collapse.

Yet, in 1994, the International Monetary Fund (IMF) and the World Bank, which are greatly responsible for the plight of the South, celebrate their fiftieth year of existence.

For the more than seventy countries that were subjected to 566 IMF and World Bank stabilization and structural adjustment programs (SAPs) in the last fourteen years, there is certainly nothing to celebrate.

These countries were told that the 'structural reforms' promoted by the SAPs were essential for sustaining growth and economic stability. Most were skeptical, suspicious, or downright opposed to these programs. But faced with the threat of a cutoff of external funds needed to service the mounting debts incurred from western private banks that had gone on a lending binge in the 1970's, these countries had no choice but to implement the painful measures demanded by the Bank and IMF. These usually included:

- Cutbacks in government expenditures, especially in social spending;
- Cutbacks in or containment of wages;
- Privatization of state enterprises and deregulation of the economy;
- Elimination or reduction of protection for the domestic market and fewer restrictions on the operations of foreign investors; and
- Devaluation of the currency.

Adjustment: Rationale and Reality

Fourteen years after the World Bank issued its first structural adjustment loan, most countries are still waiting for the market to "work its magic," to borrow a phrase from Ronald Reagan. In fact, structural adjustment has failed miserably in accomplishing what World Bank and IMF technocrats said it would do: promoting growth, stabilizing the external accounts, and reducing poverty.

Institutionalizing Economic Stagnation

Comparing countries that underwent adjustment with countries that did not, IMF economist Mohsin Khan reported the uncomfortable finding that "the growth rate is significantly reduced in program countries relative to the change in non-program countries."[1] Says Massachusetts Institute of Technology Professor Rudiger Dornbusch: "Even with major adjustment efforts in place, countries do not fall back on their feet running; they fall into a hole."[2] That is, economies under adjustment are stuck in a low-level trap, in which low investment, increased unemployment, reduced social spending, reduced consumption, and low output interact to create a vicious cycle of stagnation and decline, rather than a virtuous circle of growth, rising employment, and rising investment, as originally envisaged in World Bank theory.

Guaranteeing Debt Repayments

Despite global adjustment, the Third World's debt burden rose from $785 billion at the beginning of the debt crisis to $1.3 trillion in 1992. Thirty-six of sub-Saharan Africa's forty-seven countries have been subjected to SAPs by the IMF and World Bank, yet the total external debt of the continent is now 110 percent of its gross national product.[3]

Structural adjustment loans from the World Bank and the IMF were given to indebted countries to enable the latter to make their immediate interest payments to the western commercial banks. Having done this, the Bank and the IMF then went on to apply draconian adjustment policies that would assure a steady supply of repayments in the medium and long term. By having Third World economies focus on production for export, foreign exchange would be gained which could be channeled into servicing dollar denominated foreign debt.

The policy was immensely successful for first world banks, effecting as it did an astounding net transfer of financial resources from the Third World to the commercial banks that amounted to $178 billion between 1984 and 1990. So massive was the decapitalization of the South that a former executive director of the World Bank exclaimed: "Not since the conquistadores plundered Latin America has the world experienced a flow in the direction we see today."[4]

Intensifying Poverty

If structural adjustment has brought neither growth nor debt relief, it has certainly intensified poverty. In Latin America, according to Inter-American Development Bank president Enrique Iglesias, adjustment programs had the effect of "largely canceling out the progress of the 1960s and 1970s."[5] The numbers of people living in poverty rose from 130 million in 1980 to 180 million at the beginning of the 1990s. Structural adjustment also worsened what was already a very skewed distribution of income, with the result that today, the top twenty percent of the continent's population earn twenty times that earned by the poorest twenty percent.[6]

In Africa adjustment has been a central link in a vicious circle whose other elements are civil war, drought, and the steep decline in the international price of the region's agricultural and raw material exports. The number of people living below the poverty line now stands at 200 million of the region's 690 million people, and even the least pessimistic projection of the World Bank sees the number of poor rising by fifty percent to reach 300 million by the year 2000.[7] So devastated is Africa that Lester Thurow has commented, with cynical humor tinged with racism: "If God gave it (Africa) to you and made you its economic dictator, the only smart move would be to give it back to him."[8] So evident is the role of SAPs in the creation of this blighted landscape that the World Bank chief economist for Africa has admitted: "We did not think that the human costs of these programs could be so great, and the economic gains so slow in coming."[9]

Adjusting the Environment

IMF and Bank-supported adjustment policies have been among the major contributors to environmental destruction in the Third World. By pushing countries to increase their foreign exchange to service their foreign debt, structural adjustment programs have forced them to over-exploit their exportable resources. In Ghana, regarded as a 'star pupil' by the IMF and the World Bank, the government has moved to intensive commercial forestry, with World Bank support. Timber production more than doubled between 1984 and 1987, accelerating the destruction of the country's already much reduced forest cover, which is now twenty-five percent of its original size.[10] The country is expected to soon make the transition from being a net exporter to being a net importer of wood.[11] Indeed, economist Fantu Cheru predicts that Ghana could well be totally stripped of trees by the year 2000.[12]

Impoverishment, claims the World Bank, is one of the prime causes of environmental degradation because "land hungry farmers resort to cultivating erosion-prone hillsides and moving into tropical forest areas where crop yields on cleared fields usually drop after just a few years."[13]

What the World Bank fails to acknowledge is that its structural adjustment programs have been among the prime causes of impoverishment, and thus a central cause of this ecological degradation. In the Philippines, a World Resources Institute study claims that the sharp economic contraction triggered by Bank-imposed adjustment in the 1980s forced poor rural people to move into and super-exploit open access forests, watersheds, and artisanal fisheries.[14]

Rollback: The Strategic Objective

But if structural adjustment programs have had such a poor record, why do the World Bank and the IMF continue to impose them on much of the South?

This question is valid, only if one assumes that the Bank and IMF's intention is to assist Third World economies to develop. Then, the failure of SAPs can be laid to such things as bad conceptualization or poor implementation. However, it is becoming increasingly clear that, whatever may be the subjective intentions of the technocrats implementing them, structural adjustment programs were never meant to reduce poverty or promote development. Instead, they have functioned as key instruments in the North's effort to roll back the gains that had been made by the South from the 1950s through the 1970s.

These decades were marked by high rates of economic growth in the Third World. They also witnessed successful struggles of national liberation, and the coming together of southern states at the global level to demand a "New International Economic Order" (NIEO) that would entail a more equitable distribution of global economic power. This sense of a rising threat from the South—underlined by such events as the U.S. defeat in Vietnam, the OPEC oil embargo of 1973 and 1979, restrictions on multinationals' operations in Mexico and Brazil, and the Iran hostage crisis—contributed to the victory of Ronald Reagan in the U.S. presidential elections of 1980.

Central to the economic achievements of the South was an activist state or public sector. In some countries, the state sector was the engine of the development process. In others, state support was critical to the success of domestic businesses wishing to compete against foreign capital. It was not surprising, therefore, that when Reaganites came to power in Washington with a clear agenda to discipline the subordinate Third World, they saw as a central mission the radical reduction of the economic role of the Third World state, and structural adjustment programs by the World Bank and IMF as the principal means to accomplish this.

Not surprisingly, few southern governments were willing to accept structural adjustment loans when they were first offered. However, they had no choice but to capitulate, since at the onset of the debt crisis in 1982, Washington, notes Latin America specialist John Sheahan, took advantage of "this period of financial strain to insist that debtor countries remove the government from the economy as the price of getting credit."[15] Similarly, a survey of structural adjustment programs in Africa carried out by the United Nations Economic Commission for Africa concluded that the essence of these programs was the "reduction/removal of direct state intervention in the productive and distributive sectors of the economy."[16]

The New South

By the end of the twelve-year Reagan-Bush era in 1992, the South had been transformed: from Argentina to Ghana, state participation in the economy had been drastically curtailed. Government enterprises were passing into a few private hands in the name of efficiency. Protectionist barriers to Northern imports were being eliminated wholesale. Restrictions on foreign investment had been radically reduced. And, through export-first policies, the internal economy was more tightly integrated into the capitalist world market.

The erosion of Third World economies translated at the international level to the weakening of the formations that the South had traditionally used to attain its collective goal of bringing about a change in the global power equation: the Non-Aligned Movement, the United Nations Conference on Trade and Development (UNCTAD), and the Group of 77. The decomposition of the Third World was felt at the United Nations, where the U.S. was emboldened to once again use that body to front the interests of the North, including providing legitimacy for the U.S.-led invasion of Iraq in 1991. Rollback via structural adjustment had succeeded.

Corporate-driven structural adjustment was not, of course, limited to the South. In the U.S., for instance, Reaganomics ensured that inequality and poverty were greater in 1990 than in 1980. It was, however, the Third World that was made to shoulder the main burdens of adjustment.

In the 1950s and 1960s, the peoples of the South were optimistic that the future belonged to them, the eighty percent of the world's population that had long been treated as second or third class citizens of the world under colonialism. The illusions were gone by the beginning of the 1990s. As the South stood on the threshold of the 21st century, the South Commission captured the essence of its contemporary condition: "It may not be an exaggeration to say that the establishment of a system of international economic relations in which the South's second-class status would be institutionalized is an immediate danger."[17]

1 Khan, Mohsin. "The Macroeconomic Effects of Fund-Supported Adjustment Programs," *International Monetary Fund Staff Papers*, vol. 37 no. 2, June 1990, pg. 15.

2 Rudiger Dornbusch, quoted in Jacques Polak, "The Changing Nature of IMF Conditionality," *Essays in International Finance*, Princeton University, No. 184 (September 1991), pg. 47

3 World Bank. *World Debt Tables 1991–92*, vol. 1 (Washington, DC: World Bank, 1991), pp. 120, 124.

4 Miller, Morris. *Debt and the Environment: Convergent Crises* (New York: United Nations, 1991), pg. 64.

5 Iglesias, Enrique. *Reflections on Economic Development: Toward a New Latin American Consensus* (Washington, DC: Inter-American Development Bank, 1992), pg. 103.

6 Fidler, Stephen. "Trouble with the Neighbors," *Financial Times*, February 16, 1993, pg. 15.

7 World Bank. *Global Economic Prospects and the Developing Countries* (Washington, DC: World Bank, 1993). pg. 66.

8 Thurow, Lester. *Head to Head: The Coming Struggle Among Japan, Europe and the United States* (New York: William Morrow, 1992), pg. 216.

9 Quoted in *Debt and the Environment: Convergent Crises*, op. cit., pg. 70.

10 Development GAP. *The Other Side of Adjustment: The Real Impact of World Bank and IMF Structural Adjustment Programs* (Washington, DC: Development GAP, 1993), pg. 25.

11 French, Hillary. "Reconciling Trade and the Environment," in *State of the World* (New York: W.W. Norton, 1993), pg. 161.

12 Cheru, Fantu. "Structural Adjustment, Primary Resource Trade, and Sustainable Development in Sub-Saharan Africa," *World Development*, vol. 20, no. 4 (1992), pg. 507.

13 World Bank. *World Development Report 1992: Development and the Environment* (Washington, DC: World Bank, 1992), pg. 30.

14 Cruz, Wilfredo and Robert Repetto. *The Environmental Effects of Stabilization and Structural Adjustment* (Washington, DC: World Resources Institute, 1992), pg. 8.

15 Shahan, John. "Development Dichotomies and Economic Strategy," in Simon Teitel, ed., *Toward a New Development Strategy for Latin America* (Washington, DC: Inter-American Development Bank, 1992). pg. 33.

16 Cited in Seamus Clery, "Towards a New Adjustment in Africa," in "Beyond Adjustment," special issue of *African Environment,* vol. 7, nos. 1–4 (1990), pg. 357.

17 South Commission. *The Challenge to the South* (New York: Oxford University Press, 1990), pp. 72–73.

Part Four

The Free-Market Path

When the Market Decides Who Can Eat

The ultimate test of any economic policy is how well it provides for the basic needs of ordinary people. For two decades, Food First researchers have been examining the impacts of different development strategies on the ground—not just in theory, but in the real lives of Third World people. They have asked: beyond the rhetoric, does the concrete implementation of policies in poor nations around the world, really lead to the progress promised by economic theories?

In this section, we look at the actual results of free-market, free trade policies in several parts of the world. What does the neo-liberal economics promoted by the World Bank, the IMF, and the U.S. and other governments mean for the everyday life of peasant farmers? How do decisions taken in Washington change the lives of people in villages in Chile, Taiwan or Mexico? What happens when the market decides who can eat?

In *Needless Hunger: Voices from a Bangladesh Village*, Hartmann and Boyce found a picture very different from the 'basket-case' image spread by the international media. They discovered a fertile land with many natural resources, clearly capable of feeding its dense population. But this abundance had been siphoned off from the villages for centuries, first to British colonial rulers who 'underdeveloped' the country and eliminated the peasants' right to till the soil, and since independence to rich landowners and merchants who control the land and the agricultural markets. The wealth of the countryside is concentrated in a few hands, with poor peasants and landless workers suffering from chronic poverty and periodically, from massive famines.

The 'newly industrializing countries' of east Asia were for many years touted as an example of the success of free-market policies. Building on successful land reforms in the 1950s, countries such as Taiwan and South Korea had high rates of GDP growth for decades, and developed major industries in fields such as electronics, automobiles and steel. But as Bello and Rosenfeld point out with remarkable foresight in "Dragons in Distress; The Economic Miracle Unravels in South Korea, Taiwan, and Singapore," and demonstrated years before the 'Asian crisis' became worldwide news in 1997 and 1998, this development model would inevitably lead to crisis. Based on exportation with a competitive basis in cheap labor, its expansion to Thailand, Indonesia, Malaysia, and other southeast Asian nations insured that when those nations suffered currency crises, the effects would reverberate through markets around the world.

Even before the crises, this development model was showing clear evidence that it could not be sustained indefinitely. Unrestricted industrial growth led to major environmental problems, provoking popular reactions and raising costs. The small farm sector which had subsidized industrialization in the past, fell victim to free trade policies which opened up the economy to cheap foreign imports. The export-led growth for which east Asia was known is clearly no longer a viable model for the rest of the Third World.

When the Zapatista rebellion in Chiapas, Mexico burst onto the world scene in 1994, on the same day that NAFTA went into force, it made it clear that decades of pro-business policies and one-party rule had finally provoked armed resistance. The Zapatista struggle grew out of centuries of repression of indigenous communities, but it also represented a new response to the forces transforming the countryside—free trade in agricultural commodities, the demise of the Mexican Revolution's land reform, the boom and then collapse of the oil industry, and the financial crisis stemming from the country's enormous international debt. In *Basta! Land and the Zapatista Rebellion in Chiapas*, Collier and Quaratiello describe the political and economic crisis caused by the Zapatista uprising, which led to the collapse of the peso in December 1994 and a crisis of the ruling ideology which has not yet been resolved.

The examples in this section show that, while the free-market model may sometimes produce high growth rates, these numbers do not represent a real improvement in the life of peasants and workers. On the contrary: free trade agreements, USAID projects, and World Bank development schemes tend to aggravate rural inequality, driving small farmers out of business and concentrating wealth in the hands of large landowners, merchants, and multinational businesses. The environment and the rights of labor are sacrificed in order to attract foreign investment, and the economy becomes extremely vulnerable to boom-and-bust cycles spreading to and from faraway markets. The results are predictable: the erosion of rural society, economic and political crisis, and sometimes, as we have seen in Chiapas, armed revolution.

From *Needless Hunger: Voices from a Bangladesh* Village by Betsy Hartmann and James Boyce

Hunger in a Fertile Land: The Paradox

In U.S. news media Bangladesh is usually portrayed as an 'international basket case,' a bleak, desolate scene of hunger and despair. But when we arrived in Bangladesh in August 1974 we found a lush, green, fertile land. From the windows of buses and the decks of ferry boats, we looked over a landscape of natural abundance, everywhere shaped by the hands of men. Rice paddies carpeted the earth, and gigantic squash vines climbed over the roofs of the bamboo village houses. The rich soil, plentiful water and hot, humid climate made us feel as if we had entered a natural greenhouse.

As the autumn days grew clear and cool and the rice ripened in the fields, we saw why the Bengalis in song and verse call their land "golden Bengal." But that autumn we also came face to face with the extreme poverty for which Bangladesh has become so famous. When the price of rice soared in the lean season before the harvest, we witnessed the terrible spectacle of people dying in the streets of Dacca, the capital. Famine claimed thousands of lives throughout the country. The victims were Bangladesh's poorest people who could not afford to buy rice and had nothing left to sell.

As we tried to comprehend the contrast between the lush beauty of the land and the destitution of so many people, we sensed that we had entered a strange battleground. All around us silent struggles were being waged, struggles in which the losers met slow, bloodless deaths. In 1975 we spent nine months in the village of Katni, collecting material for a book on life in the Third World. There we learned more about the quiet violence which rages in Bangladesh.

Katni is a typical village. The majority of its 350 people are poor: most families own less than two acres of land, and a quarter of the households are completely landless. The poorest often work for landlords in neighboring villages who own over forty acres a piece. Four-fifths of the villagers are Muslims and one-fifth are Hindus. Except for two rickshaw pullers, all make a living from agriculture.

To minimize the differences between ourselves and the villagers, we lived in a small bamboo house, spoke Bengali and wore local clothing. By approaching the villagers as equals we were eventually able to win their trust. Jim spent most of his time talking with the men as they worked in the fields or went to the market, while Betsy talked with the women as they worked in and around their houses. The villagers taught us what it means to be hungry in a fertile land.

Golden Bengal

Bangladesh lies in the delta of three great rivers—the Brahmaputra, the Ganges, and Meghna—which flow through it to empty into the Bay of Bengal. The rivers and their countless tributaries meander over the flat land, constantly changing course, since most of the country lies less than one hundred feet above sea level. The waters not only wash the land, they create it; their sediments have built the delta over the centuries. The alluvial soil deposited by the rivers is among the most fertile in the world.

Abundant rainfall and warm temperatures give Bangladesh an ideal climate for agriculture. Crops can be grown twelve months a year. The surface waters and vast underground aquifers give the country a tremendous potential for irrigation in the dry winter season. The rivers, ponds, and rice paddies are alive with fish; according to a report of the United Nations Food and Agriculture Organization (FAO), "Bangladesh is possibly the richest country in the world as far as inland fishery resources are concerned."[1]

The country's dense human population bears testament to the land's fertility; historically the thick settlement of the delta, like that along the Nile River, was made possible by agricultural abundance. Today, with more than eighty million people, Bangladesh is the world's eighth most populous nation. Its population density is the highest of any country in the world except for Singapore and Hong Kong,[2] a fact which is all the more remarkable in light of the country's low level of urbanization. Nine out of ten Bangladeshis live in villages, where most make their living from the land.

Bangladesh's soil may be rich, but its people are poor. The average annual income is less than $100 per person, the life expectancy only forty-seven years, and like all averages these overstate the well-being of the poorest.[3] Twenty-five percent of Bangladesh's children die before reaching the age of five.[4] Malnutrition claims many. Over half of Bangladesh's families consume less than the minimal calorie requirement, and sixty percent suffer from protein deficiencies.[5]

Health care is poorly developed and concentrated in the urban areas. Less than a quarter of the population is literate.[6]

A United States Senate study notes that Bangladesh "is rich enough in fertile land, water, manpower, and natural gas for fertilizer not only to be self-sufficient in food, but a food exporter, even with rapidly increasing population size."[7] But despite rich soil, ideal growing conditions, and an abundant supply of labor, Bangladesh's agricultural yields are today among the lowest in the world. According to a World Bank document, "Present average yields of rice are about 1.2 metric tons per hectare, compared with 2.5 tons in Sri Lanka or 2.7 in Malaysia, which are climactically similar, or over four tons in Taiwan where labor inputs are greater."[8] Production has stagnated; today's yields are similar to those recorded fifty years ago.[9]

Why is a country with some of the world's most fertile land also the home of some of the world's hungriest people? A look at the history of Bangladesh sheds some light on this paradox. The first Europeans who visited eastern Bengal, the region which is now Bangladesh, found a thriving industry and a prosperous agriculture. It was, in the optimistic words of one Englishman, "a wonderful land, whose richness and abundance neither war, pestilence nor oppression could destroy."[10] But by 1947, when the sun finally set on the British Empire in India, eastern Bengal had been reduced to an impoverished agricultural hinterland.

The Making of Hunger: Who Owns the Land?

The pattern of landownership in Bangladesh profoundly affects both the production and distribution of food. Although Bangladesh is often called a "land of small farmers," the reality in the villages is more complex. On the one hand, many villagers own no land at all and depend upon wage labor for their livelihoods. On the other hand are landlords whose holdings, though modest by American standards, are large enough to free them from the necessity of working in the fields.

A recent study commissioned by the United States Agency for International Development (USAID) found that a "dichotomy between ownership of land and labor on it" is widespread in Bangladesh. Less than ten percent of Bangladesh's rural households own over half the country's cultivable land, while sixty percent of rural families own less than ten percent of the land. One third own no cultivable land at all, and by including those who own less than half an acre, the study concludes that forty-eight percent of the families of rural Bangladesh are "functionally landless." Pointing to the difficulties of collecting reliable data, the authors of the study note that these figures probably *underestimate* the actual extent of landlessness and the true level of concentration of landownership.[11]

Based on their different relationships to the land, the villagers of Bangladesh fall into five basic classes:

- Landlords do not work on the land themselves, except sometimes to supervise their workers. Instead they hire labor or let out land to sharecroppers.
- Rich peasants work in the fields but have more land than they can cultivate alone. They gain most of their income from lands they cultivate with hired labor or sharecroppers.
- Middle peasants come closest to our image of the self-sufficient small farmer. They earn their livings mainly by working their own land, though at times they may work for others or hire others to work for them.
- Poor peasants own a little land, but not enough to support themselves. They earn their livings mainly by working as sharecroppers or wage laborers.
- Landless laborers own no land except for their house sites, and sometimes not even that. Lacking draft animals and agricultural implements, they are seldom able to work as sharecroppers, and must depend upon wages for their livelihoods.

A villager in Katni told us, "Without land, there is no security." Indeed, without land there is often no food. An International Labor Organization study reports that landless laborers consume only seventy-eight percent as much grain as those who own over seven and one-half acres of land, despite the fact that the landless need forty percent more calories because they work harder.[12] Landownership not only determines who will have enough to eat, but also affects how much food is actually produced.

Not surprisingly, the small minority of rural families who own over half the country's farmland are, in the words of the USAID study "at the apex of the structure of power in rural Bangladesh; the political economy of the countryside is controlled by them."[13] Land is the key to their power, power which in turn brings them control over other food-producing resources such as irrigation facilities and fertilizer. Since these agricultural inputs are often highly subsidized by the government, they are all the more desirable to the rural elite.

Similarly, the large landowner is better able to receive low interest loans from government banks. His land serves as collateral, and he knows how to deal with the bank officials: how to fill out the necessary forms and when to propose a snack at the nearest tea stall. The large landowners also usually dominate village cooperatives which have access to government credit.

The rural poor meanwhile must turn to the village moneylender when they need cash, often paying interest rates of more than one hundred percent a year. Not coincidentally, the moneylender and the large landowner are often one and the same person. Since Islam, Bangladesh's main religion, condemns the taking of interest, moneylenders ease their consciences through such simple expedients as buying a peasant's crop before the harvest—at half the market rate. To get credit small farmers frequently mortgage their land, forfeiting the right to cultivate it until they repay the loan.

The large landowners' control of food-producing resources—land, inputs, and credit—allows them to appropriate much of the wealth produced in the countryside. As a result, they are able to buy out hard-pressed smaller farmers, driving them into the ever growing ranks of the landless. One study found that peasants who own less than an acre of land sell half their remaining land every year.[14] Land, the ultimate source of wealth and power in rural Bangladesh, is becoming concentrated in fewer and fewer hands.

Just as Bangladesh is often called a land of small farmers, so the country's agriculture is sometimes described as 'subsistence farming.' The implication is that the peasants grow barely enough to feed themselves, with little left over for anyone else. Once again, reality is more complex. Much of the wealth which the peasants produce in the fields is siphoned by large landowners, moneylenders and merchants. The hunger of Bangladesh's poor majority is intimately related to the ways this wealth is extracted and used.

Who Works, Who Eats?

Surplus is siphoned from poor peasants and landless laborers by the twin mechanisms of sharecropping and wage labor, production relationships which determine who works the land and who eats its fruits.

Sharecropping, according to a 1977 AID study, covers at least twenty-three percent of Bangladesh's farmland.[15] In Katni's vicinity, landlords and rich peasants generally cultivate about three-fourths of their land by means of sharecroppers and the remaining one-fourth with hired labor. The landowner and sharecropper normally split the crop equally, although in some districts the landowner often takes two-thirds.[16] The sharecropper usually must bear the costs of seed and fertilizer, so that in practice his share is really less than half the crop.

Although the rewards from sharecropping may seem meager, those of wage labor are even less. In Katni the standard wage for male laborers is about thirty-three cents U.S. per day, paid in a combination of rice, cash, and a morning meal which insures that the laborer has enough strength to work all day in the fields. Women from poor families who work processing crops in well-to-do households earn even less—about twenty cents for a day's hard labor.

The number of landless laborers in Bangladesh is rising rapidly due to population growth and the displacement of small farmers. The dramatic rise in landlessness has not been matched by a rise in employment opportunities. As a result, in 1974 real wages for agricultural laborers had fallen to less than two-thirds of their 1963 level.[17] As Dalim, a landless laborer, told us: "I earn two pounds of rice, one *taka* (about seven cents) and a meal for a day's work. With that taka I used to be able to buy two more pounds of rice, with a little left over for oil, chilis and salt. But today one taka won't even buy one pound of rice. Employers used to let their workers take a few free vegetables when they went home in the evening, but nowadays they aren't so generous. Times are getting harder for men like me."

With wages declining, it is becoming more profitable for the landowner to cultivate with hired labor than to give land to sharecroppers. The large landowners in Katni's vicinity calculate that wage labor only costs them one-fourth to one-third of the crop. They are slowly shifting more and more land to hired labor.

In some countries this shift has been associated with the 'Green Revolution'—the introduction of new crop varieties, chemical fertilizers, and irrigation—which by raising yields also makes wage labor more attractive to the landowner than sharecropping. But in Katni the main reason for the shift is not that yields are going up, but rather that wages are going down.

Poor peasants and landless laborers are caught on an economic treadmill. No matter how hard they run they keep slipping backwards. The siphoning of the surplus makes it almost impossible to save enough money to buy land of their own. Instead, illness and unemployment often force them to sell their remaining tiny plots of land and their meager household possessions. Though they devote their lives to growing and processing food, they face perpetual hunger.

What happens to the surplus once it passes into the landowners' hands? If it were used productively, the suffering of the poor might not be entirely in vain. After all, any society must generate a surplus for investment if the economy is to grow. But in Bangladesh very little of the surplus finds its way into productive investment. Luxury consumption absorbs much of the income of the rural elite. Nafis, a big landlord, bought himself a new Japanese motorcycle while we were in Katni. It cost him as much as a laborer working on his land would earn in twenty years.

Large landowners are reluctant to invest in agriculture, for farming is a difficult and risky business. They may buy more land, but this is simply a transfer of resources (usually from small farmers), which adds nothing to the nation's productive base. Even less of the surplus is mobilized for investment elsewhere in the economy through taxes or savings because the government does not want to tax the large landowners for fear of losing their political support, and the interest paid on savings deposits cannot compare with other more profitable uses to which the landowner can put his money. Trade and money lending—both of which siphon surplus from the peasants while leaving the production process untouched—offer by far the most lucrative and easy avenues for investment.

The Market

Through the exchanges of the marketplace, merchants are able to siphon surplus from Bangladesh's peasants. Receiving low prices for the crops they sell and often paying high prices for the goods they buy, the peasants lose whether they enter the market as producers or as consumers. The transfer of wealth is often hidden by the seemingly impersonal movement of prices, but in Katni we found several examples which throw the relationship between peasants and merchants into sharp relief.

Jute, the fiber used to make rope, burlap, and carpet backing, is the main cash crop of Bangladesh's peasants and provides about four-fifths of the country's export earnings. After independence jute prices stagnated, while rice prices soared. As a result, peasants grew less jute and more rice, so that by 1975 jute acreage had shrunk to two-thirds of its pre-independence level. Worried about export earnings, the government announced a floor price for jute which it hoped would check the decline in production. Government purchasing centers throughout the country were instructed to buy jute from the growers for about ninety taka per *maund* (one *maund* is about eighty pounds).

One such purchasing center was located three miles from Katni. Nevertheless, the villagers sold their jute in the local markets for sixty taka per maund, two-thirds of the government rate. At this price jute was decidedly a losing venture. As the rich peasant Kamal complained: "I sold my jute for less than half of what it cost me to grow it!"

A visit to the local jute procurement center yielded some clues as to the reasons for the striking discrepancy between the official support price and the actual market rate received by the peasants. The purchasing center consisted of a half dozen large warehouse buildings which had once belonged to a West Pakistani company. The buildings were piled high with rough bales and loose mounds of the golden fiber.

The manager, a heavyset young man dressed in Western clothing, was happy to explain how the jute is graded and how to operate the baling press. But when asked where the jute in his warehouse came from, his reply was guarded: "We buy from the growers."

"At what price?"

"We pay the government rate, 91.50 taka per maund."

"How curious. In the markets a few miles from here, the growers are selling their jute for only 60 taka."

"Oh, they must be selling to the merchants."

"Ah, the merchants. And do you buy from them too?"

"Yes, we buy from the licensed traders." The manager stressed the word 'licensed,' to emphasize the legitimacy of such transactions.

"The jute in this warehouse—did you buy most of it from growers, or from merchants?"

The manager began to look uneasy. "Well, actually we buy mostly from the merchants. You see, these growers only bring in eight or ten maunds at a time, so it is very inconvenient to buy from them. From the merchants we can buy hundreds of maunds at once."

The manager declined to elaborate as to why the growers are willing to sell so cheaply in the local markets, if they could receive the much higher government price simply by coming to the ware-house a few miles away. The villagers were less reticent. "If I bring in a cartload of jute," said one middle peasant, "the warehouse people say, 'Today we are closed—come back tomorrow.' So my time has been wasted. If I return the next day they will have another excuse: 'We've already bought our quota for the day,' or 'We have to wait for funds from Dacca.' We can never sell our jute to the government."

Why do the warehouse people turn the peasants away? Our middle peasant friend explained: "We sell our jute in the market at sixty taka. The merchants then sell it to the government at ninety taka, making a thirty taka profit on each maund. They share this profit with the warehouse manager, giving him maybe half. So of course he won't buy from us!" Besides paying kickbacks, the merchants are said to pay the warehouse manager a monthly retainer in order to ensure his cooperation. The manager buys only from those who make it 'convenient' for him to do so. Perhaps he does buy directly from a few growers—local landlords who make suitable arrangements.

In previous years jute prices in the local markets had been somewhat higher, because much jute was smuggled to India and this demand had helped to keep prices up. But in 1975, following the assassination of Sheikh Mujib, the political connections which had protected the smugglers unraveled and the illegal flow of jute across the border slowed to a trickle. In collusion with the government authorities, a few merchants were then able to strengthen their control over the local jute market, pushing prices to the abysmally low level of sixty taka per maund.

"In Bangladesh we call our jute 'the golden fiber,' " said one peasant. "But tell me, who gets the gold?"

While some merchants buy cash crops from the peasants, others sell various goods to them. No villager is entirely self-sufficient; all rely on the market to meet some of their needs. Landless laborers and poor peasants must buy food, everyone has to buy salt and cloth, and those who can afford them purchase such items as medicine and footwear. The prices the villagers pay are often high, in part because of hoarding by merchants.

Sometimes hoarding is a reaction to genuine scarcities, but other times it is a means to deliberately raise prices. Merchants not only manage at times to control the supply of a particular good within a given locality; on occasion they are able to corner a market throughout the country. In the autumn of 1974 a cartel cornered the market for salt in Bangladesh. The price rose to fifty times its normal level, and salt riots broke out in major towns. For two weeks the merchants who hoarded the nation's salt reaped tremendous profits. Then they loosened their grip, and prices returned to normal.

Most hoarding is less spectacular, and it is often hard to say where natural scarcities end and artificial ones begin. For example, rice prices are generally lowest at harvest time and highest just before the next harvest. This predictable fluctuation makes speculation in rice attractive to merchants. If they hold stocks in anticipation of rising prices, this in itself helps to lift prices. Landless laborers and poor peasants, who rely on the market for much of their food needs, must pay the price. Middle peasants often lose coming and going: they sell their rice cheaply at harvest time because they need cash for consumption, investment in the next crop and repayment of debts; a few months later they have to buy rice at inflated prices.

The rice trade was not particularly lucrative in the 1960s, but after independence this changed. In 1974 the price of rice climbed to ten times its pre-independence level. Peasants were particularly vulnerable in parts of the country where floods had damaged crops. As prices rose, many sold their animals, their land and their household possessions in order to buy rice. The poorest, with nothing left to sell, came to the towns in search of work or relief. An estimated 100,000 people starved to death.

A villager recalls: "Lalganj (a town five miles from Katni) became a town of beggars. Whole families were living, sleeping and dying in the streets. Each day there were new bodies along the roadside."

Officially, the government blamed the famine on floods, but many observers believed that hoarding by merchants and a breakdown in government administration were responsible for turning a manageable, localized shortage into a catastrophe. According to an USAID official in Dacca: "The food supply was there; it just didn't get to the right people."

The 1974 famine brought terrible suffering to many people, but to some it brought profit. The merchants who hoarded grain were not the only ones to benefit. Moneylenders did a brisk business, and large landowners were able to buy land cheaply from their poor neighbors. In the hardest hit areas land registry offices had to stay open late into the night to handle the record sales.

The villagers are well acquainted with many of the merchants who profit at their expense. The men who buy their jute, for example, are local landlords who have diversified into trade and moneylending. A pyramid of trade extends above these local merchants to regional, national and sometimes even international economic interests. At the base of this pyramid are the peasants. They sum up their situation with a simple phrase: "The merchants drink our blood."

The Trials of a Poor Peasant Family

Abu and Sharifa live with their six children in a one-room bamboo house with broken walls and a leaky straw roof. They are poor peasants, and year by year they are becoming poorer.

"I wasn't born this way," says Abu. "When I was a boy I never went hungry. My father had to sell some land during the 1943 famine, but still we had enough. We moved to Katni when he died—my mother, myself, and my three brothers. We bought an acre and a half of land. As long as none of us brothers married that was enough, but one by one we married and divided the land."

"I was young," recalls Sharifa, "and I worked very hard. I husked rice in other women's houses to earn money, and finally I saved enough for us to buy another half acre of land. But my husband's mother was old and dying, and he wanted to spend my money to buy medicines for her. He threatened to divorce me if I didn't give him the money, so I gave in. The money was wasted—she died anyway—and we were left with less than half an acre. Then the children came. Our situation grew worse and worse, and we often had to borrow to eat.

Sometimes our neighbors lent us a few taka, but many times we had to sell our rice to moneylenders before the harvest. They paid us in advance and then took the rice at half its value."

"People get rich in this country by taking interest," Abu interjects bitterly. "They have no fear of Allah—they care only for this life. When they buy our rice they say they aren't taking interest but really they are."

"No matter how hard we worked," continues Sharifa, "we never had enough money. We started selling things—our wooden bed, our cattle, our plow, our wedding gifts. Finally we began to sell the land."

1 Food and Agriculture Organization. "Bangladesh Country Development Brief, 1973," cited in Frances Moore Lappé and Joseph Collins, *Food First: Beyond the Myth of Scarcity* (New York: Ballentine Books, 1979).

2 World Bank. *Bangladesh: Development in a Rural Economy*, vol. 1: The Main Report, September 15, 1974, pg. 1.

3 According to the World Bank, *Bangladesh: Current Trends and Development Issues*, December 15, 1978, pg. iv, per capita income is $91. Life expectancy from U.S. Agency for International Development (USAID) *FY 1978, Submission to Congress: Asia Programs*, February 1977, pg. 16.

4 *Bangladesh: Development in a Rural Economy*, op. cit., pg. 2.

5 Institute of Nutrition and Food Science. *Nutrition Survey of Rural Bangladesh, 1975–1976* (Dacca, Bangladesh: University of Dacca, 1977).

6 *Bangladesh: Current Trends and Development Issues*, op. cit.

7 "World Hunger, Health, and Refugee Problems: Summary of Special Study Mission to Asia and the Middle East," report prepared for the subcommittee on labor and public welfare and the subcommittee on refugees and escapees, Senate Committee on the Judiciary, January 1976, pg. 99.

8 World Bank. *Bangladesh: Current Trends and Development Issues*, May 19, 1977, pg. 34.

9 Rice yields for 1928–32 can be found in Nafis Ahmad, An Economic Geography of East Pakistan (London, UK: Oxford University Press, 1968), pg. 129.

10 Ibid.

11 Jannuzi, F. Tomasson, and James T. Peach. *Report on the Hierarchy of Interests in Land in Bangladesh* (Washington, DC: Agency for International Development, 1977), pp. xxi, 30.

12 Khan, Azizur Rahman. "Poverty and Inequality in Rural Bangladesh," in *Poverty and Landlessness in Rural Asia* (Geneva, Switzerland: International Labor Organization, 1977), pg. 142.

13 *Report on the Hierarchy of Interests in Land in Bangladesh*, pg. 70.

14 "Poverty and Inequality in Rural Bangladesh," pg. 159.

15 *Report on the Hierarchy of Interests in Land in Bangladesh*, pp. 41–42, 81.

16 Ibid., pp. 42–43.

17 Clay, Edward J. "Institutional Change and Agricultural Wages in Bangladesh," paper presented at Agricultural Development Council Seminar on Technology and Factor Markets, Singapore, August 9–10, 1976.

From *Food First Action Alert*, 1990 by Walden Bello and Stephanie Rosenfeld

Dragons in Distress: The Economic Miracle Unravels In South Korea, Taiwan, and Singapore

"There is nothing in it for us," Lee So Sun answered without hesitation when we asked her what benefits high-speed industrialization had brought to Korean workers. Her conviction was founded on painful experience. In November 1970, her son Jeon Tae-Il, a pioneering organizer of garment workers in Seoul, set himself on fire with the cry, "Do not mistreat the young girls."

Jeon's fiery gesture touched off an extraordinarily energetic and heroic effort to organize Korea's working class. His mother picked up Jeon's banner and endured arrests and repeated jailings to become the symbol of the determination of Korean labor. When Korean workers launched over 7,000 strikes between 1987 and 1990—or 6.5 per day, probably a world record for a labor force the size of Korea's, they were carrying on Jeon's heroic tradition. When thousands of workers armed with Molotov cocktails battled 10,000 riot policemen for control of the strategic Hyundai shipyard in Ulsan in late April 1990, they were affirming Jeon's uncompromising stance.

Korea's workers are the secret of their country's so called 'economic miracle.' They work an average of fifty-four hours a week, the longest of any country surveyed by the International Labor Organization.

They also suffer the world's highest rate of industrial accidents, with an average of five workers killed daily and another 390 injured. This record of exploitation produced a sizzling nine percent GNP growth rate in the 1980s, but it also resulted in a worsening distribution of income and wealth, as the top twenty percent of the population cornered almost forty-four percent of the national income and a minuscule five percent monopolized sixty-five percent of all private land-holdings nationwide.

The wave of insurrectionary strikes since the summer of 1987, when popular pressure forced an easing of repression, has put South Korea's rulers and the huge conglomerates or *chaebol* that control the Korean economy on notice that labor demands its share of the economic miracle and will no longer tolerate a pattern of high growth achieved through harsh exploitation.

The Crisis of the NICs Model

Labor's rebellion in Korea is one of the manifestations of a many-sided crisis now overtaking the so-called "Newly Industrializing Countries" (NICs) of East Asia: Taiwan, Singapore, Hong Kong, and South Korea. The NICs model might be described as a strategy of high-speed industrialization centered on production for export based on cheap labor, and with the exception of Hong Kong, closely guided by an authoritarian regime. By the mid-eighties, the NICs model had been enshrined as the new orthodoxy in development economics. But such is the cunning of history that by that time, the NICs formula had run out of steam in Singapore, South Korea, and Taiwan. (We have not included Hong Kong in this essay since the key political and economic developments in the last few years have been influenced by one overriding future event: the colony's devolution to China in 1997). The crisis was most visible in South Korea, where social and political unrest fed off a fifty percent decline in the annual GNP growth rate from 12.2 percent in 1988 to 6.5 percent in 1989.

The Protectionist Threat

The NICs grew by attaching themselves to the U.S. market at a time when the U.S. was still the guardian of free trade in the so-called "free world." But, by the mid-eighties, free trade had given way to aggressive protectionism in official U.S. circles. Massive trade deficits with the NICs and Japan led many influential politicians and businessmen to see these countries as the leading candidates to replace the Soviet Union as the 'enemy.'

"Although the NICs may be regarded as tigers because they are strong, ferocious traders," declared a senior U.S. Treasury official in October 1987 in what was tantamount to a declaration of economic warfare, "the analogy has a darker side. Tigers live in the jungle, and by the law of the jungle. They are a shrinking population."

The U.S. has launched a two-pronged assault on the NICs. In an attempt to make NICs exports less attractive to U.S. consumers, the U.S. forced the appreciation of the Korean won and the new Taiwan dollar relative to the U.S. dollar, making NICs products more expensive in dollar terms. The U.S. has also threatened to use the infamous 'Super 301'—legislation which obligates the U.S. president to take retaliatory measures against those tagged as 'unfair traders'—to successfully force the NICs to liberalize their trade and investment restrictions on thousands of imported commodities and services, including foreign banking operations, beef, and cigarettes.

This double-barreled blast has drastically curbed further expansion of NICs exports to the U.S. market: Both Taiwanese and Korean exports to the U.S. grew by only one percent from 1987 to 1988, while imports from the U.S. to the NICs rose sharply, climbing by more than one hundred percent in the case of Taiwan. The U.S. share of NICs exports fell from thirty-nine percent in 1987 to thirty-two percent in 1988 in the case of Korea, and from forty-five percent to thirty-nine percent in the case of Taiwan.

The American shock treatment has been quite effective.

The Crisis of Legitimacy

As the NICs' external environment worsened, so did their internal environment. Alienation was widespread, as the social and political costs of the authoritarian imposition of export-oriented growth began to catch up with the NICs governments in the mid-eighties.

The People's Action Party of Singapore, led by Lee Kwan-Yew, propelled export-oriented growth as a means to strengthen its political legitimacy, while in Korea and Taiwan the state elites sought to use economic development to neutralize what was widely perceived to be their illegitimate assumption of power. Prosperity, it was hoped, would buy legitimacy and excuse the repressive policies imposed to achieve high-speed growth. By the mid-eighties, however, it was clear that this strategy of purchasing popularity was not working. Instead, both the economic model and the power structure that imposed it were being fundamentally challenged.

While the Lee Kwan-Yew regime in Singapore claimed that one-party rule was necessary to ensure economic prosperity, the real sentiment of the population was expressed in the sharp rise in the emigration rate. Proportionally, the emigration from Singapore in 1989, which was under no external threat, was not far below the exodus from Hong Kong, which faces absorption into China. Most of those leaving were members of Singapore's skilled labor force, and one of them told us his reasons for leaving: "People demand more democracy, they want honest-to-goodness opposition, and if they cannot get it, they will vote with their feet and leave."

Though less publicized than in Korea, labor in Taiwan was also mounting a formidable challenge to the repressive partnership of management and the Kuomintang (KMT) state elite. Many of the firms struck by workers were owned by foreign investors, who had long profited from the KMT's harsh control of labor. Three hundred workers of Nestle's Taiwan subsidiary on Hsinchu shut down operations and imprisoned the foreign managers in their offices. Wildcat strikes at Ford Motor Company's Taiwan subsidiary forced Ford to give workers a bonus amounting to five percent of profits. And as in Korea, labor's demands were, according to one observer, "far ranging, calling not only for wage increases but also for union autonomy, fair labor practices, reform of current labor laws, and liberalization of management's authoritarian style."

Dissatisfaction in both Taiwan and Korea, however, spread far beyond the ranks of labor. Land speculation by the rich was pricing not only the poorer classes out of homeownership, but also significant sections of the politically volatile middle class. In Korea, for example, where the average per capita income was below $5,000, the sale price of a small apartment in a middle-class area of Seoul reached $225,000 in 1989—the same level as in the San Francisco area in California! It was not surprising then that seventy percent of Koreans perceived a 'severe' gap between the haves and the have-nots, with nearly half attributing this to land speculation and other non-productive activities.

Skyrocketing land prices stemmed from speculation in real estate by the giant conglomerates or *chaebol* that dominate the Korean economy. Instead of chanelling capital into research and development (R&D) to improve Korea's technological infrastructure, the chaebol preferred to make super-profits in land speculation and the stock market. The widespread revelation of the backward state of Korean R&D coincided with revelations that the chaebol had parked about $16.5 billion in speculative investments in land, luxury hotels, and golf courses. For the public at large, the link between profits and production—on which rested the fragile legitimacy of the chaebol—had snapped. The chaebol were increasingly scorned as rentiers instead of being respected as producers. "Chaebol" became a dirty word, and suspicion of wealth became so widespread that one ruling party legislator remarked, "The mood of the country is like a people's court in a communist country."

The perception of the chaebol as rentier capitalists was joined by the popular conviction that the partnership between big business and the military-technocratic elite that had propelled the traditional growth strategy was increasingly corrupt, especially under the reign of Chun Doo-Hwan (1980–1987). Nearly all the top chaebol were found to have contributed millions of dollars to Chun's personal fund-raising front, the Ilhae Foundation, with Hyundai's head, Chung Ju-Yung, serving as bag man.

Dominated by the same politicians, soldiers and bureaucrats prominent in the Chun government, the Roh Tae-Woo administration, which came to power in 1987, continued the pro-conglomerate policies of the old regime, leading to the widespread perception of it as the 'Chaebol Republic.'

Agriculture: The Road to Extinction

Though land reform was carried out in the early 1950s in Korea and Taiwan, subsequent policies sacrificed the interests of farmers to those of urban-based export manufacturers. Policies were adopted which kept the prices of agricultural commodities low so as to keep the wages of urban workers low. Peasant income was also kept deliberately low to encourage migration of rural people to the cities, where they could be employed as workers in industrial enterprises. Not surprisingly, the agricultural population in both countries declined rapidly.

Subordinated to the demands of industry, agrarian interests were also marginalized by U.S. moves to dump surplus wheat and grain via the PL 480 program. The diversified agrarian economy was ruined, as imports of wheat, soybeans, and cotton resulted in the almost complete disappearance of Taiwanese and Korean farms growing those crops.

With renewed U.S. pressure to eliminate barriers to other non-rice agricultural commodities in the mid-eighties, Taiwanese and Korean farmers realized that the extinction of the non-rice sectors of agriculture was the price that would be exacted for the continuing access to the U.S. market of the NICs manufactured exports. Farmers did not relish being led to extinction, and in 1988, they battled police in the streets of Taipei and Seoul as they launched demonstrations to protest the sacrifice of agriculture to the trade relationship with the United States.

Industrialization and Toxification

Like agriculture, the environment has been a prime victim of the NICs high-speed industrial growth. With the technocrats assuming that some degree of ecological destabilization was the price of economic growth, export-oriented industrialization telescoped into three decades of environmental destruction that took many more years to unfold in earlier industrializing countries.

With anti-pollution laws hardly implemented, it is not surprising that the lower reaches of most of Taiwan's major rivers are heavily polluted, or that Korea's tap water is unsuitable for drinking, carrying heavy metals like iron, manganese and cadmium at nearly twice the official tolerance level, and containing ammoniacal nitrogen—generated by human and animal waste—at nearly ten times the maximum tolerable level.

In both countries, technocrats persist in trying to build more nuclear plants, backed by U.S. firms like Westinghouse and Combustion Engineering which stand to profit from future contracts. Yet, in 1985, Taiwan's Number One Nuclear Power Plant is reported to have set a world record of fifty-six days of continued radiation leakage outside the plant. Since Korea's first nuclear reactor went into operation in 1978, its plants have experienced 185 accidents, including leakage of heavy water at the Woolsung plant. Had the leakage not been quickly brought under control, the resulting accident would have been one on the scale of Three-Mile Island. Undeterred, Taiwan's technocrats hired the U.K. public relations firm

Ogilvy and Mather to convince an increasingly critical public to accept a fourth nuclear plant. In Korea, energy technocrats continued to support the establishment of fifty-five nuclear plants by the year 2031 in a country smaller than the state of Ohio.

Aroused by high-speed destruction of the environment, citizens in both countries have formed grassroots groups aimed at opposing ecologically destructive industrial activities. This movement is most advanced in Taiwan, where it has, among other things, stopped the building of more naptha crackers for the highly polluting petrochemical industry, halted the construction of a $160 million titanium dioxide plant by the Dupont Corporation, and forced the closing of a petrochemical plant owned by the British ICI Corporation that fishermen had accused of dumping acid waste on their fishing grounds. A 1985 poll showed that fifty-nine percent of Taiwanese favored environmental protection over economic growth.

The Structural Squeeze

But perhaps the most potent threat to the continued viability of the NICs model is the worsening 'structural squeeze' in the key manufacturing sector of the economy—that is, a situation in which the NICs have lost their 'competitive advantage' in cheap labor, without being able to successfully make the transition to more value-added, high-technology production.

One of labor's greatest weaknesses is its inability to present a common front worldwide. And one of capital's greatest advantages is its ability to leap nimbly over national boundaries. By the late seventies and eighties, the depletion of the reserves of rural migrants in the NICs exerted strong upward pressures on wages despite the repression of labor. With the explosion of labor organizing, wages in Taiwan and Korea rose by sixty percent between 1986 and 1989. By 1989, the average hourly cost of a textile operator was $3.56 in Taiwan and $2.87 in Korea; in China, the rate was $0.40 and in Indonesia, $0.23.

Taiwan's response to this competitive crisis has focused on finding new sources of cheap labor. Manufacturers continue to transfer labor intensive operations to China and other low wage countries in Southeast Asia and the Caribbean. Investors have flocked across the Taiwan Straits to invest an estimated $1 billion in China, disregarding the formal political hostility that still marks the relationship between Taiwan and China. A huge chunk of Taiwan's famed shoemaking industry—some 225 enterprises—have, in fact, relocated their production facilities across the straits.

The other path to a cheap labor supply followed by Taiwan is importing foreign workers, a method pioneered by Singapore, where about 12.5 percent of the 1.2 million work force is foreign. An estimated 50,000—80,000 foreign workers are now working in Taiwan without papers, with another 10,000 enjoying legal status. Most of them are Filipinos, Thais, Indonesians, and Malaysians who are paid less than half the average monthly wage of Taiwanese workers.

In both Taiwan and Singapore, there is developing a labor system similar to that which emerged in Europe in the 1960s and 1970s: a two-tier labor force composed of poorly paid, unorganized 'guest workers' and better-paid local workers. The presence of the foreign workers dampens the wage demands of the local labor force, while chauvinism is encouraged to keep foreign workers in their place. This arrangement may well slow down the loss of competitiveness in wage costs, but only at the price of destabilizing social conflicts in the medium term.

Unlike Taiwan, Korea's response to the loss of labor competitiveness has been to attempt to create a technology and skill-intensive domestic industrial base to support a high-tech export economy. Korean technocrats have laid out grandiose plans to compete with Japan and the U.S. on the frontiers of high technology by developing sixty-four megabit memory chips, artificial intelligence computers, video disc players, rockets to launch satellites, robots for seabed mining, and superspeed linear electric trains.

There are some very serious obstacles to the ambitions of Korea's technocrats. One is simply the massive financial resources demanded by this effort. In 1988 alone, Japanese manufacturers of microchips invested $4.5 billion in semiconductor-related plant and equipment, more than the total South Korean investment to that date. Korea's industrial giants are already heavily in debt, with liabilities outweighing assets by five to one or more.

A second bottleneck is dependence on a chronically high level of imports of basic materials, parts and components, and general machinery. Eighty-five percent of the value of a Korean color TV is made up of components from Japan. The engine and transmission of the bestselling Excel sub-compact car produced by Hyundai Motors are designed in Japan. And as many as ninety percent of the components of a Korean-made laptop computer come from foreign sources, mainly Japan. Indeed, imports from Japan alone now make up a third of the value of Korean exports, prompting some analysts to remark that Korean manufacturing is for the most part the assembly of Japanese components with Japanese technology.

The third obstacle is Korea's low capability for self-sustaining technological innovation. Almost all the key advanced technologies are licensed, purchased, or copied from Japan and the U.S. To take just one example, the three top Korean manufacturers of VCRs—Samsung, Goldstar, and Daewoo—all acquired their technology from one source, the Matsushita subsidiary JVC, to which they forked over six percent of their export earnings in 1987.

Government spending on R&D is much lower than the figure for industrialized countries and, at 0.4 percent of GNP in 1988, even lower than that of Taiwan. At the same time, the chaebol prefer to place their funds, not in risky R&D, but in the stock market and in real estate, where quick profits can be made.

Another barrier to high tech success is Korea's still very limited pool of engineers and scientists. While this pool is proportionally larger than that of most developing countries, it is rather small in comparison with that of Japan and the United States, Korea's main competitors in high tech. This problem is compounded by the continuing drain of skilled personnel to the U.S., where an estimated 6,000 highly trained Korean scientists and engineers decided to reside after studying in U.S. institutions.

The barriers to the high tech option, however, are not only technical but also political. The decision to go high tech was made in the old technocratic style typical of command capitalism. However, as one top technocrat put it, "old formulas for keeping the economy on track—usually technocratic solutions developed in a political vacuum—are no longer appropriate given the new socio-political environment."

People now demand a greater voice in the making of economic policy; and in a period of slower economic growth, the massive allocations for high tech are likely to clash with popular demands for more investment in social welfare, improving the quality of life, and cleaning up the environment. With limited capital resources, the choice may well boil down to either having clean air to breathe and clean water to drink or mass producing the megabit memory chip and building more nuclear reactors.

The strategy of phasing out or automating labor-intensive industries will also certainly provoke the opposition of labor, especially the twenty percent of the manufacturing work force that is in textiles and garments, which is already one of the most militant sectors. In general, focusing on high tech will lead to a declining capacity to absorb unskilled and semi-skilled labor—that is, to a rise in structural unemployment. Already, there are disturbing signs of this trend. According to one source, if one were to add the underemployed, the 'precariously employed,' and the seasonally unemployed to the figure for the formally unemployed, the current unemployment rate would reach twenty to twenty-five percent of the work force. A survey of recent graduates of selected Korean universities showed that forty percent were unemployed in 1989, compared to about eighteen percent in 1985. "Factory and office automation is leaving a lot of people out in the cold," notes one account, and there are increasing cases of suicides because of failure to find employment.

Trapped in the model of export-oriented growth, the technocrat-big business coalition in Korea is likely to discover that the high tech option is not an escape route but the path to a techno-economic debacle and even sharper social conflict.

Transforming the NICs

As the NICs enter into crisis, there is active discussion of economic reform in all three countries. But while the debate has not yet yielded a comprehensive and coherent vision of an alternative model of development, its contours can be discerned.

- With export markets unlikely to grow in this time of increasing protectionism, the domestic market must become the engine of economic growth. In fact, in Korea, the workers' push for higher wages in the past three years created a boom in domestic demand that compensated for the severe slowdown of export growth in 1989.

  Growth will not, however, be sustained unless measures are taken to correct the worsening distribution of income in all three NICs. These policies will undoubtedly hurt the now dominant groups. For instance, in Singapore, it will require keeping more of the value of Singapore's production in the country, in the form of income for its population, rather than as profits to be repatriated by the multinational firms. In Korea, income redistribution will necessarily have to be tied to dismantling the chaebol and ending their stranglehold on national resources. And in both Taiwan and Korea, a dynamic internal market cannot emerge unless there is a massive 'Marshall Plan' to save the agrarian economy and U.S. moves to dismantle the mechanisms protecting local farmers from extinction are stopped.

- Development has to be environmentally sustainable. Making environmental preservation a priority, however, need not mean an end to growth. Restricting the growth of the petrochemical industry can be counterbalanced by large-scale investment in a program to save the environment, which may spawn a whole manufacturing subsector devoted to making anti-pollution equipment, waste-treatment facilities, and other environmental protection systems. Also a shift from chemically intensive agriculture to organic farming would mean turning to more labor-intensive yet higher value-added production. So strong has the fear of toxics become in Taiwan that consumers would probably be willing to pay more for clean grain, produce, and meat. Organic farming, with its high demand for labor-intensive care, could help reverse the depopulation of the countryside and spark a resurrection of agriculture.

- Making the domestic market the engine of growth does not imply an end to exports as a key component of economic strategy. But instead of chasing evanescent markets all over the world, Korean, Taiwanese, and Singaporean entrepreneurs could focus on selected markets like Southeast Asia. And instead of competing with the Japanese on a whole range of commodities, from low tech to medium tech to high tech products, they could focus on a narrower band of medium tech exports for which they can build a more solid base of intermediate and basic industries. Such products might include energy-efficient and non-polluting public transportation vehicles, electro-mechanical implements for labor-intensive organic farming, and simple, user-friendly personal computers for use in Third World educational and work settings.

- With Japan rising quickly as the dominant power in the Pacific, the future of the Asia-Pacific region depends greatly on the NICs. They can either remain economic and technological dependencies of

Japan, or they can work with their less developed neighbors in the Association of Southeast Asian Nations (ASEAN), Indochina, and the Pacific to create a techno-trading association that would serve as a regional counterweight to Japanese economic colonization.

Such a regional association can be built on very different relationships than those characteristic of both U.S. and Japanese economic expansion, which has traditionally been marked by a lopsided division of labor, technological monopoly, unequal transfers of capital, chronic trade deficits, and environmental degradation. In contrast, the NICs, ASEAN, Indochina, and the Pacific countries can fashion institutions and agreements that would, among other things, ensure that divisions of labor which facilitate trade do not congeal into permanent cleavages, technological know-how is spread around systematically, and foreign investment develops an economy integrally instead of simply creating cheap-labor enclaves.

- To be successful, strategies for growth can no longer be imposed from above by state technocratic elites. Command economics, whether of the capitalist or socialist variety, is obsolete: that is the message of the concurrent social upheavals in Eastern Europe, Korea, and Taiwan. That is also the meaning of the spectacular rise in the numbers of people leaving Singapore. More than ever, people want an active say in economic policy, and unless it is based on a rough consensus forged by democratic means, that policy is likely to founder. Democracy, one might say, has become a factor of production.

The currently dominant economic and political groups, however, have developed a very strong stake in the old, discredited export-oriented growth model. A break with the past appears to require the coming to power of a new coalition of forces—a coalition based on those sectors of society that have the greatest stake in a new strategy of democratic development: the farmers, small and medium businesses, workers, and environmentalists.

The Demise of the Small Farmer

Taiwan and South Korea are often cited as examples of successful land reform. True, decisive land reform was enacted in both countries in the 1950s which radically reduced the number of tenant-farmers and installed small landowners as the dominant force in the countryside.

But since then it was downhill for the agrarian sectors of both countries, as they were systematically exploited to serve export-oriented industrialization.

Low grain price policies translated into low wages for urban workers, which in turn translated into low prices for Taiwanese and Korean exports. Low grain prices also served another purpose, as Lee Teng-Hui admitted before he became president of Taiwan: "Government policy has intentionally held down the peasants' income, so as to transfer the people who originally engaged in agriculture into industries."

As a result of such policies, the agricultural work force has declined from a majority to less than twenty percent of the total work force in both Taiwan and Korea. In Korea, the rate of migration from the countryside to the cities comes to an average of 400,000 yearly. With the vast majority of migrants being young men and women, the rural work force has 'aged': the portion of the agricultural work force age fifty or over leaped from nineteen percent in the early eighties to almost thirty-three percent by the end of 1988.

So difficult had it become to earn a living from agriculture that by the early eighties, most of Taiwan's farmers derived the bulk of their income from non-farm work, prompting one observer to write that "farming has become the real 'sideline' for most farm families." In Korea, going into heavy debt became the only means to keep the farm afloat: between 1975 and 1985, rural household debt rose ten times faster than household income.

Today agriculture in Taiwan is plagued by high costs and low productivity. But instead of seeing the cause as the subordination of agriculture to the needs of export-oriented industry, Taiwanese and Korean technocrats blame what one Taiwanese bureaucrat described as "the overprotection of tenants and the strict restrictions on the transfer of farmland after land reform, (which) have resulted in a rigid land tenure system which cannot provide an effective mechanism for enlarging farm size."

Seeing American agriculture as the model, Taiwanese and Korean technocrats want a reverse land reform which would buy out marginal farmers, concentrate landholdings in the hands of fewer families, and promote efficient mechanization.

Because they wish to reduce the farm population, the technocrats are hardly resisting the U.S. drive to open up Taiwan and Korea's markets to cheap American fruits, poultry, beef, and tobacco. The consequences of trade liberalization on the agrarian community are grimly portrayed by one Korean writer:

> As the planting season loomed in the spring of 1989, farmers were confused: "Hey," they asked each other, "what are we going to plant? We have to plant rice no matter what, but wheat has been a lost cause for a long time— imported soybeans have grabbed more than eighty percent of that market... and western cigarette imports have already forced reductions in tobacco acreage. We just don't have a crop to plant." In the end, the farmers reached the inescapable conclusion that they were caught in a vicious cycle that runs like this:
>
> "First, a crop is opened to imports, which causes collapse of that crop. That triggers, in turn, mass flight to another crop, which causes over-production, which causes the bottom to drop out of prices. Which causes penniless farmers."

In short, in the name of 'efficiency' and 'free trade,' the tragedy of the American family farm is being replayed in Korea and Taiwan.

From *Basta! Land and the Zapatista Rebellion* in Chiapas by George A. Collier with Elizabeth Lowery Quaratiello

---

Basta!

When a housemaid in the Chiapas state capital, Tuxtla Gutierrez, suddenly quit her job, why had she just used her entire Christmas bonus to buy hundreds of bandages?

Why did a man purchase an itinerant merchant's entire stock of rubber boots in March 1993 at the entrance to Palenque, Chiapas's famous classic Mayan ruin?

What made peasants eking out a precarious existence in the rain forests of eastern Chiapas cite 'war' as a threat more dangerous to the world than 'poverty,' 'disease,' 'deforestation,' or 'pollution' in a 1992 survey of attitudes about global change?[1]

In the summer of 1993, Tucson writer Leslie Marmon Silko's *Almanac of the Dead*, a novel predicting native American rebellion from Chiapas to Arizona, suddenly captured an audience of readers in Chiapas. Was there a special reason for such fascination?

The answers to many such puzzles suggested themselves on January 1, 1994, when the EZLN (the *Ejercito Zapatista de Liberación Nacional* or Zapatista Army of National Liberation), equipped with rubber boots, homemade army uniforms, bandanas, ski masks, and weapons ranging from handmade wooden rifles to Uzi machine guns, seized towns in eastern and central Chiapas, proclaiming a revolution on the inaugural day of the North American Free Trade Agreement (NAFTA).

Taking advantage of the New Year holiday to catch security forces off guard, the Zapatistas—a force of young, disciplined, and mostly indigenous men and women soldiers—ransacked the town halls of Altamirano, Chanal, Huistan, Las Margaritas, Oxchuc, Ocosingo, and San Cristobal de las Casas—once the colonial seat of government of Chiapas and today an important commercial and tourist center. Some burned district attorney, judicial, and police records (but spared archives in San Cristobal that a local scholar told them had historic value). Others fanned out into the mountains to seek recruits from among the indigenous and other peasants of the region. Treating startled tourists and civilians with courtesy, the EZLN pronounced itself in rebellion against the government, the army, and the police. In printed circulars and broadcasts from captured Ocosingo radio station XOECH, the Zapatistas declared:

> *¡Hoy Decimos Basta!* Today we say enough is enough! To the people of Mexico: Mexican brothers and sisters: We are a product of five hundred years of struggle: first against slavery, then during the War of Independence against Spain led by insurgents, then to promulgate our constitution and expel the French empire from our soil, and later (when) the dictatorship of Porfirio Díaz denied us the just application of the Reform laws and the people rebelled and leaders like Villa and Zapata emerged, poor men just like us. We have been denied the most elemental education so that others can use us as cannon fodder and pillage the

> wealth of our country. They don't care that we have nothing, absolutely nothing, not even a roof over our heads, no land, no work, no health care, no food, and no education. Nor are we able freely and democratically to elect our political representatives, nor is there independence from foreigners, nor is there peace nor justice for ourselves and our children.[2]

Invoking Article 39 of Mexico's 1917 Constitution, which invests national sovereignty and the right to modify government in the people of Mexico, they called on other Mexicans to help them depose the 'illegal dictatorship' of President Carlos Salinas de Gortari's government and party. They declared war on the Mexican armed forces and called on international organizations and the Red Cross to monitor under the Geneva Conventions of War. They appealed to other Mexicans to join their insurgency.

Within twenty-four hours, the EZLN launched an attack on the Rancho Nuevo army base about six miles southeast of San Cristobal and freed 179 prisoners from a nearby penitentiary. They kidnapped Absalon Castellanos Dominguez, governor of Chiapas from 1982 to 1988, announcing he would be tried summarily and shot for crimes of repression. Instead, he was 'sentenced' to a life term of hard peasant labor.

The Mexican government quickly moved 12,000 troops and equipment into the region. Within days, and after two pitched battles, the Zapatistas retreated east and southward out of the central highlands into rugged and inaccessible strongholds in the tropical forests of the eastern lowlands. Backed by air strikes, federal troops pursued the EZLN in armored vehicles to where roads give way to wilderness.

By that time, journalists from around the world had arrived on the scene to chronicle the Zapatista's exploits and explore their motives, which struck chords of sympathy with the Mexican public. Reporters portrayed the rebels as Maya Indians upset over years of poverty and discrimination as they began to write poignant articles describing the abuses heaped upon the Indians of Mexico's southernmost state. Human rights organizations that had for some time been trying to alert the world to the plight of the region's poor suddenly found an avid audience as they began to document federal army abuses against Zapatista prisoners and civilians caught up in the warfare.

The rebels, disguised with bandanas and ski masks, became instant icons, their images replicated on everything from cloth dolls sold in Mexico's outdoor marketplaces to cartoons in Mexico City's daily newspapers. Zapatista images even appeared on condom wrappers. Guessing the identity of "Subcomandante Marcos," the shadowy, green-eyed ideologue and military director of the uprising, suddenly became a popular obsession.

The Mexican government sought at first to paint the Zapatista rank and file as guileless but gullible natives inspired and led by foreign subversives. But public opinion, international scrutiny, and widespread dissension within the ranks of the ruling party soon led Salinas de Gortari to acknowledge that the Zapatistas had justifiable grievances based on Chiapas's and Mexico's internal affairs. Salinas fired Interior Minister José Patroncinio Gonzalez Garrido, who as governor of Chiapas from 1988 to 1993 had been accused of repression and abuses against peasants. The President declared a unilateral cease-fire on January 12, calling on the Zapatistas to disarm and negotiate with a specially designated Commission for Peace and Reconciliation.

In the weeks that followed, tensions rose. Citizens seized dozens of town halls in Chiapas and neighboring states to protest that incumbent mayors, most of whom were pawns of the governing party, had stolen office through fraudulent elections. In eastern Chiapas, peasants invaded private ranches; some landowners fled; others counterattacked with hired gunmen. A broad coalition of peasant and indigenous organizations denounced the government for past neglect and abuse and proclaimed themselves in favor of Zapatista demands for reforms.

Peace talks finally began on February 21 in San Cristobal de las Casas, under the aegis of the Catholic diocese and noted liberation theologian Bishop Samuel Ruiz Garcia. Salinas designated Manuel Camacho Solis as the government's negotiator, a man respected for the integrity and negotiating skills he had demonstrated as former mayor of Mexico City and as foreign affairs minister. As a gesture of good will, the Zapatistas released hostage and former governor Absalon Castellanos Dominguez. With Subcomandante Marcos as their public relations and military spokesperson, the EZLN's general command, the Committee of Clandestine Indigenous Revolution,

FIGURE 1: Chiapas and the area of Zapatista Rebellion

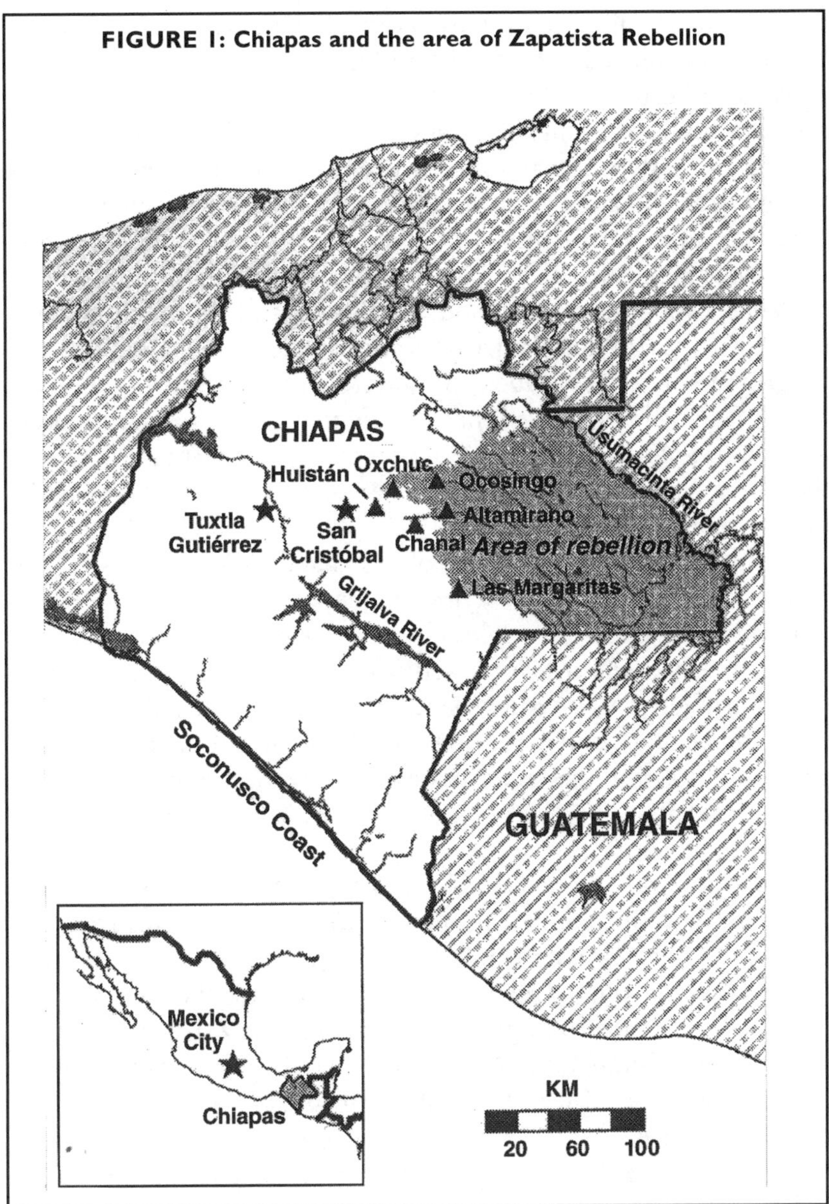

made up of eighteen representatives of four Mayan-language ethnic groups, began the talks and demanded attention to thirty-four broad-ranging issues of political, economic, and social reform. Two weeks later, negotiators announced thirty-two tentative accords, and the talks recessed to allow both sides to consult with their constituencies. Mexicans breathed a sigh of relief and hope.

Their optimism proved ephemeral. The uprising had thrown Mexico into crisis, calling into question President Salinas de Gortari's program to restructure and modernize the Mexican economy. Deep fissures in the Mexican political system, papered over during Salinas's regime by a facade of 'consensus' building, had revealed themselves to international scrutiny as the highly publicized rebellion evoked public debate over government fraud, corruption, and dereliction of duty to Mexico's peasant and indigenous poor. It became apparent that the government had known about the guerrilla army in Chiapas for over a year. Rumors flew that Salinas had done nothing for fear of jeopardizing U.S. congressional approval of NAFTA, or that hard-liners in the ruling party and in the military had lent support to the rebels in protest against Salinas's policies.

The presidential candidacy of Salinas's hand-picked successor, Luis Donaldo Colosio Murrieta, seemed in jeopardy as speculation flared that Salinas might replace Colosio with peace negotiator Camacho, or that Camacho, whom Salinas had initially passed over in favor of Colosio, might declare a renegade candidacy. For the first time in more than half a century, it began to seem as though Mexico's ruling party, the Institutional Revolutionary Party (PRI), might actually have to relinquish power in the presidential elections slated for August 21, 1994.

On March 22, Camacho declared that he would not run for President, but the next day PRI presidential candidate Luis Donaldo Colosio fell to an assassin's bullet, shocking the nation and provoking charges that the killing had been plotted within the PRI itself.

The violence polarized public debates over the Zapatistas and the government. Although the PRI hastily replaced Colosio with Ernesto Zedillo Ponce de Leon in the hopes of capitalizing on the reaction against violence, Zedillo, a Yale-trained economist who had been Colosio's election advisor, ran a lackluster campaign. But so did candidates of the opposition parties, Cuauhtemoc Cardenas of the Party of the Democratic Revolution (PRD) and Diego Fernandez de Cevallos of the National Action Party (PAN).

Then, on June 12, the Zapatista rank and file rejected the tentative peace accord their own representatives had negotiated with the government, calling for a national convention to revamp the political system while refusing to lay down their arms. Zedillo deemed the earlier negotiations with the EZLN a failure, blaming Camacho, who then resigned as the government's peace negotiator. As leading Mexicans began to suggest publicly that Mexico might become ungovernable, regardless of the outcome of the August elections, Mexico's Jorge Carpizo MacGregor, a jurist and former human rights commissioner who had agreed in January to step in as interior minister to oversee the elections, submitted his resignation, asserting that an unnamed party was making fair elections impossible. Salinas persuaded Carpizo to withdraw his resignation, but uncertainties continued to grow as the Zapatistas called for a convention prior to the elections to rewrite Mexico's Constitution.

Despite widespread fascination with the Zapatista rebellion, little is understood about its root causes. I want to look behind the romantic images, instant analyses,[3] and outrage prompted by the rebellion, to explore the circumstances and forces—both domestic and international—that created a situation ripe for rebellion.

The region behind the headlines is a complicated one. Chiapas is sometimes described as a picturesque backwater—a quaint stop on the tourist circuit where time has stood still and Maya Indians can be observed performing their age-old crafts and rituals. But beneath he surface seen by the casual visitor, Chiapas is filled with paradoxes that defy easy categorization. In addition to peasants who weave and wear the traditional *huipul* (tunic) and carry loads on *tumplines*, there are peasants who dress in jeans and drive trucks. Within even the tiniest Indian hamlets there are wealthy entrepreneurs who own such modern luxury items as televisions and videocassette recorders; poor, marginalized farm workers; and opportunistic political bosses. The state has plenty of wealthy *ladinos* (nonindigenous Mexicans of the region) who sympathize with the Zapatistas and poor peasants who do not. There are disaffected intellectuals, grassroots organizers, elite colonial families, and ranchers, each with their own political agenda. And in this place where Catholics and Protestants have clashed bitterly, where women are praised for passivity rather than activism, a nonsectarian rebellion has arisen, and mothers, wives and daughters are among those who make up the ranks of the Zapatista army.

What is it that unites those who took up arms with the Zapatistas? In contrast to some analysts, I posit that it is primarily a peasant rebellion, not an exclusively Indian rebellion, because although the

Zapatistas are demanding rights for indigenous peoples, they are first and foremost calling attention to the plight of Mexico's rural poor and peasants, both indigenous and non-indigenous. By *peasants,* I mean rural people who produce their own food or who are closely connected to others who produce for subsistence, as contrasted with those who farm commercial crops primarily for sale and profit. In southern Mexico, many peasants, but not all, are *indigenous* people, descendants of those who were conquered and subordinated by the Spanish during the period of colonial rule.[4]

Some may wonder why the rebellion was instigated by peasants and not by the urban poor or those who toil in the *maquiladoras,* people who have certainly suffered greatly in recent years. Although there is no clear answer to this question, one possible explanation is that on top of the severe hardships peasants have had to endure during the past decade of economic restructuring, they were also disappointed by a number of broken promises from the government: land reform that never occurred; price supports guaranteed, then taken away; and credits extended, then withdrawn. When, in 1992, the government of President Salinas de Gortari brought land reform—the issue on which his party had originally risen to power—to a halt, he signaled an abrupt end to a traditional government covenant with the peasantry and deprived many peasants not just of the possibility of improving their livelihoods, but of their power as a constituency. The Zapatistas are trying to reclaim that constituency.

The Zapatista rebellion bears some resemblance to the revolutionary movements of Central America, in that those movements also involved peasants who had no safety net to help cope with marginalization. Mexico's government, however, has been much more stable than any of Central America's regimes and has steadfastly—and successfully—fought off involvement in the geo-politics of the Cold War.[5] Until the 1982 debt crisis, Mexico also catered to peasants as a distinct constituency—something most Central American governments never even pretended to do. One of the paradoxes of the rebellion is that the Zapatistas have responded to the adversity of eastern Chiapas more as Mexican nationals than as doctrinaire revolutionaries. While they are demanding changes for their region and for the rural poor in other areas, they are also holding Mexico's ruling party responsible for undemocratic politics and for betraying historic commitments to social welfare when it opened Mexico up to free trade and foreign investment.

As economic and political forces have transformed peasant farming, peasant communities have become less egalitarian, demarcated by class and by national political affiliation. My observations are drawn from research I performed as an anthropologist during three decades spent studying agrarian change in highland Chiapas, including ways in which Mexico's oil boom of the 1970s and the resulting debt crisis of the 1980s redefined the lives and roles of the region's peasants. Among the indigenous peasants I know, I saw a gap grow ever wider between the wealthy, who were able to infuse their farming with cash derived from wage work near the oil fields, on dam projects, and in urban construction projects, and the poor, who are finding it increasingly impossible to be able to afford to farm even their own land.

Some peasants in Chiapas have been able to weather the changes wrought by Mexico's economic restructuring by diversifying their farming activities, becoming produce and flower merchants, or starting up transport businesses. But many have not. Their successes and failures have often resulted from and contributed to politics, as can be seen by tracing the rise of local political bosses who have used their ties with the Institutional Revolutionary Party (PRI) to establish monopolies on small businesses and quell opposition in their towns. Economic restructuring brought particular suffering to the inhabitants of the eastern part of Chiapas, from which the rebels draw the majority of their members, and where cultural isolation, political exclusion, and economic depression have combined to leave people in what is commonly called Mexico's 'last frontier' without hope and without even the most basic necessities of life.

Because the situation of peasants in Mexico's countryside seems so bleak, the Zapatista rebellion inspired enormous sympathy from people throughout the world who read about the uprising in their newspapers and watched reports about it on television. Journalists tended to paint an image of the poor, honest peasants on one side and the greedy ranchers and corrupt politicians on the other.

This idealization of peasants is inaccurate, however, because some of the inequalities in the countryside are the result of stratification within peasant communities, not merely the result of injustices heaped upon them from outside. Understanding indigenous politics in this way necessarily complicates the sympathies one might hold toward peasants, but I view this as salutary. I think we misrepresent peasants if we allow ourselves to view. them in simplistic terms—as either the passive victims of the state or as 'noble savages' who can reinvigorate modern society with egalitarian and collective values. By acknowledging tensions and differences in peasant communities, we face up to both the virtue and the vice inherent in peasants' exercise of power over one another, and we integrate individual agency into our understanding of peasant communities. We also arrive at an appreciation of why not all peasant and indigenous groups welcomed the Zapatistas. In some highland communities, some people referred to the Zapatistas in native Tzotzil as 'trouble makers' or as 'thieves'—a reference to marauders who roamed the countryside in the 1910–1920 decade of the Mexican Revolution. Others told me that when the Zapatistas fanned into the mountains to look for recruits, a giant snake and a whirlwind—ancestor deities—rose in their paths to block off the entrances to Zinacantan and Chamula, communities whose leaders are loyal to the PRI.

The Zapatista uprising has forced a public debate about Mexico's priorities and has galvanized many impoverished Mexicans to demand better lives, even as it has to some extent polarized relations among indigenous peoples' organizations in Chiapas.

1 Arizpe, Lourdes and Fernanda Paz and Margarita Velázquez. *Cultura y Cambio Global: Percepciones Sociales sobre la Deforestación en la Selva Lacandona* (México, DF: UNAM-Centro Regional de Investigaciones Multidisciplinarias/Grupo Editorial Miguel Angel Porrúa, SA, 1993)

2 From *Declaracion de la Selva Lacandona* emitted by the *Comandancia General del EZLN* from the Selva Lacandona, Chiapas, Mexico on December 31, 1993.

3 Arvide, Isabel. *Crónica de una Guerra Anunciada* (México, DF: Grupo Editorial Siete, SA, 1994); Julio César Guillén Trujillo, *¿La guerra o la pas?* (Tuxtla Gutiérrez: Editorial Diálogo, 1994); Luis Méndez Asensio and Antonio Cano Gimeno, *La guerra contra el tiempo: Viaje a la selva alzada* (México, DF: Espasa Calpa Mexicana, SA, 1994); César Romero, Marcos: *¿Un professional de la esperanza?* (México, DF: Grupo Editorial Planeta, 1994).

4 Indigenous people generally speak one of the native American languages as their first tongue, though many also learn some Spanish. Many indigenous people identify themselves members of ethnic communities that were classified as 'Indian' rather than 'Spanish' or *mestizo* (mixed race) under the period of colonial rule. There are also indigenous people who lack such links to specific communities because they have been diasporic throughout history. Indigenous people often experience discrimination and subordination at the hands of nonindigenous Mexicans.

5 See Robert G. Williams, *Export Agriculture and the Crisis in Central America* (Chapel Hill, NC: University of North Carolina Press, 1986) on Central American radical movements as responding to peasants' displacement into marginal lands by developing commercial agriculture and ranching oriented to exports.

Part Five
Alternatives

The Human Right to Survive

Although much of the Food First critique of the world food system has come to be accepted even by mainstream institutions, progress toward changing that system is inhibited by the lack of alternative visions. Even in its heyday, Soviet agriculture shared many of the same defects as its Western counterparts, with its large farms, push for mechanization, and valuing of production more than the environment. In practice, it was chronically plagued by shortages, breadlines, and rationing. Even before its disappearance, there seemed to be no other way to structure the world food system except along the current free-market lines. So, despite widespread discontent with the failure of our present system, change has been blocked by the conservative argument that, in Margaret Thatcher's words, "There Is No Alternative."

This final section shows that there are in fact alternatives, and that they offer real visions of how food production could be reoriented to serve human needs rather than corporate profits. They range in size from small community projects up to national food policies, and from Third World peasant villages to giant American cities. Nor must they wait until 'after the revolution' to be implemented; most are already functioning, right now, within capitalist countries and despite the pressure of the world market.

The state of Kerala in India has shown for years that even poor lands can provide a decent living for all their people if there is a social commitment to do so (see *Kerala: Radical Reform as Development in an Indian State*). Although it has a large population and low per-capita income, Kerala has achieved very impressive levels of literacy, nutrition, life expectancy, and infant survival. Under a succession of leftist governments, the state carried out a far-reaching land reform and established public food programs through schools, nurseries and 'fair price shops' to guarantee adequate food for all, even in times of crisis. Among the results have been a low rate of population growth and the repeated democratic re-election of Marxist state governments, despite the frequent opposition of India's central government.

During the first few years of the Sandinista revolution, Nicaragua's food system was oriented away from its traditional focus on export crops toward producing the food grains which are the basic diet of the poor majority. In *Nicaragua: What Difference Could a Revolution Make?*, Collins, Lappé, Allen, and Rice show how broad land reform, redirection of credit toward the small farmers rather than the rich landowners, and raising the prices paid to farmers formed the basis of the strategy of food self-sufficiency. Banning dangerous pesticides such as DDT and replacing them with integrated pest management, the country was able to reduce pesticide poisoning and at the same time save millions of dollars. However, these new policies provoked violent opposition from wealthy Nicaraguans and the Reagan Administration; the authors describe the effects on food production of the first years of the Contra war, which eventually led to the Sandinista revolution's downfall.

Cuba, for several decades after its revolution, had based its agricultural strategy on large state farms, intensive use of Green Revolution technology and prioritization a single export crop, sugar cane. Although successful at providing food for all and achieving the highest level of health care in Latin America, this model was thrown into crisis with the disappearance of the Soviet Union. Rosset and Cunningham in "The Greening of Cuba: Organic Farming Offers Hope in the Midst of Crisis" tell how Cuba has turned to organic farming as an alternative. Using animal power instead of petroleum, biological control instead of pesticides, and home-grown grain production instead of paying for imported food with sugar exports, the country was able to weather the storm and turn the corner, resuming economic growth based on a new model of agriculture.

Green Revolution approaches to increasing agricultural production in the Third World have been based on the assumption that peasant farmers are ignorant, and must be shown how to use 'modern' technologies by outside experts. Eric Holt-Gimenez's "The Campesino a Campesino Movement: Farmer-led, Sustainable Agriculture in Central America and Mexico" describes how the Campesino a Campesino movement has rejected that assumption, and builds on peasant knowledge and popular education instead of outside extension agents. Blending political organization with indigenous traditions of mutual aid, the movement has spread widely in Mexico and Central America. It presents a sustainable agriculture alternative to failed Green Revolution schemes, and shows how the social and environmental traditions of Latin America peasants can incorporate scientific knowledge in a way that adapts it to their own needs.

In the United States, communities are also working to build locally-based alternatives to the world food market (see "Community Food Security: A Growing Movement"). The community food security movement, which has spread to cities across the country, provides a grassroots alternative to depending on agribusiness for our food. It demonstrates how urban dwellers can grow much of their own food on small parcels of land, using safe alternatives to chemical pesticides and keeping costs under control. Borrowing from Third World examples, the movement has served as a basis for economic development in neighborhoods that business considers too poor to invest in profitably. It shows that even in the homeland of the free-market system, alternatives based on the idea of food self-sufficiency can take root and grow.

The transformation of the world food system is not a change that will happen merely through good will and intellectual arguments. Political mobilization by those without access to land and resources, as described by Langevin and Rosset in "Land Reform From Below: The Landless Workers' Movement in Brazil," is fundamental to making progress. In Brazil, the Landless Workers' Movement is not waiting for the government to act; they are taking over idle land and planting crops on it—a land reform "from below." The Movement's actions directly challenge a food system that will not produce for people when producing for export, or not use the land at all. Despite government indifference and the murders of hundreds of activists, Brazil's landless workers have shown that through organization, the poor can force the system to change.

The alternatives described in this section are just a few of the many challenges worldwide to the ideology that profits should take precedence over people's basic needs. They show that to build a new world food system which is sustainable and just, we must recognize the right to survive as a fundamental right of all humanity. Such a new system will based on mobilizing the skills and energies of people, rather than replacing them with dangerous pesticides and costly machinery. Instead of exhaustible energy resources, it will use both traditional and scientific knowledge of the complex workings of nature to maintain the productivity of agriculture. Rather than considering people as simply more mouths to feed, it will see them as hands to work and minds to think. And above all, it will find its guidance not in the dictates of the market or the opinions of the experts, but in the recognition that the earth's bounty is given to be shared by us all.

From *Kerala: Radical Reform as Development in an Indian State* by Richard W. Franke and Barbara H. Chasin

Lessons from Kerala

What can we learn from Kerala's development experience so far? Kerala has many unique ecological and historical features that do not exist or cannot be repeated in other places. A thorough set of lessons can only be gathered through a careful comparative study with special attention to the details and conditions under which each nation or region is attempting to bring the benefits of modern life to all its people. Kerala cannot offer a model to be copied; it provides lessons that may be used to evaluate the development efforts of other societies or—for those directly involved in development work either from the developed or from within the developing countries themselves—to suggest strategies for improving their work. In this spirit, we offer the following tentative lessons from the Kerala experience.

1. Radical reforms deliver effective and mutually reinforcing benefits to the poor even when per capita incomes remain low.

Kerala's choice of radical reforms has produced benefits to all the state's people that no other Indian state and few other Third World nations have accomplished. Many Third World countries with far higher per capita incomes stand way below Kerala in education, life expectancy, and infant mortality.

Kerala's achievements are more than the sum of a list of reforms. The reforms are mutually reinforcing and produce greater overall effects than might appear from simply adding up the list we made in order to examine each in detail. The popular organizations and structures that have produced the reforms are themselves further strengthened by them. The land reform removed the threat of eviction of tenants, making possible their greater political participation without fear. Political organization and struggle, redistribution, democracy, participation, basic services, empowerment of the poor: these are the combined features of Kerala's recent history.

2. Popular movements and militant progressive organizations with dedicated leaders arc necessary to initiate and sustain the process of reform.

In terms of action, Kerala's most important lesson is that the poor must be organized to ensure that their needs are met. The poor cannot depend on benevolent rulers or outside development agencies. Just as in the developed countries, the poor need the strength of their numbers in organizations that truly represent them in legislative bodies and on the ground to press for their demands. They must be organized to agitate when necessary to protect themselves against the interests of the wealthy, whose power is built right into the daily workings of the social and political structure. The success of popular movements in Kerala has resulted in the following additional lessons, both positive and negative.

3. Despite their beneficial consequences in many areas, radical reforms cannot necessarily create employment or raise general levels of per capita income.

Kerala's reforms have not raised agricultural production or alleviated unemployment—both chronic problems in many Third World economies. Whether reforms can stimulate production and employment opportunities is hotly debated and must be studied in greater comparative detail. Kerala's experience in this regard, however, is sufficient to indicate that it is not a cheap alternative to the need for income growth and production increases.[1]

4. Local reformers are restricted by national politics.

Kerala's reforms have been carried out despite frequent opposition from the more powerful Indian central government. Kerala's minister for finance recently complained that the state is receiving thirty percent less than the all-India average in development funds. He has also charged the central government with freezing state bank funds and providing only about half the per capita investment in industry deserved by the state.[2] If such allegations are correct, Kerala's left-wing governments face difficult challenges in administering their development approach.

5. Public distribution of food is a rational and highly effective policy choice in very poor agrarian economies.

Kerala's school and nursery feeding programs and especially its thorough and full-coverage fair price shops ensure at least a minimum food package to nearly all people in the state. Although the shops do not offer credit during periods of greatest food shortage, through low prices they nonetheless help reduce the plight of the most potentially undernourished groups and help them remain free of private moneylenders whose practices are a source of exploitation and misery in many Third World rural areas.

6. Devoting significant resources to public health and health care can bring about low infant mortality, high life expectancy, and low birth rates even when incomes are not high or increasing.

Sanitation, safe water, housing, and regularly staffed and accessible health care facilities play a major role in reducing the incidence and effects of disease. Kerala proves there is no need to wait for economic growth before installing these crucial services even in the poorest parts of the world. Although some countries face greater costs in providing appropriate facilities, the rewards in terms of meeting basic needs and producing a population that can make use of other development programs is surely greater than the costs when measured against the cumulative benefits of health programs and other reforms.

7. Widespread literacy and expanding educational opportunities can help to break down traditional social barriers and create a more just and open social order.

Despite widespread recognition of the need for education, many Third World countries devote far too little to both basic literacy and raising the levels of education in rural areas. Illiterate people cannot compete for modern employment and may be intimidated into fear and passivity in the face of bureaucracies that demand they fill in forms or display awareness of laws they cannot read. Education can give people confidence as well as skills.

8. Meaningful land reform can reduce economic and social-political inequality and put important productive resources into the hands of a large portion of the poor.

Each agrarian society has its own particular land ownership structure and Kerala's experience can only be seen in the most general terms. The kind of land reform needed will vary, but to the extent that wealth and income are derived from land, only the breaking of the hold of the landed elite over productive resources can liberate the poorest small farmers from the effects of highly unequal land ownership.

9. Protection of farm workers through laws covering wages and working conditions can help distribute more widely the existing resources of even a very poor economy.

Many of the features of the Kerala Agricultural Workers Act would directly and immediately benefit millions of Third World workers who would be left out of even the most thoroughgoing land reform. Mere passage of such laws will not ensure their enforcement, however, and would require organization and agitation.

10. Greater social and economic equality in combination with strong organizations representing the poorest groups can lead to lower levels of violence and a generally healthier social and political environment.

The near absence of violence against lower castes in Kerala is one of the most valuable lessons the state has to offer other regions of India. Kerala's success in achieving peaceful intercaste relations may also be instructive in other parts of the Third World where much routine violence is meted out to groups at the bottom of the society, even where no caste system is present.

11. Women can benefit substantially from radical reforms even when these are not aimed directly at their problems, but such reforms must eventually be supplemented by special attention to women's needs.

When widely distributed across all groups, land reform, education, and health improvements are bound to help women in the process. Kerala's achievements for women show this. Each society, however, has its own special forms of gender oppression, and for Kerala, certain forms of violence and unemployment are still substantial problems. Kerala's militant organizations and leftist governing parties now face the difficult task of mobilizing support for types of reforms that have not been on the social agenda for very long.

12. Progressive forces including Communists and Communist parties can play a major and positive role in bringing benefits of development to very poor Third World farmers and workers.

Many people in the developed countries regard Communist organizers and Communist parties primarily as outside agents of the Soviet Union or some other antagonistic power. The stereotype of Third World Communists in the United States is that they are corrupt puppets engaging in an international conspiracy against more 'democratic' regimes. Members of the Communist Party-Marxist (CPM) in Kerala, the party currently in power along with its coalition partners, and the party with the widest or second-widest following, are independent of any other country. While they try to learn from the experiences of the socialist countries, they do not follow an agenda set in Moscow, Peking, or any other Communist capital. CPM leaders and cadres in Kerala have a reputation, even among their opponents, for being relatively honest and noncorrupt. This means they are not involved in politics for their own immediate political gains, but are trying to implement their vision of a more just society. Although communism and Communists are far more acceptable in Kerala than in the United States, there has been and could again be repression against them, including brutal beatings and killings. There are many risks and few material rewards for choosing the life of the radical organizer. Communists in Kerala follow the electoral rules. They run candidates in elections, and when they win, they do not institute undemocratic practices such as censorship of the media or repression of their opponents. An election won by the CPM has not meant an end to the Kerala electoral process. When voted out of office, they have continued to play by the official rules of Indian democracy as spelled out in that country's constitution. Anticommunism is particularly strong in the United States. Kerala offers an excellent opportunity for us to rethink our outmoded and harmful approach. Kerala does not threaten perceived United States interests even in the wildest geopolitical imagination. Kerala's history and development experiment allow us to examine thoughtfully the role of Communist organizers and structures in bringing many of the very benefits that most Americans would favor to people who might not otherwise have been able to get them. Kerala dramatically suggests the need for a reevaluation of our ingrained hostility to Communists.

13. Radical reforms can shield the poor against recessions.

As we were drafting this report, an international news item drew our attention to one of Kerala's most important lessons and one that is easily hidden from view. *This is the capacity of radical reforms to shield the poor from recessions.*

During the 1980s we in the United States have witnessed the spectacle of feverish growth and collapse on the stock market, rising and falling interest rates, luxury living and homelessness. In the Third World, the effects of the capitalist economic roller coaster have been far greater. Since 1983 the underdeveloped countries have transferred a net of over $30 billion in wealth to the already-rich countries. This anti-development flow of funds comes from the combination of interest payments on the staggering Third World debt along with a decline in rich-country support for development efforts and a drop in the prices of the major goods sold by the Third World.[3] Many countries have suffered zero or negative economic growth in the 1980s. These countries represent over 700 million people, most of whom have seen their standards of living deteriorate. Of all the population groups, young children of poor parents are the most vulnerable to economic decline.[4] In Brazil, one study estimates that 60,000 'extra' child deaths occurred because of the recession of the 1980s.[5] In the Third World generally, malnutrition is rising and more than 500,000 more children died in 1988 alone than might have been expected. War-related deaths are not included in the estimate.[6]

What do all these grim statistics have in common? They are all consequences of the market system, of capitalism's ability to provide a stunning array of products to those who can afford them, along with its dramatic inability to distribute resources and products equitably among the peoples of the world.

It is precisely this contradictory nature of capitalism and the market that Kerala's reforms help to overcome. Land redistribution, effective food rationing, and pensions insulate the poorest groups from the negative consequences of the capital business cycle, in particular, from dropping through the bottom during recessions. Certainly Kerala's people, like people everywhere, are affected by the world recession. But the reforms already in place give them protection that many others do not have. The Left Front government elected in Kerala in 1987 has actually expanded access to school lunches and pensions. In addition, the government increased per capita expenditures on education by sixteen percent in 1987 and five percent in 1988 and it increased expenditures on health by twenty percent in 1987 and ten percent in 1988—continuing its tradition of expanding public services as the cornerstone of its development strategy.[7] While many Third World governments are cutting education and medical services in order to make payments on their debts to rich country banks—including the World Bank, which is supposed to be supporting development—Kerala has structures and policies in place to buffer its people against the worst effects of this situation.

More importantly, Kerala's government, more so than most national or local governments in the Third World, is directly responsible to political parties, labor unions, women's groups, and peasant associations that are militant, well organized, well led, and ready if necessary to go into action if their members' interests are threatened. Because of the strength of the people's organizations, many of the reforms are continued and expanded even under conservative governments, though they prosper more when leftist parties are in power.

At the press conference announcing its 1989 report, UNICEF Director James Grant stated: "In the 1960s and 1970s, tremendous emphasis was put on how you get better GNP growth rates. But GNP growth rates can hide mass maldistribution of income." He continued: "In the 1990s, the target ought to be meeting more tangible human targets: assuring safe water, assuring access to health services, assuring basic education."[8]

Kerala State, India, he might have added, offers us lessons in how to meet the target.

1 This point has been forcefully argued in Joan Mencher, "The Lessons and Non-Lessons of Kerala," *Economic and Political Weekly* vol. 15, no. 41–43, 1980: pp. 1781–1802.

2 Menon, V. Viswanatha. *Budget Speech: Revised Budget 1987–88* (Trivandrum: Government of Kerala, 1987), pp. 5, 8–11.

3 Grant, James P. *The State of the World's Children: 1988* (London, UK: Oxford University Press, 1988), pp. 28–29.

4 Ibid., pg. 25.

5 bid, pg. 28.

6 *The New York Times,* December 20, 1988, pg. 1; James P. Grant, *The State of the World's Children: 1989* (London, UK: Oxford University Press, 1989), pg. 1.

7 Government of Kerala. *Economic Review: 1987,* (Trivandrum: State Planning Board, 1988), pp. 7, 75. Government of Kerala, *Economic Review: 1988,* (Trivandrum: State Planning Board, 1989), pp. 81, 87.

8 *The New York Times,* December 20, 1988, pg. 6.

From *Nicaragua: What Difference Could a Revolution Make?* by Joseph Collins with Frances Moore Lappé, Nick Allen, and Paul Rice

Getting Off the Pesticide Treadmill

The extensive monocultivation of cotton in Nicaragua during the 1950s and 1960s created an environment ideal for pests, including cotton's most notable nemesis, the infamous boll weevil. Initially the growers got reassuring results using chemical pesticides, especially DDT and methyl parathion (a byproduct of chemical warfare research during World War II). Soon, however, they found themselves caught up in a vicious cycle of dependency on more and more chemicals: Pests developed resistances to the chemicals and new pests emerged as the pesticides killed off their natural predators. Once on this 'pesticide treadmill,' growers wound up scheduling so many aerial sprayings of pesticide 'cocktails' that the profitability of cotton production was threatened. By the late 1960s, Nicaragua had the dubious distinction of holding the world's record for the number of applications of pesticides on a single crop.

The environmental consequences—and the impact on human health and food production—were shocking. In the 1969–1970 harvest season alone, 383 pesticide fatalities were reported. More than three thousand acute poisonings a year were reported from 1962 to 1972. *Reported* figures, however, drastically understated reality. In Somoza's Nicaragua many poisoning victims would not seek medical attention because they were too poor or feared being fired if they missed work. Immediate fatalities and illnesses from pesticide exposure are only the tip of the iceberg. No one really knows the long-term consequences of decades of heavy pesticide use, especially since DDT and other pesticides in the organochlorine family remain in soil and water for years. What is known is certainly ominous: Hundreds of thousands of Nicaraguans have been found to carry extraordinary levels—sixteen times the world average—of DDT and other cancer-causing chemicals in their fatty tissues. Samples of mothers milk were found to have forty-five times the DDT maximum considered permissible by the World Health Organization. Pesticide 'drift' has polluted water supplies and wreaked havoc on nearby food crops; it also set the stage for the resurgence of malaria, once the mosquitoes who carried the disease developed resistance to DDT.

In 1977 and 1978, cotton production costs exceeded export revenues. But the big growers were powerful enough to arrange for production subsidies by public borrowing from American banks. The 'hidden' costs were even greater: a 1977 United Nations investigation estimated that annual environmental and human health damage from pesticide use in Nicaragua added up to $200 million, while the foreign exchange generated from cotton never exceeded $141 million a year. On economic, environmental, and health grounds, the pesticide treadmill was monstrously irrational. But given the powerful economic and political interests behind it, it had literally taken a revolution to create the context to get Nicaragua off the treadmill.

From the start, the Sandinistas have understood agrarian reform to include improving working conditions in the countryside. Immediate steps were taken to reduce human poisonings and environmental contamination caused by excessive and improper application of pesticides. At the same time, the government moved toward the longer-range goal of sharply reducing the use of highly toxic and expensive pesticides without sacrificing production.

Since all but one pesticide used in Nicaragua was imported, the government used its control over imports to reduce and eliminate (by 1981–1982) the import of DDT, endrin, and dieldrin. (All three of these organochlorines had been banned or severely restricted in the United States for over a decade.) Pesticide-formulating companies (licensees of multinationals) were pressured to substitute pesticides considered environmentally less harmful and less hazardous to humans. In 1980, the pesticide DBCP was banned after the chemical was found to cause sterility in production workers in California and to be a potent cancer-causing agent in laboratory tests. The pesticide is a favorite in banana production as well as in many other crops throughout the Third World.

The government promoted educational programs to illustrate the dangers of pesticides to factory and rural workers. The Ministry of Labor organized over four thousand safety classes since 1979. Safety inspectors were been trained to investigate every workplace in the country. A national commission, created in 1982, sought to coordinate the efforts of the various ministries, labor unions, and producer associations (like the association of cotton growers) in addressing pesticide problems. In response to a proposal by the Sandinista-led farmworkers union (ATC), the commission developed new regulations. Included is a system of simply worded, illustrated, and color-coded pesticide labels to provide information on hazards, personal protective measures, poisoning symptoms, first aid, and medical treatment. The national literacy campaign has helped make such measures effective among rural workers.

The progress achieved in the revolution's first three years was encouraging. A visitor still can come upon shockingly nonchalant handling of highly toxic chemicals, but many farm managers and workers, as well as peasant producers, now seem aware of pesticide hazards. Once again, however, the discouraging reality was Washington's war on Nicaragua.

As the contras stepped up their attack in 1983 and the U. S. invasion of Grenada in October set off nationwide alarm in Nicaragua, many projects came to a standstill. Mobilizations drew off many of those trained as health and safety inspectors. (Not surprisingly, a high proportion of volunteers for the defense mobilizations were most actively involved in the revolution's new programs—they were Nicaragua's 'best and brightest.') By the end of 1983, of Nicaragua's forty inspectors, thirteen had been mobilized to armed combat and five others to harvest coffee in regions where Contra attacks contributed to the shortage of workers. Training courses had frequently been interrupted by defense mobilizations. Review of new pesticide regulations by the national commission was held up for nearly six months as key members were mobilized. Safety inspections and courses slowed down by war-related transport problems. Imported tires and spare parts go on a priority basis to national defense. By the end of 1983, there were only two functioning vehicles for the forty workplace inspectors. The inspectors who managed to get to the field, like other agricultural technicians, became prime targets of the Contras.

U.S. aggression also hampered Nicaragua's efforts to eliminate certain pesticides from use. Launched from a CIA ship off the coast, the October 1983 assault on the port of Corinto destroyed a major new shipment of pesticides on the docks. The attack coincided with the height of the boll weevil season. With the shipment of pesticides destroyed and shipping disrupted well into December, the government had no recourse but to reintroduce a number of banned pesticides locked away in warehouses.

Nonchemical Pest Control

The long-term reduction of pesticide use promises major economic as well as health benefits. Pesticides make up a third of the production costs in cotton growing, which accounts for seventy-five to eighty percent of all pesticide use in Nicaragua. Any significant reduction in their use would save millions of dollars. Despite subversion and sabotage, striking advances have been made toward the goal of sharply reducing pesticide use in cotton production. At the Institute for Food and Development Policy, we are pleased to have been associated with this work.

Practitioners of 'integrated pest management' (IPM) worked with the agricultural ministry on pesticide reduction. Advocates of reduced pesticide use invariably ran up against the widely held assumption that environmental protection is at odds with production. IPM workers in Nicaragua reduced pesticide use by developing alternative means of pest control while demonstrating that output can be maintained and perhaps increased. IPM maximizes the use of naturally available nonchemical insect controls. In Nicaragua natural predators of pests were introduced into cotton fields (with the help of technical assistance from the West German government). IPM also worked to see that chemical pesticides are used only when careful scouting of the pest populations indicates they are needed. (Contrast that with the practice promoted by the multinational chemical companies of scheduled sprayings—"If it's Tuesday, it must be time to spray.")

But IPM's principal technique in Nicaragua has been 'trap cropping.' Trap cropping involves leaving four rows of cotton plants standing for every two acres of land when the rest of the plants have been plowed under after the harvest. These plants are the 'trap' between seasons to which the boll weevils are attracted. Workers trained in safe techniques treat the trap crops daily with organophosphate pesticides which, although extremely toxic, unlike the organochlorines such as DDT leave no lasting residues in the environment. Before planting the commercial crop, another trap is planted next to what remains of the existing one and treated until the commercial crop sets fruit. Trap cropping has been shown to postpone the need for pesticide treatments on the commercial cotton crop until much later in the season, greatly reducing the total amount applied.

In 1982 a pilot program in trap cropping was carried out on forty-one thousand acres, approximately one-sixth the area planted that year in cotton. The private and state farms in the program used only one-third the pesticides applied on farms outside the program, and produced slightly higher yields. Pest control costs were cut sixty-three percent, for a net savings of $2.14 million, even after figuring in additional labor and other costs. (To put that in perspective, the total earnings from cotton exports that year came to $87 million.)

This success went a long way toward combating the skepticism that prevailed in the agricultural ministry and among private growers. The government expanded the program to cover three times the acreage in 1983. The national university in León offered a three-year masters program in integrated pest management mainly for technicians already employed on state and private cotton farms. At the university, IPM techniques were also being developed for controlling such health problems as malaria-bearing mosquitoes. This was of worldwide importance in the face of the resurgence throughout the Third World of malaria from mosquitoes resistant to DDT and other chemical insecticides.

IPM practitioners emphasized that agrarian reform made it possible to ensure all farms in a given area participate in the program. This coordinated approach is essential to successful implementation of biological pest controls. It also reinforced the Sandinistas' emphasis on local control and human resources development. Integrated pest management cannot be implemented by sweeping centralized decisions, but requires training numerous local managers and workers.

The Ministry of Agriculture coordinated local weekly meetings of government and private pest-control technicians to evaluate area ecological conditions and to devise coordinated control tactics. In many cotton-growing areas bean and corn production had been curtailed by the ecological disruption caused by the pesticide treadmill. With IPM in greater use, those staples could be reintroduced into the area; successfully reducing pesticide use in cotton production ultimately favors the agrarian reforms' goal of food self-reliance for Nicaragua.

From *Food First Action Alert*, Spring 1994 by Peter Rosset with Shea Cunningham

---

The Greening of Cuba: Organic Farming Offers Hope in the Midst of Crisis

Times are hard in Cuba. The collapse of the socialist bloc has led to an estimated eighty-five percent drop in total external economic relations—that is exports, imports, and foreign aid. The U.S. trade embargo has recently been strengthened through the passage of the Torricelli Act, making it impossible for the West to make up for the bulk of the loss in trade with the socialist bloc. From less than twenty consumer items that were rationed in the mid-eighties, shortages have led to the rationing of everything. There is virtually no item that a Cuban can buy with pesos that does not require a ration card. Food intake by the population may have dropped by as much as thirty percent since 1989—moving Cuba from the top five Latin America countries for both average caloric and average protein intake, to the bottom five, though Bolivia and Haiti are still worse off. Prostitution and petty theft are at their highest point since the 1959 revolution.

Amidst the suffering of the Cuban people, however, there have been some remarkable innovations that have not been widely reported outside of Cuba. It is far too early to say whether these developments will be sufficient to help Cuba weather the present storm, but they do offer some hope in contrast to the generally bleak outlook. These changes run from the legalization of small-scale private enterprise, to the privatization of the state farm sector in the form of worker's cooperatives, both within the past six months. This article focuses on another recent development: the technological transformation of Cuban agriculture in response to a massive drop in pesticide and fertilizer imports. Cuba is presently in the third year of the largest conversion of any nation in history from conventional modern agriculture to large scale organic farming.

While this is a calculated risk—and possibly a life and death gamble—for the Cuban people, it is also a critically important experiment for the rest of the world. Whether we are citizens of the United States, Mexico, Germany, or Thailand, we must all confront the declining productivity and environmental destructiveness of what passes for modern agriculture.

As soils are progressively eroded because of their exposure to the elements, compacted by heavy machinery, salinized by excessive irrigation and sterilized with methyl bromide, and as pests become ever more resistant to pesticides, crops yields are in decline, even as aquifers and estuaries are contaminated with agrichemical run-off. Organic farming and other alternative technologies are intensively studied in laboratories and experimental plots worldwide, but examples of implementation by farmers remain scattered and isolated. Cuba offers us the very first large scale test of these alternatives, perhaps our only chance before we are all forced to make this transformation, to see what works and what doesn't, what problems and which solutions will come up along the way.

Cuba Before and After the Collapse of the Socialist Bloc

From the Cuban revolution in 1959 through the collapse of trading relations with the socialist bloc at the end of the 1980s, Cuba's economic development was characterized by rapid modernization, a high degree of social equity and welfare, and strong external dependency. While most quality of life indicators were in the high positive range, Cuba depended upon its socialist trading partners for petroleum, industrial equipment and supplies, agricultural inputs such as fertilizer and pesticides, and foodstuffs—possibly as much as fifty-seven percent of the total calories consumed by the population. Cuban agriculture was based on large-scale, capital-intensive monoculture, more similar in many ways to the Central Valley of California than to the typical Latin American *minifundio* or small-scale farm. More than ninety percent of fertilizers and pesticides, or the ingredients to make them, were imported from abroad. This demonstrates the degree of dependency exhibited by this style of farming, and the vulnerability of the island's economy to international market forces. When trade relations with the socialist bloc collapsed in 1990, pesticide and fertilizer imports dropped by about eighty percent, and the availability of petroleum for agriculture dropped by a half. Food imports also fell by more than a half. Suddenly, an agricultural system almost as modern and industrialized as that of California was faced with a dual challenge: the need to essentially double food production while more than halving inputs—and at the same time maintaining export crop production so as not to further erode the country's desperate foreign exchange position.

Mobilizing Science and Technology to Respond to the Crisis

In some ways Cuba was uniquely prepared to face this challenge. With only two percent of Latin America's population but eleven percent of its scientists and a well developed research infrastructure, the government was able to call for 'knowledge-intensive' technological innovation to substitute for the now unavailable inputs. Luckily an 'alternative agriculture' movement had taken hold among Cuban researchers as early as 1982, and many promising research results—which had previously remained relatively unused—were available for immediate and widespread implementation.

The Alternative Model versus the Classical Model

Though the technological changes in agriculture might be viewed pessimistically as short-term responses to crisis, Cubans are quick to claim that this is a long overdue structural transformation. Planning authorities within the Agriculture Ministry have officially declared that all new development of agriculture be based on what they call the "Alternative Model," which they contrast with the "Classical Model" of conventional modern agriculture. They say that the

BOX 1: Quality of Life Indicators (Cuban Rank in Latin America in 1989)

CATEGORY	RANK
Average daily caloric intake	2
Average daily protein intake	4
Infant mortality	lowest
Life expectancy	2
Doctors per person	1
Scientists per person	1
Housing per family	1
Tractors per unit of farm land	1
Yields of grain crops	2
Theater attendance	1
Museum attendance	1
Movie attendance	1
Average of health & education indicators	1

Source: Peter Rosset and Medea Benjamin, *The Greening of Cuba: A National Experiment in Organic Agriculture* (London, UK: Ocean Press, 1994)

Classical Model was always inappropriate for Cuban conditions, having been imposed by European socialist bloc technicians. In this conceptual framework, the Classical Model is based on extensive monoculture of foreign crop species, primarily for export. It is highly mechanized, and requires a continuous supply of imported technology and inputs. It promotes dependence on international markets and, through mechanization, drives migration of people from rural areas to the city. Finally, it rapidly degrades the basis for continued productivity, through the erosion, compaction and salinization of soils, and the development of pesticide resistance among insect pests and crop diseases.

The Alternative Model, on the other hand, seeks to promote ecologically sustainable production by replacing the dependence on heavy farm machinery and chemical inputs with animal traction, crop and pasture rotations, soil conservation, organic soil inputs, biological pest control, and what the Cubans call biofertilizers and biopesticides—microbial pesticides and fertilizers that are non-toxic to humans. The Alternative Model requires the reincorporation of rural populations into agriculture—through both their labor as well as their knowledge of traditional farming techniques and their active participation in the generation of new, more appropriate technologies. This model is designed to stem the rural-urban flood of migrants, and to provide food security for the nation's population. It is virtually identical to alternatives proposed in the U.S., Latin America, Europe and elsewhere—differing only in one key respect. While it represents a utopian vision for the rest of us, it is now government policy and agricultural practice in Cuba.

A Cuban NGO

A rare phenomenon in Cuba—a non- governmental organization (NGO) is playing a pivotal role in what might be called the institutionalization of the alternative model. The Cuban Association for Organic Farming is composed of ecological agriculture activists ranging from university professors and students to mid-level government functionaries, farmers and farm managers. It is struggling on a shoestring budget to carry out an educational campaign on the virtues and indeed the necessity of the alternative model. Food First is working with the Association and with a Cuban university—the Advanced Institute for Agricultural Sciences of Havana (ISCAH)—on a project to document the transformation of agriculture, with particular emphasis on the evaluation of the efficacy of the new technologies, in terms of economic productivity as well as environmental and social indicators.

Conversion from Conventional Agriculture to Organic Farming

Cuba is undergoing large scale conversion from conventional agriculture to organic or semi-organic farming. Empirical evidence from the U.S. and elsewhere demonstrates that it can take anywhere from three to five years from the initiation of the conversion process to achieve the levels of productivity that prevailed beforehand. That is because it takes time to restore lost soil fertility and to re-establish natural controls of insect and disease populations. Yet Cuba does not have three to five years—its population must be fed in the short term. Cuban scientists and planners are shortening this process by bringing

BOX 2: Translation of a chart circulated to all planning personnel by the Cuban Ministry of Agriculture (originating fundamentally in developed countries)

Classical Model vs. Alternative Model

Classical Model

External dependence
- of the country on other countries
- of provinces on the country
- of localities on the province & the country

Cutting edge technology
- imported raw materials for animal feed
- widespread utilization of chemical pesticides and fertilizers
- utilization of modern irrigation systems
- consumption of fuel and lubricants

Tight relationship between bank credit and production; high interest rates

Priority given to mechanization as a production technology

Introduction of new crops at the expense of autochthonous crops and production systems

Search for efficiency through intensification and mechanization

Real possibility of investing in production and commercialization

Accelerated rural exodus

To satisfy ever increasing needs has ever more ecological or environmental consequences, such as soil erosion, salinization, and waterlogging

Alternative Model

Maximum advantage taken of:
- the land
- human resources of the zone or locality
- broad community participation
- cutting edge technology but appropriate to the zone where it is used
- organic fertilizers and crop rotation
- biological control of pests
- biological cycles and seasonality of crops and animals
- natural energy sources
  hydro (rivers, dams, etc.)
  wind
  solar
  slopes, biomass, etc.
- animal traction
- rational use of pastures and forage for both grazing & feedlots, search for locally supplied animal nutrition

Diversification of crop and autochthonous production systems based on accumulated knowledge

Introduction of scientific practices that correspond to the particulars of each zone; new varieties of crops and animals, planting densities, seed treatments, post-harvest storage, etc.

Preservation of the environment and the ecosystem

Need for systematic training (management, nutritional, technical)

Systematic technical assistance

Promote cooperation among producers, within and between communities

Obstacles to overcome:
- difficulties in the commercialization of agricultural products because of the number of intermediaries. Control over the market and its particulars.
- poverty among the peasantry
- the distances to markets and urban centers (lack of sufficient roads and means of transport)
- illiteracy

Source: Peter Rosset and Medea Benjamin, *The Greening of Cuba: A National Experiment in Organic Agriculture* (London, UK: Ocean Press, 1994)

sophisticated, 'cutting edge' biotechnology to bear on development of new organic farming practices. This is not the environmentally dangerous genetic engineering version of biotechnology that we see in U.S. agriculture, but rather a locally controlled variety based on the mass production of naturally occurring organisms to be used as biopesticides and biofertilizers. Cuba is demystifying biotechnology for developing countries—showing that it does not have to rely on multi-million dollar infrastructure and super-specialized scientists, but rather can be grasped and put into production even on peasant cooperatives.

Elements of the Alternative Model

During several trips to Cuba myself and other agricultural scientists have been able to document the development and implementation of alternatives in the areas of pest and soil management, labor mobilization, and participatory methods for generating new technology.

Management of Crop Pests

Among the alternative tactics being used to offer insect control, the most important are conventional biological control based on mass releases of parasitic and predatory insects, and the use of biopesticides. In the latter area, Cuba is substantially more advanced than other Latin American countries and compares favorably to the U.S. Cubans produce numerous formulations of bacterial and fungal diseases of insect pests which are applied to crops in lieu of chemical insecticides. A total of 213 artesanal biotechnology centers located on agricultural cooperatives produce these products of cutting edge technology for local use. They are typically produced by people in their twenties, born on the cooperative, who have received some university-level training. While industrial production of biopesticides will soon be underway for use in larger scale farming operations that produce for export, it remains most remarkable that the sons and daughters of campesinos can make the products of biotechnology in remote rural areas.

Furthermore, Cuban use of biofertilizers in commercial agriculture is unrivaled in the world, including not only standard *Rhizobium* inoculants for luguminous crops, but also free living bacteria that make atmospheric nitrogen available for other crops. Perhaps of greatest importance for other developing countries, Cubans are mass producing solubilizing bacteria which make phosphorous, which in many tropical areas is bound to soil particles, available for uptake by crop plants.

An Experiment that the World Should Be Watching

It is unclear whether the widespread implementation of an alternative model of agricultural development will, in conjunction with other government policies, allow Cuba to emerge from the crisis wrought by the collapse of the socialist bloc. As agricultural scientists, environmentalists, and concerned citizens however, we can say that the experiment in agricultural alternatives currently underway in Cuba is unprecedented, with potentially enormous implications for other countries suffering from the declining sustainability of conventional agricultural production.

From *Food First Development Report* no. 10, June 1996, by Eric Holt-Gimenez

---

The Campesino a Campesino Movement: Farmer-Led, Sustainable Agriculture in Central America and Mexico

> "Farmers helping their brothers so that they can help themselves to find solutions and not be dependent on a technician or on the bank. That is Campesino a Campesino."

This is the simple definition provided by a farmer for the grassroots movement in sustainable agriculture that has swept across Mexico and Central America. El Movimiento Campesino a Campesino, or the 'Farmer to Farmer Movement,' is one of the most extensive and successful efforts for sustainable agriculture to appear in Latin America in decades.

Led by campesinos, the subsistence farmers of the ecologically fragile hillsides and forest perimeters of the Mesoamerican tropics, Campesino a Campesino (CAC) involves hundreds of volunteer and part-time campesino 'promotores' with the support of dozens of technicians, professionals, and local development organizations. Participants number in the incalculable thousands. Incalculable because in essence, Campesino a Campesino is an extensive, expanding, and somewhat amorphous cultural process for agricultural transformation. Programatically elusive and annoyingly asystematic, Campesino a Campesino has nonetheless succeeded in regenerating tens of thousands of hectares of exhausted soils in the tropics. Simultaneously, it has significantly raised and stabilized campesino agricultural production, opening the door to crop diversification, marketing, processing, commercialization, and other alternatives.

How have campesinos helped each other develop sustainable agriculture? On the one hand, they have used relatively simple methods of small-scale experimentation, combined with horizontal (farmer to farmer) workshops in basic ecology, agronomy, soil and water conservation, soil building, seed selection, crop diversification, integrated pest management, and biological weed control. These approaches have provided campesinos with sufficient technical and ecological knowledge to reverse degenerative agroecological processes and overcome the basic limiting factors in farm production.

On the other hand, campesino promotores, leading by example, have inspired their peers to innovate and try new alternatives. This inspiration has given them the conviction, enthusiasm and pride to teach others using hands-on learning methods, and to share with others through farmer to farmer cross-visits and conferences. By forging their own alternatives for agriculture, the promotores have gained a new-found belief in themselves and have formulated a fresh, campesino vision for the future of agriculture. Campesino a Campesino is not a program or project looking for peasant 'participation;' it is a cultural phenomenon, a broad-based movement with campesinos as the main actors. As protagonists, they are foremost and directly involved in all stages of generation and transfer of technologies for sustainable agriculture including adoption and adaptation of appropriate modern techniques as well as traditional practices.

This approach is fundamentally different from the conventional 'generation and transfer' strategies for agricultural development applied by governments and research centers for nearly half a century in the developing world. Granted, the 'old' conventional approach has evolved from the top-down extension of technological packages to 'passive' farmers, to sophisticated 'new' conventional approaches that use 'participatory' techniques to elicit farmer involvement. What remains the same is the conventional chain of command and control over research and development, generation and transfer process: researchers and governments ultimately control the agenda and resources, whereas farmers 'participate.'

While this approach was, by its own criteria, effective for the once-touted 'Green Revolution,' it has fallen flat for sustainable agriculture. With a few exceptions, campesinos are simply not adopting techniques pushed by international research centers and state extension programs. In the face of the growing agroecological crisis sweeping Latin America, solving the 'problem' of farmer participation has become a critical issue for both technicians and researchers.

The Campesino a Campesino movement does not presume to solve the problems of governments or research centers for them. It simply turns the fundamental question on its head. Rather than asking how to get farmers to participate, it poses the question: How do those interested in the development of sustainable agriculture participate in farmer-led development? A closer look at the movement may help to understand the question and formulate possible answers.

Sustainable Agriculture, Campesinos, the Green Revolution, and the Lost Decade

In the 1960s and 1970s Central America experienced dramatic gains in economic growth. But the concentration and uneven distribution of wealth and resources provoked social and political unrest, rebellion, revolutions and eventually, (with the help of the United States), civil war. In the conflict-ridden 'Lost Decade' of the 1980s, these gains disappeared just as dramatically as they had appeared.

In the Central American countryside, the negative socio-economic effects of uneven growth were exacerbated by the negative ecological effects of the Green Revolution: To allow for the concentration of capital-intensive, high-external input, monocrop agriculture on prime agricultural land, Central America's peasants had been driven onto hillsides and into the reaches of the agricultural frontier. Later, they were extended credit and technology packages in an effort to 'modernize' the sector.

This massive displacement brought with it the depletion of the tropical forest cover as poor farmers cleared land for crops using traditional slash and burn, bush-fallow techniques. Because newly-available fertilizers temporarily maintained nutrient levels, fallows were virtually abandoned. With the high mineralization rate of soil organic matter in the tropics, lush vegetation is essential for maintaining soil cover and replenishing tropical nutrient cycles. Once these cycles were interrupted, soils, which in the tropics are extremely fragile, ecologically, were destroyed.

Loss of forests meant loss of habitat for predators and the ensuing ecological imbalance unleashed recurrent explosions of pest populations and plant diseases and choked the fields with weeds, resulting in higher and higher application of fertilizers, pesticides and herbicides, provoking a vicious cycle of falling productivity and ecological degradation through Latin America. Application rates rose almost as fast as yields dropped. Long before the regional economic recession dried up agricultural credit in the 1990s, for most campesinos, production costs had already dwarfed profits, making agricultural chemicals too risky and expensive to use. They cut back to subsistence production to keep from falling into debt.

The aftermath of the Green Revolution, along with the upheaval of the Lost Decade, has thrown Central America into a chronic state of economic crisis and ecological degradation. Half of the population now lives in abject poverty. Deforestation advances at over three percent a year, now the highest rate in the hemisphere. Countries like Nicaragua, which bore the brunt of the region's wars and natural disasters, lost over forty years of economic development in less than ten years.

In the 1990s, when the devastation of Central America's soils and tropical forests passed the threshold levels for even subsistence production, major research institutions had to admit two critically important facts:

1 Although there had been massive adoption of Green Revolution 'technology packets' over twenty years, the increases in overall grain production were not due to adoption but to the expansion of grain cultivated areas by poor farmers (campesinos).[1]

2 The Green Revolution's research and extension programs held few answers to the agroecological crisis of deforestation, degraded soils, diminishing yields, and negative real economic returns to Central America's poor majority.[2]

Not until the Central America countryside was driven to the brink of economic and ecological disaster, however, did low-input, sustainable agriculture with campesinos become accepted as a legitimate strategy for rural development programs.[3]

Once a minor budget item of integrated rural development programs, sustainable agriculture was quickly promoted to major program status as development agencies attempted to deal with the crisis. Sustainable agriculture technology 'packages' were developed and the rich body of information available through Farm systems research was revisited in an effort to retrain extensionists and reconvince skeptical farmers of the next agricultural 'miracle' being offered to overcome the crisis produced by the first one.

But precisely when they are most needed, 'privatization' of government agencies resulting from International Monetary Fund-imposed structural adjustment programs has led to the virtual disappearance of government agricultural services, leaving the region's five million campesinos to fend for themselves.

It is true that non-governmental agricultural development programs—financed by foreign agencies—were implemented en masse in the 1980s as governments sought to palliate the crisis. This financing spawned a plethora of non-governmental organizations (NGOs) and local development agencies (LDOs), whose job it was to mitigate the negative after-effects of uneven growth among the hardest-hit: the campesinos.

But NGO and LDO efforts have been hampered by the pre-existent development paradigms established as early as the 1960s by the politically-motivated 'Alliance for Progress' and the capital-driven Green Revolution. Apart from equating development with technology transfer from North to South, the general view was that campesinos were an obstacle to development. The transformation—or the elimination—of campesinos in favor of capital-intensive producers was necessary in order to "get on with the business of development."

But even though it is now acceptable program fare for developing agencies and research institutions, 'adapting' sustainable agriculture to existing agricultural extension systems has provided few success stories on the ground. Without the lure of credit that made the Green Revolution's seed and fertilizer packages so attractive, selling sustainable agriculture to farmers largely depends on the ability of the extensionist to convince farmers of medium-to long-range benefits that the extensionist cannot guarantee.

If sustainable agriculture is hard to sell, it is even harder to teach. Few extensionists in Central America have ecological training and even fewer have any real farm experience. Unlike the quick fix offered by the *paquete tecnológico*, sustainable agriculture is more a process—or an art—than a recipe. The simple 'credit cycle' farm visits that characterized extensionist-to-farmer relationships during the last thirty years are hopelessly inadequate to the complex task of helping farmers transform their degraded agroecosystems into sustainable farms.[4] Further, while sustainable agriculture benefits from over fifty years of research, documentation and legitimate adoption of alternative and organic techniques in northern, developed countries, tropical organic and/or sustainable agriculture in the Third World has gone largely unnoticed. With a few exceptions, transfer of northern/developed sustainable techniques to tropical/underdeveloped regions attempts have been difficult, often inappropriate and more often than not, unsuccessful.

Luckily, these difficulties do not mean that sustainable agriculture in Central America is impossible or that it is nonexistent. The technology packages offered by the Green Revolution were frequently altered during the process of adoption, often because poor farmers could not afford the high levels of required chemical inputs. Further, as grain prices and chemically fertilized yields dropped, soil eroded and pest outbreaks became harder to control with chemicals, farmers began to look for ways to cut costs and preserve natural fertility. In fact, aside from many sustainable traditional practices, over the last twenty years campesinos and a handful of professionals—largely financed by NGOs—have been patiently and empirically developing locally based, sustainable agriculture systems "from the ground up." When credit all but dried up for poor farmers in the early 1990s, green manures, composts, and natural pesticides and herbicides were further developed and/or adapted in an effort to make ends meet. There now exists a significant body of knowledge and practice increasingly recognized as economically, socially and ecologically appropriate.

CAC: Innovation, Solidarity, and a Little Help

The tremendous political, economic, and social upheavals of the eighties not only redistributed millions of acres of farmland, they also uprooted hundreds of thousands of campesinos, exposing both land recipients and refugees to different agroecological systems. *But just as importantly, the upheavals also thrust the region's campesinos into contact with each other.* The result was both a tremendous exchange of cultural and agroeconomic information and the development of a widespread, informal 'system' of mass communication for a traditionally isolated, provincial sector.

Agroecological crisis and the social explosions of the eighties may have been the catalyst for what up to that time had been a locally-based farmer-to-farmer initiative once limited to a few disperse NGO projects in Guatemala and Mexico: Campesino a Campesino.

Supported almost entirely by local initiative and non-governmental organizations, the Campesino a Campesino is based on farmer innovation and solidarity. Trans-institutional, decentralized, horizontally and informally organized, Campesino a Campesino functions largely as a local, national, and regional cultural exchange for technical and methodological innovations developed by individual farmers and groups of farmer promotores both with and without professional assistance. In this sense, it is more than just a program or a method: its persistent, changing and amorphous nature suggests that it might best be understood as a broad-based, farmer-led movement for sustainable agriculture.

The rapid spread of Campesino a Campesino is not simply conjunctural, nor is it the result of a sophisticated combination of methods. In essence, it has built upon its own successes—success in raising production and protecting the natural resources of campesino agroecosystems, and success in empowering campesinos.

Not without its critics—particularly post Green Revolution extensionists—nor without its failures, Campesino a Campesino is a distinctly Latin American phenomenon which merits a closer look. The lessons learned may well pertain to sustainable agriculture far beyond the borders of Latin America.

A Partial History of the Campesino a Campesino Movement

POPULAR EDUCATION AND ALTERNATIVE AGRICULTURE

The development of 'Popular Education' in the 1970s and 1980s in Latin America coincided with the wave of social unrest sweeping the region. Horizontal communication between 'learner and teacher,' combined with 'reflection-action-reflection' methodologies for social transformation and local empowerment, effectively put teaching into the hands of local people and produced a highly motivated, broad-based movement of social activists. In Central America, especially, thousands of promotores sprang up to teach literacy, health, and community organization.

Agriculture, however, was never the forte of Popular Education. The educated professionals involved with Central America's political and social movements tended to have studied political science, sociology, or law. They were virtually bereft of technical or agricultural knowledge. Agronomists from the technical schools tended to work for large plantations, and fertilizer, machinery, and seed companies. Those in the ministries of agriculture were primarily involved with distributing credit and fertilizers, and did very little teaching.

The gap between Popular Education's horizontal and deductive learning techniques and agricultural extension was bridged by sustainable agriculture. As early as the late seventies, appropriate technology advocates found some receptiveness for their alternative approaches to agriculture among campesinos not benefiting from the Green Revolution. Campesinos in particularly dry, mountainous, or frontier areas were somewhat receptive, basically because they had no other alternatives. Adoption, even among this forgotten sector, was slow, in part because many poor farmers simply could not risk change or were unwilling to invest their labor in something untried. Further, many of the technologies extended by alternative agriculture advocates had been developed in northern climates in more technified agricultural systems and simply didn't fit the agroecosystems of poor campesinos in the tropics.

In general, until the 1990s most campesinos had relatively easy access to chemical inputs. The degradation of the fragile tropical soils under the new Green Revolution practices was not readily apparent to farmers, who could still compensate for the drop in soil fertility and pest resistance by adding more chemicals. Sustainable agriculture, aside from being isolated to small, village-level projects was not particularly attractive. Also, compared to the virtual army of credit and chemical-toting extensionists employed by the Green Revolution in state agricultural ministries and local seed and fertilizer companies, the handful of organic and alternative agriculture advocates found it quite impossible to diffuse their extended technologies to the remote agroecosystems farmed by Mesoamerica's campesinos.

It took a small NGO program and a small group of Cachikel Mayan Indians to hit upon the combination of small-scale campesino experimentation, mutual assistance groups, and farmer exchanges to give sustainable agriculture an extension methodology appropriate for campesino farming systems.

TWO EARS OF CORN: THE GUATEMALAN CASE

Don Marcos Orozco, a retired soil conservationist employed by World Neighbors, had a problem. He had successfully demonstrated the effectiveness of contour ditches and organic soil amendments on his small, backyard plot. His corn plants were visibly higher and the ears bigger than the control he left alongside using the conventional methods of the Mayan farmers whose small farms dotted the hillsides. An extensionist with over forty years experience, he could easily have explained the new techniques to the surrounding farmers except for one thing: he spoke no Cachikel and the Mayans spoke little or no Spanish. Ironically, this was the beginning of a wonderful relationship—a new one for 'técnicos' and campesinos.

Soil erosion and low fertility were major factors on their steep, eroded hillsides. Yields were low and parcels small. Most farmers had difficulty paying back credit for hybrid seeds and fertilizers and were forced to work off their debts as laborers. The vicious credit-debt cycle was held in place by company loan sharks and extension agents working for the plantations and the fertilizer companies.

Following the example of promoter-led health services pioneered by the Bearhorse Clinic in Guatemala, World Neighbors encouraged Don Marcos to work along with a few Spanish-speaking, Cachikel campesinos to reduce erosion, improve natural fertility, diversify production, and improve local seed. Don Marcos helped the small group to design experiments on their own land. When their own experiments demonstrated the value of the alternatives, the farmers applied them to their entire fields. Because Don Marcos could not speak the local Mayan language the only way to extend the knowledge effectively was through those campesinos working closely with him. The Mayan farmers discovered that the best way to convince their neighbors was by demonstrating the techniques incorporated in their own plot and helping interested farmers test the alternatives through small-scale experimentation.

Because soil and water tended to be the major limiting factors in Cachikel production in the highlands of Chimaltenango, contour ditches, bunds, terraces, and heavy applications of compost tended to provide quick and visible results. The heavy labor requirements were assimilated by the Cachikel 'kuchubal' or mutual assistance groups traditionally used by the Mayans. Groups of three to ten men took turns working as a group on each others' land until all had implemented the basic conservation and fertility practices.

Since yields were extremely low to begin with (less than one ton per hectare), and the limiting factors were basically soil and water, increases of one hundred to two hundred percent within a year or two of implementing the conservation practices were not uncommon.[5] These increases resulted in enthusiasm on the part of the participating farmers and widespread interest on the part of neighbors. Soon, the knowledge and skills of the small group working with Don Marcos were in high demand.

Demands upon the small group of innovators soon outstripped their capacity to respond through the traditional Cachikel systems for mutual assistance. As small-scale experimentation reached higher and higher levels and limiting factors became more complex, the knowledge and skills employed also become more complex. The need for more theory as well as practical ability required understanding of abstract agronomic concepts. The transfer of technology from farmer to farmer began to require more time, effort and different techniques for understanding and sharing what was essentially a locally developed—if embryonic—sustainable agriculture.

World Neighbors and OXFAM-UK must be credited as the first development organizations to recognize the potential of the Chimaltenango experience. By employing technicians to support local farmers in the development of their own agriculture—rather than simple technology transfer—they set the guidelines for a new farmer-to-technician relationship that was later to characterize the Campesino a Campesino movement. These institutions supported the local efforts by helping to sponsor farmer to farmer workshops, cross-visits, technology 'fairs' and provide further organizational and technical training and seed money for agricultural and commercial cooperatives.

Eventually, this farmer-led development program was so successful that the Cachikels were able to establish 'Kato-Ki,' a 900-member cooperative, which bought supplies, sold the farmers' basic grains, provided farmer-to-farmer training in basic conservation and fertility techniques and even established a farmer-run land bank.

The theoretical and methodological lessons from Chimaltenango are articulately described in Roland Bunch's *Two Ears of Corn*, published by World Neighbors. *Two Ears of Corn* became a programmatic classic for locally-based alternative agriculture projects from as early as the late seventies. To this day it is regarded as the primary source for Campesino a Campesino methodology, even though at the time of its writing, the Movement as such was yet to be recognized.[6]

The secret of Chimaltenango's success was, of course, success, though this was also the cause of its eventual destruction. As Cachikel campesinos gained more and more economic autonomy, they began to arouse the ire of the local landowners. Sustainable agriculture and cooperatives meant higher yields and higher incomes for poor campesino families, and subsequently fewer agricultural laborers willing to work at starvation wages on the plantations. Cooperatives meant local organization and economic autonomy from the traditional system of debt peonage. But the crowning blow to plantation owners came when Kato-Ki began to buy up eroded coffee plantations and redistribute them among their members who, by implementing sustainable agriculture, not only grew their traditional basic grains, but began to compete by selling coffee through their cooperative outlet. Suddenly the coffee oligarchy realized that the Cachikels were, in effect, organizing, financing, and providing technical assistance for their own, locally based land reform. Action was swift and brutal. According to its members, the Kato-Ki and its campesino extensionists were designated communists and the Guatemalan army was called in to 'disappear' the cooperative. Those campesinos who could, fled, and the cooperative and much of the land was left abandoned. After a decade of patient training, painstaking organization, and much backbreaking work, the Chimaltenango experience appeared aborted.[7]

The political events of the early 1980s in Guatemala led to the unfortunate disbanding of the cooperative, but not before they had trained visiting campesinos from Mexico, Honduras, and other areas of Guatemala. The basic knowledge and principles for farmer-led development passed, orally and experientially from farmer-to-farmer, (campesino a campesino) where they continued to grow and spread with the help of local development organizations (LDOs).

Cachikels in Exile: From Refugees to Campesino Consultants

The handful of Cachikel campesinos who left the Kato-Ki cooperative in the early 1980s were instrumental in setting up dozens of farmer-led agricultural development projects in Mexico and Honduras. Far from their homes and cultures, they learned Spanish, became familiar with a wide range of humid-tropical and semi-humid tropical agroecosystems and integrated themselves into the social and cultural milieu of the Mesoamerican campesino.

Keeping a low political and organizational profile, they concentrated on promoting local campesino innovations to overcome the limiting factors in production. Further, they looked for ways to stabilize the diverse and fragile campesino agroecosystems on the Pacific plains, central mountains, and wet Atlantic basins. Because they owned no land, and had no other means of support, they were paid modestly by World Neighbors as full-time 'extensionistas.'

As news of the Cachikels work spread, this time the demand for their professional services grew. The campesino extensionistas split up, hired out to other NGOs, coordinated rural projects, and directed sustainable agriculture programs. One man even started his own school. Then several of the Cachikels formed an agricultural consulting firm with Roland Bunch, the Association of Consultants for a Sustainable Ecological, People-Centered Agriculture (COSECHA). COSECHA offered international consulting services worldwide, and soon the Cachikels were traveling to South America, Africa, and Asia. They attended to clients, participated in regional and international seminars, presented their work at symposia, and helped organize forums. They began to carry out research. They published.

Their mobility and full-time working status allowed them to vastly improve their theoretical and working knowledge of agroecosystems. Their superior agronomic skills became renowned. Inspiring other campesinos throughout the isthmus to develop sustainable agriculture on their own distinguished them from the great majority of professional agronomists working in region, causing conflicts with many professionals, who felt threatened by the Cachikels' success. Many argued that these extensionistas and their villager-led model for development were not only taking work away from other professional agronomists, they were also eliminating the very need for agronomists by training local villagers to fend for themselves. But both the lack of real conflict with other professionals and the key to their effectiveness at motivating villagers were best stated by one extensionista who wrote:

> We do not see village extensionists as some sort of second-class agronomists—people with the same role as agronomists, but with less education. They are not substitute agronomists—sometimes preferred because they cost less. Nor should the objective of their training be that of teaching them everything about agriculture.
>
> The role of the village agricultural extensionist, like that of the village health workers before them, is not of having a general, over-all knowledge of their particular subject area—of becoming village doctors or agronomists. Their role is, rather, *to learn the solutions to the most common problems in the village…* (emphasis added)[8]

Campesino Principles for Sustainable Agriculture in the Tropics

Countless observations and direct, hands-on experience discovering local solutions to local problems in the diverse agroecosystems throughout the world brought the Cachikels to the conclusion that common limiting factors and critical ecological points in tropical, campesino agroecosystems had common solutions. They developed a set of principles for sustainable agriculture in the humid tropics:

1 The vast majority of soils in the tropics, though often poor, can be made highly fertile by maximizing the production of organic matter.

2 Migratory agriculture is frequently motivated by decreasing fertility, increased weed problems, or both. Mulches of crop residues and fast-growing green manure/cover crops drastically reduce the weed problem by keeping the soil covered, also controlling erosion and solation.

3 Shallow soils and/or steep slopes that contribute to erosion in the tropics can also be significantly controlled by zero tillage. The secret to zero tillage in the tropics is the application of massive amounts of organic matter to the soil—in this case through cover crop/green manures.

4 The prevalence of plant diseases and insect pests in tropical agroecosystems is managed by maintaining biological diversity. There are over sixty green manure/cover crop species in use by campesinos developing sustainable agriculture today.

5 Plant growth can overcome the 'hostile'—acidic, low fertility—soils of the tropics when plants are fed through mulch.

Studies in northern Honduras indicated that agroecosystems of maize which applied these principles were thirty percent more profitable than the high-input maize systems nearby.

And so, after over thirty years of developing and sharing the campesino innovations, campesinos themselves have hit upon an astounding concept:

> In order for tropical agriculture to be both highly productive and sustainable, it must imitate the highly productive, millions-of-years-old, humid tropical forest.[9]

It is highly notable that campesinos, when given the support, were not only able to develop appropriate techniques, they were able to extract the underlying principles for sustainable agriculture in the tropics from their own empirical experience. This is no different from what campesinos have done—slowly—since the beginning of agriculture. What is highly significant is the time frame: less than thirty years.

Lessons Learned

What can Campesino a Campesino teach us that might be useful in promoting sustainable agriculture?

It seems highly important to note that campesinos have succeeded where agricultural science has failed: by developing basic agroecological and methodological tenets for the development of sustainable agriculture in the tropics. This is no easy point for many to admit, but it is critical in establishing the importance of campesinos culture in the future paradigms for sustainable development. Without returning to any romanticized notion of the peasantry or idyllic village life, it is essential that scientists and development agencies recognize that campesinos are not just 'clients' or 'target groups.' Nor are they simply a disadvantaged social class in need of resources to "pull themselves up by their bootstraps," thereby solving the "campesino problem." In fact, modern day campesino culture holds the crucial keys for the development of sustainable agriculture. Campesinos—the oldest social class on the planet—are not a problem to be solved. The world needs campesinos in order to solve the present agroecological crisis affecting all of us. Campesinos, in this sense are a socially desirable sector and a legitimate partner in development.

The task then becomes not how do we change campesinos into capital-intensive farmers or Third World yeomen in order to develop agriculture, but rather: how can we help make campesino culture stronger and more dynamic?

Short of suggesting that every country or region have a farmer-to-farmer movement, the first major orientation might be to insure that farmers and campesinos be genuinely involved in the entire process of agricultural development, from analysis of the problems to generation of alternatives, including their validation, diffusion, evaluation and back again.

There are no quick fixes. The bottom line is: sharing the development process with farmers, means sharing power with them: power to set experimental agendas, use extension resources, determine diffusion strategies, and plan and execute production and conservation plans. The question then is not whether farmers are capable of developing sustainable agriculture, but are the rest of us capable of letting them?

If we accept the notion that today's campesinos have more social, scientific and intellectual resources at their disposal than they did a generation ago (illiteracy notwithstanding), and if we accept the legitimacy of farmers' endogenous development processes, then we may be able to honestly conceive of participating in a partnership rather than a patronship for agricultural development.

Participation, in this context means our participation in their process. Can we help? Can we open lines of communication? Clear up conceptual voids? Provide methods for discovery, generation and sharing of knowledge? Can we support farmer organization for developing and sharing innovations? Can we stand back and allow campesinos to become the protagonists in agricultural development?

If we *are* interested in a partnership with the farmers, what does this mean, institutionally, programmatically, or politically? The Campesino a Campesino experience suggests the following guidelines may be helpful:

1 Promote small-scale experimentation and farmer to farmer training and exchanges through local development and farmers' organizations.

2 Provide local, regional, and national opportunities for farmer gatherings, symposia and conferences.

3 Actively promote the development of sustainable, low-external input agriculture.

4 Provide direct and indirect technical and financial support for groups of farmer-promotores.

5 Train and orient professionals technically and methodologically in their new support roles.

6 Provide formal and informal opportunities for farmers and promoters to share their experiences, their doubts and express their demands to formal research and extension facilities.

7 Document and make available videos, radio spots, and pictorial articles based on direct farmer experiences and testimony in sustainable agriculture.

Clearly, some of us are already capable of supporting farmer-led development of sustainable agriculture. It has implied fundamental changes in our development paradigms. But as agricultural development turns toward sustainable agriculture for solutions to the agroecological crisis, and as peasants continue to exert their influence in the countryside, one wonders what will happen to those institutions not capable of making the change. The sustainability of agriculture itself may depend not on whether the state ministries, the agricultural universities, the technical training schools and the agricultural centers can convince farmers to adopt technologies, but rather, whether these institutions can change their traditional role of provider to that of facilitator. This will be no easy task, and institutions will probably resort to a myriad of new 'participatory' approaches before they concede any real power sharing with their former 'clients.' But then, 'wresting' subsidized power from these institutions may not be the best strategy anyway. Rather, it may be more important to do just what promotores have been doing: developing sustainable alternatives from the ground up. Campesinos have proven it can be one. We hope the rest of the development community catches on.

1 Dr. Jorge Bolanos, Center for Research on the Improvement of Corn and Wheat, CIMMYT, Mexico. Presentation to PCCMA on "Generation and Transfer," Managua, Nicaragua, 1992.

2 Dr. David Kaimowitz, IICA-Interamerican Institute of Central American Research, San Jose, Costa Rica. Presentation to FUNDESCA-EC, Program for the Agricultural Frontier, Panama, May 1994.

3 Prior to this time, the Green Revolution 'took care' of agricultural development while rural development programs 'helped the poor.' Problems arose when rural development programs could no longer attend to the poor as fast as the Green Revolution could produce them.

4 Even the concept of sustainability is onerous. Some project extensionists maintain that 'sustainability' means that the project rather than the agriculture itself, will continue on for generations.

5 Bunch, Roland and Gabino V. Lopez. "Soil Recuperation in Central America: Measuring the Impact Three to Forty Years After Intervention," paper presented at the International Institute for Environment and Development's International Policy Workshop in Bangalore, India, November-December 1994.

6 Bunch, Roland. *Two Ears of Corn: A Guide to People-Centered Agricultural Improvement* (Oklahoma City, OK: World Neighbors, 1982).

7 Author interviews with Guatemalan promotores from Chimaltenango, 1987, Mexico, 1994, and Honduras, 1995. Names of promotores withheld on request.

8 Lopez, Gabino V. "The Village Extensionist in Developing Nations," paper presented at the International Workshop on Farmer-Led Approaches to Agricultural Extension, July 17–22, 1995, IIRR, ODI, World Neighbors, Silang, Cavite, Philippines.

9 Bunch, Roland. "An Odyssey of Discovery: Principles of Agriculture for the Humid Tropics," COSECHA, Tegucigalpa, Honduras, 1995.

From *Food First Backgrounder*, Spring 1996 by Christopher D. Cook and John Rodgers

---

Community Food Security: A Growing Movement

Agribusiness produce typically travels over one thousand miles before it reaches the consumer, yet this food has a far easier time reaching the supermarket than do low-income residents.[1] Due to the food industry consolidation and supermarket relining of low-income areas, among other developments, supermarket chains have become a meeting point for agriculture conglomerates and middle class consumers—leaving small farmers and rural and urban low-income groups hungry and out in the cold.

Leaders of an emergent national grassroots movement are turning this common plight into common ground, challenging a U.S. food system which they say lacks two vital ingredients—community and security. Forging local solutions to make food more healthful, equitable, and sustainable, the Community Food Security Coalition has captured attention throughout the country and support in several areas including Congress.

Stressing local-based economic development, ecological restoration, hunger prevention, and sustainable agriculture, community food security is an innovative approach with far-reaching potential. Urban subsistence and market farming is taking hold internationally, as poor communities attempt to redevelop land for both economic and dietary nourishment.[2] Environmental sustainability and economic opportunity are two ideas that community groups are literally fertilizing with long-neglected urban soils.

Reviving Communities

One of the poorest sections of San Francisco is home to one particularly promising effort to redefine urban ecology and empowerment. On any given day, the office of the San Francisco League of Urban Farmers (SLUG) bustles with energy and optimism. Teenage gardening interns drop by to pick up paychecks and brag about their vegetable crops, or to encourage the group's executive director, Mohammed ("Mo") Nuru, to give them more farming work. Once strictly a gardening group, SLUG now promotes, with help from city government and foundation grants, small-scale urban agriculture projects as a means to help develop economically depressed communities. SLUG's aim, Nuru explains, is not only to give summer jobs to kids or provide inexpensive produce to the poor, but to restore vitality to the community by directly involving its members in its welfare. "It's a whole cycle we're addressing," Nuru says, "not just one issue."

The broader struggle for community food security could be described in much the same way. Merging food sufficiency and sustainability with grassroots economic development, activists from New York to Austin, Texas, to Los Angeles[3] are forging an alternative U.S. food chain of urban farms, farmers' markets, and joint urban-rural agriculture ventures—making good nutritious food more economically viable in poor communities.

With rows of organic vegetables, flowers, and a greenhouse, SLUG's four-acre Alemany Youth Farm is a meeting ground for teen jobs, community pride, and fresh, nutritious food. In newly landscaped backyards, Alemany residents tend six-foot-square planter boxes filled with tomatoes, collard greens, red chard, and string beans—produce often unavailable in low-income areas. Instead of sacrificing their tight food stamp budgets to high-priced convenience stores or busing across town in search of a supermarket, Alemany residents are gaining direct access to affordable, nutritious food by growing their own.

"We're looking at really reviving communities," says Nuru. "There needs to be a local base." The local base Nuru envisions rests, in part, on his childhood memories of growing up in Nigeria, where "everybody farmed," and where food security was a strong measure of a community's vitality.

While projects such as the Alemany Farm encourage people to grow their own nutritious food, other grassroots groups—sustainable agriculture advocates, urban redevelopment organizers, and environmental activists—are working to open up consumer markets in the U.S. for small-scale, diversified agriculture. These groups argue that the extreme competitive pressure exerted by multinational agri-food corporations has made it difficult to preserve community-oriented and ecologically conscious agriculture.[4]

"Our food system—from production through consumption—has been experiencing major restructuring in the last fifteen years," says Kathleen Fitzgerald, director of the Sustainable Food Center in Austin, Texas. "And small farmers and low-income consumers have been on the losing end of these developments."

Fitzgerald and other activists describe a chaotic and vastly inequitable U.S. food system where small farmers battle for dwindling market opportunities, and many low-income consumers face prodigious barriers to obtaining healthy food. Twelve million children go hungry, reports Fitzgerald, while some 500,000 farms and ranches have gone under in the past fifteen years—a rate of eighty-seven per day.

Under the banner of community food security—the notion that all people should have access to a nutritious diet from ecologically sound, local, non-emergency sources—these groups have joined together to promote what co-founder Andy Fisher calls "a more democratic food system." Linking 125 groups—food banks, family farm networks, anti-poverty organizations, and others—the coalition infuses grassroots projects with a structural approach aimed at building alternative systems of food production and consumption.[5]

Emergence of a National Network

The concept of a food security coalition emerged almost simultaneously on the East and West Coasts in the early 1990s, as researchers and activists began creating new working relationships between anti-hunger organizations and sustainable-agriculture groups. Mark Winne, director of Hartford Food System, a community gardening and food policy organization in Hartford, Connecticut, was building a national network of local food security projects that shared resources and ideas. Meanwhile, UCLA urban planning professor Robert Gottlieb and researcher Andy Fisher were documenting the shortcomings of U.S. food policy—as well as the disjointed efforts by sustainable-agriculture and anti-hunger groups to change it.

In a policy paper that would later become a cornerstone of the movement, Fisher called for comprehensive reform of traditional policies to alleviate hunger. "Food security represents a community need, rather an individual's plight," Fisher wrote. "It is an essential building block in enabling people to better control the conditions of their lives." Reorienting the U.S. food economy towards communities, Fisher insisted, begins with a systemic analysis incorporating issues such as food industry consolidation, supermarket redlining, and excessive reliance on long-distance transportation of processed commodity crops. Fisher and Gottlieb identified the need to restructure federal policy incentives away from market and production concentration; create new sources of funding for innovative projects at the grassroots level; and improve the coordination of the myriad federal agencies involved in agricultural production and food distribution policy.[6]

The coalition draws on several movements with strong, wide-ranging constituencies. Prominent among these is the sustainable agriculture movement, which has been fighting to protect small family farmers while promoting environmentally sound farming techniques. The community food security concept has helped inject hunger issues into the sustainable agriculture movement, redirecting attention to programs that benefit lower income people. The California Sustainable Agriculture Working Group (CSAWG) has played a major role in developing and implementing community food security plans. Among their collaborative projects, CSAWG and the California Alliance for Family Farms are working to create direct marketing relationships between small-scale diversified farmers and local schools and food banks.[7]

Another key player in the coalition, the community development movement, has focused on expanding housing and jobs and ensuring more accessible and empowering social services for the poor. Community development programs add to their arsenal of approaches by creatively pursuing combinations of job-development, community-based food production, marketing, and other initiatives to bring the fruits of sustainable agriculture to the doorsteps of the inner city.[8]

The U.S. community food security movement comes on the heels of similar efforts in developing countries, especially those in the South, where community-based urban agriculture is one response to burgeoning unemployment, ecological decay, and malnutrition. "Globally, about 200 million urban dwellers are now urban farmers, providing food and income to about 700 million people," reports the Ottawa, Ontario-based International Development Research Center (IDRC). "In some Latin American centers a third of the vegetable demand is met by urban production," the IRDC reported in a 1993 publication.[9] In recent years, urban gardening in Cuba has been particularly successful in meeting the food crisis brought on by the collapse of trade relations with the socialist bloc.[10]

With Third World conditions of poverty, unemployment, and malnourishment mounting in U.S. cities, Andy Fisher's proposals for long-term, community-driven food security programs caught on. "It was an issue that was ripe to happen," says Gottlieb. Thirty organizations representing groups as disparate as small farmers and food bank operators, policy researchers and environmentalists, attended a 1994 conference in Chicago cosponsored by Hartford Food System and UCLA's Pollution Prevention Education and Research Center. During discussions of local pilot efforts, Winne says, "there was a fairly spontaneous recognition of mutual interest."

In a time when many progressive groups are retrenching, food security organizers are literally growing a movement from the grassroots. Directly challenging a food system dominated by agribusiness conglomerates and their powerful lobbies, the Community Food Security Coalition is trying to wrest promising, if small-scale, legislation from a conservative U.S. Congress. The Community Food Security Act (CFSA), a coalition-authored bill, would fund local projects "designed to meet the food needs of low-income people, increase the self-reliance of communities in providing for their own food needs, and promote comprehensive, inclusive, and future-oriented solutions to local food, farm, and nutrition problems."[11]

Grassroots collaborative programs would receive one-time seed grants to build self-sustaining economic relationships between farmers, food stores, and low-income consumers. While a version in the House of Representatives requests $2.5 million, a Senate measure calls for $4 million—a meager allowance—but organizers point out that the bill strengthens alternative concepts and institutions to address the structural economic and nutritional problems faced by people who are locked out of the system. "We're already investing time and money in making these things happen. If you're talking about putting just a little more money to give it that extra push, I think it could go a long way. If your mission is to ensure food and fiber for the nation," says Winne, "why not look at communities as the basis for change?"

Similar government initiatives—better funded but narrower in scope—have already proved successful. In 1992, after decades of state level experimentation, the U.S. Department of Agriculture authorized the Farmers' Market Nutrition Program (FMNP) to supply fresh produce vouchers to recipients of Women, Infants, and Children (WIC) assistance. In three years, the program has issued vouchers to some 800,000 WIC recipients in twenty-four states, the District of Columbia, and the Cherokee Nation. These vouchers have been redeemed at hundreds of farmers' markets that participate in the program. Although impressive, these aggregate numbers belie the minimal funding committed to the FMNP: each WIC participant gets just twenty dollars a year in farmers' market coupons as a supplement to food stamps.[12]

Nevertheless, according to Kai Siedenberg, coordinator of the California Sustainable Agriculture Working Group, eighty-seven percent of 6,615 farmers nationwide who have participated in the FMNP have increased their overall sales. And for many WIC recipients, the program is their first entree into the fresh produce market. Tom Haller, head of the California Alliance for Family Farms, a group supporting sustainable agriculture and small farmers, says a survey by the alliance found that more than half of the WIC recipients involved in the FMNP had never been to a farmers' market. "Most WIC recipients didn't know about it or didn't think it was for them," says Haller.

Since entering the program, Haller says, participants have doubled their intake of fresh fruits and vegetables.

More qualitative surveys show the program has been successful for both farmers and WIC participants. "When the farmers see a woman with children coming to buy produce, it looks like a little bit different than just anybody coming in to use food stamps," says Haller. "They feel that the people getting the food are the ones that really need it." One woman using the WIC program responded, "My applause one hundred times for this magnificent program... It was great being able to buy produce with no pesticides and no preservatives."[13]

Just as the WIC farmers' market program has begun to take root, however, congressional Republicans are threatening to cut it. In fact, while the CFSA appears headed for approval, other federal food-access and nutrition programs are under attack. Republicans are pushing hard to cut as much as $200 million from the WIC nutrition program.[14]

With welfare entitlements under attack, food security activists and others are working to fill the void left by a fraying federal safety net and over-stretched public and private funds. Most government food aid programs, like the traditional stopgap homeless shelter approach, were never designed as a long-term solution to poverty and hunger, but rather as a plan of last resort. "We've gone a little bit too far down the line in relying on federal food assistance systems and emergency food networks," says Hartford Food System's Winne. "They don't provide a real opportunity for people to participate in solving their own problems."

New Links, New Twists

As promising alternatives, food security programs across the country are building new links in local food networks, creating partnerships among small rural and urban farmers, food banks, soup kitchens, and low-income communities. Many are giving a new twist to some old models such as Community Supported Agriculture (CSA), arrangements in which communities fund local farmers in exchange for a season's worth of crops. In Los Angeles, the Southland Farmers' Market and UCLA began collaborating late 1995 on a CSA project that delivers fresh farm produce weekly to low-income neighborhoods.[15]

In the heart of Eastside, the poorest neighborhood in Austin, Texas, organizers have set up a community garden and farmers' market in a previously unused vacant lot. More than forty percent of Eastside's families live below the poverty line, hunger and nutrition problems are widespread, and the supermarkets are few.[16] The community garden serves as a hub for nutrition education for families and grows enough produce to provide each resident with one vegetable or piece of fruit every day of the year. Every weekend the Eastside Community Farmers' Market sets up shop alongside the neighborhood garden, featuring more than twenty local farmers, many of them certified organic growers.[17]

Other food security efforts focus on creating direct marketing relationships between small farmers are larger food service organizations. A pilot effort in Oakland, California has sought to bring farm produce to food banks and residential rehabilitation centers by brokering produce sales from local farmers to food service managers. The idea, says Leslie Mikkelsen, a nutritionist heading the project, is to create new markets for small farmers and to improve nutrition. The approach works "if you can get some big programs on board and can bring the prices down without shafting the farmers."[18]

The difficulty, says Mikkelsen, is that small farmers can't provide the variety and volume of produce that these centers have come to expect year-round. Large-scale food servers generally seek a premium of convenience at the lowest possible cost, and typically follow menus of canned and frozen food (and very little fresh produce) provided by major distributors, such as Allied Sysco. "The fact that one farmer does not grow everything that people need is going to be a real barrier," Mikkelsen says.

In Austin and other cities, activists are also promoting municipal food policy councils, and linking food security and access to reliable transportation. A special food shopping bus route in Austin provides low-income consumers from Eastside with transportation to markets offering cheaper, better produce than can be found in neighborhood convenience stores.[19] Coalition co-founder Andy Fisher cites a nascent "groundswell in municipalities to look at food policy issues. They deal with water, utilities, and housing, but they don't deal with food as a municipal issue."

The food security movement has set its sights beyond the five-year lifespan of the CFSA. In the process of campaigning for the bill, organizers have established national information-sharing networks and reinvigorated efforts to strengthen food system planning. Communities long isolated from the U.S. food economy are beginning to procure their own food and fiber.

It is also evident that grassroots groups by themselves lack the time and resources to challenge the daunting inequities of the national food system and international forces of industry consolidation. Movement organizers do not expect their coalition, nor any congressional legislation, to be a panacea for such mammoth trends. But they are redefining progressive food aid approaches, broadening and diversifying policy debates, and creating a dynamic coalition with far-reaching potential—providing what Robert Gottlieb calls "a language that puts together environmental protection, economic development, and the needs of the poor."

1 Campaign for Sustainable Agriculture's Greg Watson, quoted in *The Neighborhood Works*, April/May 1995.

2 International Research Center's October 1993 report documented and profiled urban agriculture trends and projects in Asia, Africa, and Latin America.

3 *Food Security News*, published by the Community Food Security Coalition.

4 Interviews with Kathleen Fitzgerald (Sustainable Food Center, Austin, Texas), Andy Fisher (Community Food Security Coalition co-founder), Robert Gottlieb (UCLA), and Kai Siedenberg (California Sustainable Agriculture Working Group).

5 Kathleen Fitzgerald's testimony before U.S. House of Representatives, Committee on Agriculture, June 8, 1995.

6 "Community Food Security: A Food Systems Approach to the 1995 Farm Bill and Beyond," policy paper by Andy Fisher, presented to Working Meeting on Community Food Security, August 25, 1994; "The Community Food Security Empowerment Act," the Community Food Security Coaltion's original legislative proposal, which later became the Community Food Security Act.

7 California Sustainable Agriculture Working Group materials; interview with Kai Siendenberg, CSAWG coordinator.

8 Interviews with Andy Fisher and the San Francisco League of Urban Gardeners staff, and numerous press reports. Examples of entrepreneurial urban agriculture include Food from the Hood and the Homeless Gardening Project in Santa Cruz, California.

9 International Development Research Center, October 1993 Report.

10 Rosset, Peter. "The Greening of Cuba," *NACLA Report on the Americas*, November/December 1994.

11 HR 2003, The Community Food Security Act, pending before U.S. House of Representatives.

12 Interviews with Kai Siedenberg and California Food Policy Advocates policy analyst Ed Bolen; FMNP fact sheets, by California Food Policy Advocates.

13 Interview with Kai Siedenberg.

14 California Food Policy Advocates.

15 Interview with Robert Gottlieb, professor at UCLA.

16 "Access Denied," 1995 report by Sustainable Food Center, Austin, Texas.

17 Interview with Kathleen Fitzgerald, and Sustainable Food Center materials.

18 Interview with Leslie Mikkelsen, nutritionist at Alameda County Community Food Bank.

19 Sustainable Food Center, Austin, Texas.

From *Food First Backgrounder*, Fall 1997 by Mark S. Langevin and Peter Rosset

Land Reform from Below: The Landless Worker's Movement in Brazil

Cícero Lourenço da Silva Neto and eight other military police officers rode their motorcycles into Brasilia, Brazil's capital, around noon on April 17, 1997. Cícero, the son of landless rural workers from the state of Rio Grande do Norte, and his fellow officers entered the avenue of the Esplanada dos Ministerios where government buildings form a corridor leading to the National Congress, Presidential Palace, and Supreme Court. Marching behind Cícero were nearly five thousand landless rural workers, their families, and supporters.[1] They came to demand land reform. Cícero was leading them into the very heart of Brazil's body politic, a year to the day after police forces carried out the country's largest massacre of landless rural families.[2] Frustrated by government inaction, the landless in Brazil are today carrying out land reform "from below."

This March to Brasilia was organized by the Movimento dos Trabalhadores Rurais Sem Terra (MST) or the Landless Workers

Movement, founded in 1985. Many of the marchers walked for two months to reach Brasilia and galvanized the largest demonstration of opposition to the government of President Fernando Henrique Cardoso. For Brazil's 4.8 million landless families, the march to Brazilia showed the nation how far the landless rural worker's struggle has come.

The MST, Land Reform, and Brazilian Democracy

Since 1985 the MST has been organizing Brazil's rural poor to include them in the economic and political life of the nation. During the past six years the MST has organized 151,427 landless families for the occupation of well over twenty-one million hectares of idle land.[3]

Operating on a shoestring budget and despite government repression the Landless Workers Movement now organizes more landless families to occupy and produce on idle farmland than the government's land reform measures. Landless workers are carrying out land reform from below and challenging the Brazilian elite's domination of so-called democratic rule. Gilmar Mauro of the MST's National Directorate explains the role of the movement:

"There is a great and urgent need to restructure Brazil's land tenure system in order to guarantee access to land, promote equitable social and economic development, and insure the citizenship of the rural population. We believe that our struggle for land reform, occupying and cultivating large tracts of idle farmlands, democratizes access to land as well as to our society and government."[4]

The MST offers the rural poor an alternative, ensuring their welfare and participation in economic development and democracy. The MST is providing important health care and education to landless families. The MST's National Confederation of Brazilian Land Reform Cooperatives is providing agricultural extension services. They assist in organizing production and facilitate marketing the surplus produce of the MST's squatter settlements. This has transformed MST land occupations into productive agricultural cooperatives providing ample food, cash income, and basic services for thousands of member families. Moreover, this social movement has created small industries among the most advanced cooperatives, including a clothing factory in Rio Grande do Sul, a tea processing plant in Paraná, and a dairy processing operation in Santa Catarina.

The MST's alternative rural development strategy is challenging the political and policy limitations of the Cardoso government by providing a more just and productive alternative to the dominant system's preferential austerity for the poor.

According to João Pedro Stédile of the MST, "the struggle for land reform unfolds in the countryside, but it will eventually be resolved in the city where there is the political power for structural change."[5] Since its formation in 1985 the MST has worked closely with the Workers Party, many of whose leaders and elected officials come from the ranks of landless workers.

Today, the MST's struggle for land reform is supported by a majority of Brazilians and threatens to turn Brazilian politics on its head. A March 1997 public opinion poll sponsored by Brazil's elite National Confederation of Industry, reported that seventy-seven percent of respondents approve of the MST and eighty-five percent approve of the non-violent occupation of idle farmland.[6] Even the conservative president of the Brazilian National Conference of Catholic Bishops, the Reverend Lucas Moreira Neves, recently met with the Minister of Land Policy, Raul Jungmann, to request that the government work with the MST to solve the problem of rural poverty.[7] On March 20, 1997 the Brazilian Association of Journalists honored the MST and sponsored a declaration of support signed by more than 200 journalists, artists, and renowned intellectuals. The day before the MST received Belgium's prestigious King Boudouin Foundation Award, given every two years to recognize outstanding contributions to development worldwide. President Cardoso's own political party, the Brazilian Social Democratic Party, is split over the MST and land reform. Many of the party's elected officials, from federal deputies to city mayors, openly support the MST and its demands for a sweeping national land reform.

Land Ownership

Most of Brazil's rich agricultural land is increasingly concentrated in a few wealthy hands after decades of monocrop export agriculture and successive waves of government sponsored repression against rural workers and their organizations. According to Brazil's new Super Ministry of Land Policy, created immediately after the Eldorado dos Carajás massacre, small family farmers with ten hectares of land or less comprise 30.4 percent of all Brazilian farmers, but together hold only 1.5 percent of all agricultural lands.[8] Since 1985 the number of small farms has sharply decreased from just over three million to under one million.[9]

In contrast the country's largest farms, of 1,000 hectares or more, comprise only 1.6 percent of all farms, but hold 53.2 percent of all agricultural land.[10] The largest seventy-five farms, with 100,000 hectares or more, control over five times the combined total area of all small farms.[11] The consolidation of farmland increased agricultural exports and provided an effective hedge against inflation for the wealthy. However, the major impact of land concentration has been inescapable poverty and the spread of chronic malnutrition.

Further aggravating rural poverty and hunger is the pervasive use of agricultural lands for pasture and the high proportion of idle land among the country's largest landholdings. 42.6 percent of agricultural land is not cultivated, and among Brazil's largest landholdings of 1,000 hectares or more 88.7 percent of arable land is permanently idle.[12] Today, idle farmland may be the most important cause of both rural and urban poverty and hunger in Brazil.

The control and use of Brazil's vast and rich agricultural landholdings is a national problem, challenging the country's decade old democracy. For Dr. Ladislau Dowbor, Professor of Economics at São Paulo's Catholic University, "to maintain this situation when millions of agriculturists want to cultivate, but do not have access to land, while millions of people go hungry in the cities, demonstrates the level of absurdity reached in the absence of true participatory democracy. In the context of rising tensions in our cities we can only conclude the obvious; land reform is not just a rural problem, but a key question for urban society. We will all have to subsidize the poor management of our rich agricultural soils if our agrarian structure is not reformed."[13]

Reform and Repression

The transition from military dictatorship to civilian democracy in 1985 promised a sweeping national land reform. Months after the MST was founded to advocate land reform under democracy, the new civilian government announced the National Land Reform Plan.

The Plan was originally designed to redistribute farmlands to 1.4 million landless rural families during President Sarney's tenure from

1985–1989.[14] However, the land reform plan drew strident opposition from large landowner organizations which effectively stalled efforts to redistribute idle lands to rural workers. Since 1985 only a small fraction of the proposed beneficiary landless families have received land through government measures.

This slow pace of reform was matched with violence and repression against the MST and those struggling for social justice in the Brazilian countryside. From 1985 to 1996 there were 969 assassinations of rural workers and MST activists.[15] Between 1985–1995 there were 820 documented assassination attempts and 2,412 rural workers, family members, and MST leaders were threatened with death because of their support for land reform.[16] Since 1985 Brazilian government authorities have convicted only five people of crimes associated with the violence against the landless and the MST.

In 1994 the Minister of the Economy and world renowned sociologist, Fernando Henrique Cardoso, promised economic stabilization and land reform if Brazil would elect him president. He promised to redistribute land to 280,000 families over four years. Since taking office in 1995 President Cardoso's land reform record has been greatly tarnished by the slow pace of reform, questionable government claims, the brutal massacres of landless rural families, and the continued impunity of those responsible for the violence against those who struggle for land reform.

The Cardoso government reported that 42,912 families were settled by the official program in 1995 and 50,238 in 1996.[17] However, these claims have been called into question by both the MST and the National Confederation of Professional Associations of INCRA (representing the employees of the Ministry of Land Policy) and the National Institute of Resettlement and Land Reform (known by the acronym INCRA).[18] Moreover, President Cardoso has repeatedly cut the budgets of INCRA and the Ministry of Land Policy to 'fight inflation.' False claims and budget cuts aside, the Cardoso administration does not appear willing or able to fulfill its campaign promise of redistributing land to 280,000 landless families in four years.

Not only has the current administration raised and then frustrated expectations for land reform, it has also governed over the horrific massacres at Corumbiara and Eldorado dos Carajás. During the first two years of Cardoso's term in office at least eighty-six rural workers, family members, and MST activists were assassinated, most by the military police.[19] In 1997, violence, sponsored or condoned by the government, rages on against those who struggle for land and defend democracy. Yet, the government's brutality against the rural poor is now challenged by the MST's national campaign to cultivate democracy in the countryside, to occupy idle lands, resist repression, and produce food for the nation.

Brazil's land reform from below now plays an important role in shaping the emerging challenge to the global economic and political order imposed by the World Bank, the IMF and the World Trade Organization. These efforts are central to the MST's push to replace rural poverty with equitable access to land and participatory democracy.

The struggle for land, social justice, and participatory democracy, from the MST in Brazil to the Chiapas land takeovers in the wake of the Zapatista uprising in Mexico,[20] now depend on our global efforts to guarantee the human rights of those who struggle against hunger, disease, and poverty at the margins of the global order.

1 Gaspari, Elio. *O Globo,* April 21, 1997.

2 On Aril 17, 1996, over 200 military police troops attacked approximately 1,500 landless workers and their children as they blocked Highway 150 just outside Eldorado dos Carajás. Within minutes nineteen landless were dead and fifty-one severely injured. The landless were engaged in non-violent civil disobedience to draw attention to their struggle for land and protest the government's failed promise of land reform.

3 *MST Informa,* no. 15, January 1997.

4 *Estado De São Paulo,* November 3, 1995.

5 *Veja,* no. 74, August 28, 1996.

6 *Noticias da Terra,* no. 1, March 21, 1997.

7 *Istoé,* no. 35, March 5, 1997.

8 Ministerio Extraordinário de Pol'tica Fundiária. *Atlas Fundiário Brasileiro,* 1996.

9 Comparison based on the 1985 IBGE Agricultural Census and the *Atlas Fundiário Brasileiro.*

10 *Atlas Fundiário Brasileiro.*

11 Ibid.

12 IBGE's Agricultural Census.

13 Dowbor, Ladislau. "Reforma Agrária—dados básicos," *Estado De São Paulo,* October 3, 1995.

14 Graziano, Francisco. *A Tragédia Da Terra* (São Paulo, Brazil: Iglu Editora, 1991), pg. 17.

15 *Boletim da Commissio Pastoral da Terra-CPT,* no. 136, August 1996, with 1996 data provided by the documentation sector of the Commissio Pastoral da Terra.

16 Ibid.

17 *Atlas Fundiário Brasileiro.*

18 *Jornal dos Trabalhadores Rurais Sem Terra,* no. 162, 1996.

19 *Boletim da Commissio Pastoral da Terra-CPT,* no. 136, August 1996, with 1996 data provided by the documnetation sector of the Commissio Pastoral da Terra.

20 White, Peter. "A New Kind of Mexican Land Reform," *In These Times,* May 2, 1994.

Afterword by Peter Rosset

Food and Justice in the New Millennium: Changing How We Think About Hunger

Hunger occurs in the midst of plenty. In Latin America and in Asia recent decades have seen food production increases far outstrip population growth, with more food available per person than ever before. Yet there are also more people who are going hungry.[1] In the United States soup lines have grown during a decade of economic recovery and abundance.[2] In the Third World, seventy-eight percent of all malnourished children live in countries with food surpluses. There is enough food available worldwide to provide every person with more than four pounds of food every day, more than enough to make everyone fat.[3] With so much food available, hunger is profoundly needless.

The preceding should alert us to falling into the trap set by agribusiness publicists, biotechnology companies, and lending agencies like the World Bank. They would have us believe that people go hungry because there isn't enough food to go around. They mislead us as to the true causes of hunger in the hope that we will throw all our efforts into boosting food production at any cost, whether the cost be the health effects of pesticides, the risk of 'genetic pollution' associated with genetically engineered crop varieties, or further corporate domination of our food system.[4]

Inequality Causes Hunger, Social Justice Movements Can End It

Only growing inequality can explain why more people are hungry. Inequality leaves too many people too poor to buy the abundant food that is available, or leaves them without sufficient land and other resources to produce it for themselves.

This inequality cannot be addressed by supporting biotech giants like Monsanto in their drive to stop governments from regulating genetic engineering, nor can it be addressed by cheering on the World Bank and agribusiness in their push for a 'second' Green Revolution.

If anything was learned from the first Green Revolution it is that increases in food production that depend on expensive technology increase inequality, even as they boost food production, leading to more hunger.[5] The only way to truly address inequality—and hunger—is through broad structural changes in access to productive resources like land, and jobs that pay living wages.

These broad structural changes can only happen through one of two mechanisms. Undesirable changes toward greater concentration of wealth and power in the hands of a few come about when those at the top exert their influence over societies and policy makers. The kind of changes needed to end hunger—toward more democratic access to resources, jobs, and political power—can only happen when ordinary people join together in movements that are strong enough and numerous enough to challenge entrenched power. Effective movement building requires dispelling the myths that immobilize us.

Changing How We Think About Hunger

The most difficult task in movement building is encouraging people to join in. *The way we think about hunger is the principle obstacle to overcoming it.* As long as it is believed that hunger is due to forces beyond our control, like nature, we feel helpless. To believe the agribusiness myth that the solution lies with a technological 'fix' will direct any effort down the wrong path. If it is accepted that there are no real alternatives, people will consider action to be pointless. To believe that alternatives can work, but that negative economic and political conditions are so strong that change is impossible, leaves one feeling powerless. Finally, if by becoming an activist means a large and sudden change in someone's way of living—a commitment they are unable to make—they will be reluctant to take the first step. If we can change the way people think about each of these points, large numbers of people can be moved to effective action. That is the precisely the task ahead.

Is Nature to Blame? It's too easy to blame nature whenever we hear of famine caused by drought. Human forces make people vulnerable to natural forces.[6] Examples include the way commercial producers of export crops have displaced Africa's poor into drought-prone regions of sporadic rainfall, and Central American peasants onto the steep slopes that washed away during Hurricane Mitch in 1998. Closer to home, hundreds of homeless people, America's hungriest citizens, die of the cold every year in the U.S. In each case the weather is not the root cause of disaster, but just the final push that shoves people over the brink. The real culprits are economies that fail to offer opportunities to all and societies that place economic efficiency over justice and compassion. Hunger is not caused by natural forces beyond our control, but by very real human decisions over economic policies, land use, housing, and job creation. That is good news, for while we can do little to change nature, policies fall squarely into realm of things we can change by building movements.

Do We Need a Technological Fix? If hunger isn't caused by a scarcity of food, then 'magic bullets' like pesticides and genetic engineering cannot end hunger, and there is no reason to use them on such a grand scale. Even if food production needs to be increased in the future, there are better ways to do so that are less risky for the environment and human health, and which provide inequality-reducing opportunities for small farmers to play a key role in feeding the world.[7] Only be improving access to food, land to grow it, and jobs with wages adequate to buy it, can we truly end hunger. Technology cannot substitute for social justice

Are There Real Alternatives? Kerala has achieved long life expectancy, low infant mortality and fertility rates, and virtually ended hunger, despite being one of the poorest states in India. The landless workers movement (MST) in Brazil has shown that is indeed possible to force 'land reform from below' and create real structural change in land holdings, while building successful enterprises run by the poor.[8] The Community Food Security movement demonstrates how poor communities in the U.S. are taking over abandoned spaces in their neighborhoods and turning them into productive urban gardens that generate jobs and provide food for low income families. Cuba's new emphasis on agriculture shows that a nation's population can be fed by small farmers and urban gardeners relying on sustainable agriculture rather than agrochemicals and genetically engineered crops.[9] Thousands of small farmers around the world have increased their production and incomes through agroecological alternatives,[10] in many cases developing and spreading these alternatives through their own initiative and organizations. The alternatives show more promise for ending hunger than does 'business as usual.'

Is Change Possible? We, the people, have built powerful and successful movements for change before. That is how American women won suffrage, how slavery was ended, how the pesticide DDT was banned, how civil rights legislation was passed, and how the war in Vietnam was ended. The time has come to build national and international movements to attack inequality and eliminate hunger. Alone we are powerless, but together we are powerful. A real objective basis exists for building cross-class and cross-border movements for change. The slashing of social safety nets and runaway free trade policies have placed working people everywhere—whether small farmers or computer programmers—into a global 'race to the bottom.' As transnational corporations relocate production to where the lowest wages and weakest environmental regulations are found, the standards of living and job security for everyone are placed in danger. The same forces that perpetuate poverty in the countries of the South are driving job insecurity, poverty, hunger, and homelessness in the North. The majority of the world's population have faced declining living standards since the onset of Reaganomics, structural adjustment, and trade treaties like NAFTA and GATT. In a global economy what American workers have achieved in employment, wage levels, and working conditions can be protected only when working people in every country are freed from economic desperation. We truly have interests in common in reversing these policies, and we are the majority of the world's population. That gives us renewed hope—that we can realize the strength implicit in our numbers—and highlights the importance of realizing what we have in common is greater than what divides us.

Food is a Human Right

1998 was the fiftieth Anniversary of the Universal Declaration of Human Rights (UDHR), with major contributions by Eleanor and Franklin Delano Roosevelt and adopted by the General Assembly of the United Nations. The UDHR set a universal standard under which human rights are indivisible and universal. Indivisibility means that one cannot separate civil and political rights from economic and social human rights. Food is one of the fundamental economic humans rights guaranteed under the UDHR. Article 25 of the UDHR states:

> Everyone has the right to a standard of living adequate for the health and well-being of himself [or herself] and of his [or her] family, including food, clothing, housing, and medical care and necessary social services, and the right to security in the event of unemployment, sickness, disability, widow-hood, old age, or other lack of livelihood in circumstances beyond his control.[11]

This does not mean that governments themselves need provide these goods and services, or that they must feed people directly. It means that governments must insure policy options that protect, respect, facilitate, and fulfill these obligations.

Economic efficiency tells us to see hunger in terms of numbers. Thinking about hunger in terms of numbers leads to thinking about solutions in terms of numbers, yet the root causes of hunger and the solutions to it often have nothing to do with numbers, but with structures of power. If arguments and proposals are restricted to numbers, we will never get at the root causes.

At the 1996 World Food Summit, governments agreed on the goal of reducing hunger by fifty percent by the year 2015. By stating a numerical goal—the product of negotiations among parties with different viewpoints, using cost/benefit arguments—'success' in the fight against hunger was defined as a world where 'only' 400 million people continue to go without sufficient food. Cuban President Fidel Castro was the only world leader to point out that 400 million hungry people is a disgraceful goal for humanity to set. But once the idea of numerical targets is accepted, you implicitly accept a morally questionable position.

As an example, take hunger in America. If we say that food is a human right, that all have the inalienable right to be able to feed themselves and families, then to have even one hungry family in the richest nation on Earth constitutes a human rights violation, and must be addressed with all that implies.

We reject out of hand any political system in which even a small number of prisoners of conscience are tortured by the authorities. The same is true for genocide—killing 100,000 and killing one million are both totally unacceptable. Why should we view hunger in a different light? It is after all a quiet violence, one in which hunger-related diseases take the lives of 34,000 of the world's children every day. And it is preventable.

Food Is a Window

Food is a window which allows us to look into any society, anywhere in the world, and determine critically important things about its structure, especially with regard to social justice and the distribution of power and wealth. We can ask who eats and who doesn't, who benefits from the food system and who is marginalized by it. Food is a window that can illuminate a broad variety of forces acting within a society. When Kerala—one of the poorest states in India, yet there are few hungry people—is compared with the Mexico of NAFTA and structural adjustment, or with the United States of 'welfare reform,' we see how a commitment to a minimum of standard of living for all, rather than wealth, is the path to ending hunger. A window is also an entry point. By attacking the roots of hunger we work for change towards greater fairness, equity and, ultimately, true democracy. For who can argue in favor of hunger? Yet truly ending hunger requires changing the undemocratic structures of power and wealth that perpetuate it.

We *can* end hunger. Hunger is not caused by forces beyond our control. It is caused by policies put into place by human beings. The same policies can be changed by human beings. We can bring those changes by building movements, just as we did to end the war in Vietnam. We can begin we small steps that lead to larger ones. And we can arm ourselves with the powerful message of human rights.

1 Except in China, where the number of hungry people dropped dramatically. Frances Moore Lappé, Joseph Collins and Peter Rosset, with Luis Esparza, *World Hunger: Twelve Myths*, second edition (New York: Grove Press, 1998), pg 61.

2 Mittal, Anuradha, Peter Rosset and Marilyn Borchardt. "Shredding the Safety Net: Welfare Reform as We Know It," *Food First Backgrounder*, Winter 1998, pp. 1–8.

3 *World Hunger: Twelve Myths*, op. cit., pp. 8–9.

4 Weiss, Robert and Justin Gillis. "U.S. Battles to Halt World Trade Accord: Many Countries Want to Restrict the Import of Genetically-Engineered Goods Made in America," *San Francisco Examiner*, February 14, 1999, pg. A-22.

5 *World Hunger: Twelve Myths*, op. cit., chapter 5.

6 Ibid., chapter 2.

7 Ibid., chapter 5.

8 "Brazil's Landless Rebels," *Time Magazine*, vol. 151 no. 2, January 19, 1998.

9 Rosset, Peter M. "Alternative Agriculture Works: The Case of Cuba," *Monthly Review* vol. 50, no. 3, 1998, pp. 137–146.

10 Altieri, Miguel, and Peter Rosset and Lori Ann Thrupp. "The Potential of Agroecology to Combat Hunger in the Developing World," *Food First Policy Brief* no. 3, 1998.

11 United Nations Department of Public Information. *Universal Declaration of Human Rights* (New York: United Nations, 1993).

Epilogue

From *Don't be Afraid, Gringo: A Honduran Woman Speaks from the Heart* by Elvia Alvarado, translated and edited by Medea Benjamin

Turn Your Tears into Strength

When I hear that all this military buildup in Honduras is just trying to maintain peace in our country, I ask myself what peace they're talking about. Maybe it's peaceful for the politicians. The congressmen make $3,000 a month; their bellies are full of food and drink; they've got a wad of bills in their pockets. So for them there's peace.

But not for the campesinos. Do you think a mother who can't send her children to school because she doesn't have any clothes to put on their backs feels at peace? Do you think a mother who watches her child die because she doesn't have a penny to take her to the doctor feels at peace?

To protect this great peace we have, the politicians have sold our country off to the United States. They've made us a colony of the United States.

They're only doing it, they say, to protect our national security. What national security? The national security they're protecting is that of their own big stomachs. They're protecting the fat checks that come pouring in from the United States.

If I had a chance to talk to Reagan, which of course I wouldn't since Reagan is only interested in talking to the rich, I'd tell him to take all the money he's sending to Honduras—all the guns, all the tanks, all the helicopters, all the bases, all the big, expensive projects he's financing—and get the hell out of our country.

We don't need the U.S. money. We never get to see any of it anyway. What do you think that money goes for? To the foreign bank accounts of the rich, to line the pockets of our corrupt politicians, to give the military more power to repress the poor.

It's the rich who need the U.S. aid, not the poor. We've lived for years with only our beans and tortillas, and we'll go on living with our beans and tortillas. If the U.S. stopped sending money, it would be the rich who'd be hurt, not us. They're the ones who live off the dollars.

All that money does for the campesinos is divide us. USAID dangles some bills in front of the campesino groups to try to buy them off, to corrupt the leaders. It started this land titling program to say to some of the campesinos, "Stick with us and you'll get a piece of land. Don't worry about the others who have none." But the worst thing the U.S. money does is strengthen the Honduran military. For us campesinos this just means more repression, more human rights abuses, more disappeared.

We see the U.S. policy as very dangerous. The reason we haven't had a civil war here in Honduras is that we campesinos have had an alternative—our campesino movement. For us the political parties—the Liberals and the Nationalists—are all the same. As soon as they get into power they give all their friends jobs, and they start changing the laws as they please.

No, our only hope lies in the campesino movement itself. Any gains we've made have been thanks to our organizations, thanks to the fact that we work together. But if the United States is determined to break up the campesino movement, we'll be left with no alternative than to take up arms just like our neighbors have done.

And I hate to say it, but that's what I see happening. The United States is trying to draw us into a war with Nicaragua, but will end up drawing us into a civil war.

I must admit that sometimes I get so overwhelmed by the odds against us that I break down and cry. I see our children dying of hunger, and the ones that live have no jobs, no education, no future. I see the military getting more and more repressive. I see us being persecuted, jailed, tortured. I get exhausted by all the internal problems between the campesino organizations. And I see all of Central America going up in flames.

I start to wonder if it's worth it. I start to think maybe I should just stay home making tortillas.

But whenever I have these doubts, whenever I start to cry, I put my hands into fists and say to myself, "Make your tears turn into anger, make your tears turn into strength." As soon as I stop crying, I feel a sense of power go through my body. And I get back to work with even more enthusiasm, with more conviction than ever.

When I see some of my other compañeros get depressed, I say to them, "Snap out of it. Get back to work. We have too much to do to waste our time getting depressed." And they do the same to me.

One thing that gives us a great boost is when we hear that there are other people in other countries who are on our side.

Not long ago I was in a meeting with a group of Hondurans working for peace in Central America. Two gringos were visiting and joined the meeting. They weren't gringos from the United States. They were gringos from other countries I'd never heard of, some countries in Europe, they said. And they were here to show support for our struggle.

They asked me to write a message to the people in their country. I picked up the pen and I don't know how I did it—because I really don't write very well—but I wrote something and they understood it.

I wrote that I was just a poor Honduran, but that we were fighting for justice in our country. I told them how happy I was that there were people from other countries who were working for peace in Central America. I said I might not know what they look like, what language they speak, or even the names of the countries they come from, but that we were all brothers and sisters. I said that if we were both fighting for justice, then we were part of the same family.

These two gringo men were so touched by the message that they got up and hugged me and started to cry. Can you imagine gringo men crying? I never saw that in my life! So I started crying, too.

And I said to them, "Thank God there are people on our side. Thank God there are people on our side. Now we're really going to raise hell."

I later learned that there are also gringos in the United States who don't agree with their government's policies in Central America. It's amazing that Reagan has so much power and he still hasn't been able to conquer all the people in the United States. It shows he's not as powerful as we thought.

You can't imagine how much courage and hope it gives me to know that we have friends in the United States. Imagine that! Friends in the United States! Who would've ever believed it!

It's hard to think of change taking place in Central America without there first being changes in the United States. As we say in Honduras, "Sin el perro, no hay rabia"—without the dog, there wouldn't be rabies.

So you Americans who really want to help the poor have to change your own government first. You Americans who want to see an end to hunger and poverty have to take a stand. You have to fight just like we're fighting—even harder. You have to be ready to be jailed, to be abused, to be repressed. And you have to have the character, the courage, the morale, and the spirit to confront whatever comes your way.

If you say, "Oh, the United States is so big and powerful, there's nothing we can do to change it," then why bother talking about solidarity? If you think like that, you start to feel insignificant and your spirit dies. That's very dangerous. For as long as we keep our spirits high, we continue to struggle.

We campesinos are used to planting seeds and waiting to see if the seeds bear fruit. We're used to working on harsh soil. And when our crops don't grow, we're used to planting again and again until they take hold. Like us, you must learn to persist.

You also have to be fearless. If you begin with fear—fear of being persecuted or of going to jail or of being criticized—you might as well not start.

I don't know if it's the same in the United States, but here people are terrified that they'll be called communist. But if they call us communists, we have to tell them that that's a bunch of bull. We're not fighting for theories. We're not fighting for communism or Marxism. We're fighting for justice.

I'm always being criticized—that I'm a communist, that I'm a subversive, that I'm a whore, that I go around sleeping with all the campesinos, that I'm a bad mother, that I leave my children home alone, that I left my husband because I want to be free. Whatever. But we can't be afraid of criticism. We have to answer that we know where we're going and why we're going there, and if anyone wants to follow us, we'll be glad to show them the way.

You also have to be clear about your objectives, about why you're struggling. You can't struggle just because someone else tells you it's a good idea. No, you've got to feel the struggle. You've got to be completely convinced that what you're struggling for is just.

And then you have to have a plan. What are you trying to achieve? What methods will you use? How many people do you have? Who can you count on for help? How much money do you have? How long will it take you to reach a certain number of people? What will you ask them to do?

You have to begin educating people, telling them the truth about what's happening in the world. Because if the press in the United States is anything like it is in Honduras, the people aren't well informed. You have to teach them what's really happening in the United States, what your government is really doing. And once you've educated people, then get them organized.

Start out forming small groups, first in your own house, then with your neighbors. You might have to start out with just a handful of people—three women or three men. It doesn't matter if you start out small. Things that start out small get bigger and bigger. One group becomes two groups, two groups become four; and before you know it, you have a lot of well-organized people.

Then you start dividing up the tasks, and you make up your committees—the education committee, the women's committee, the youth committee. And soon you branch out to other neighborhoods and other villages and cities.

The other thing you have to do is make allies. I used to think you had to be poor to be part of this struggle. But there are people in Honduras who aren't poor, yet they're on our side. They're well-educated people—doctors, lawyers, teachers, engineers—who identify with the poor. I suppose it's the same in the United States. So don't only organize the poor and working people. You can also look for middle-class people, or even rich people who want to help change things.

But if you sit around thinking what to do and end up not doing anything, why bother even thinking about it? You're better off going out on the town and having a good time. No, we have to think and act. That's what we're doing here, and that's what you have to do.

I hate to offend you, but we won't get anywhere by just writing and reading books. I know that books are important, and I hope this book will be important for the people who read it. But we can't just read it and say, "Those poor campesinos. What a miserable life they have."

Or others might say, "What a nice book. That woman Elvia sounds like a nice woman." I imagine there'll be others who say, "That Elvia is a foul-mouthed, uppity campesina." But the important thing is not what you think of me; the important thing is for you to do something.

We're not asking for food or clothing or money. We want you with us in the struggle. We want you to educate your people. We want you to organize your people. We want you to denounce what your government is doing in Central America.

From those of you who feel the pain of the poor, who feel the pain of the murdered, the disappeared, the tortured, we need more than sympathy. We need you to join the struggle. Don't be afraid, gringos. Keep your spirits high. And remember, we're right there with you!

FOOD FIRST

BOOKS OF RELATED INTEREST

BASTA! Land and the Zapatista Rebellion in Chiapas
George Collier with Elizabeth Lowery Quaratiello
Examines the root causes of the Zapatista uprising in southern Mexico. Paperback, $12.95

Benedita da Silva: An Afro-Brazilian Woman's Story of Politics and Love, As told to Medea Benjamin and Maisa Mendonça
With a foreword by Jesse Jackson
Afro-Brazilian Senator Benedita da Silva shares the inspiring story of her life as an advocate for the rights of women and the poor.
Paperback, $15.95

Breakfast of Biodiversity: The Truth about Rain Forest Destruction
John Vandermeer and Ivette Perfecto
Analyzes deforestation from both an environmental and social justice perspective. Paperback, $16.95

Dark Victory: The U.S. and Global Poverty
Walden Bello, with Shea Cunningham and Bill Rau
Second edition, with a new epilogue by the author
Offers an understanding of why poverty has deepened in many countries, and analyzes the impact of U.S. economic policies.
Paperback, $14.95

Dragons in Distress: Asia's Miracle Economies in Crisis
Walden Bello and Stephanie Rosenfeld
After three decades of rapid growth, the economies of South Korea, Taiwan, and Singapore are in crisis. The authors offer policy recommendations to break these countries from their unhealthy dependence on Japan and the U.S. Paperback, $12.95

Education for Action: Graduate Studies with a Focus on Social Change
Edited by Sean Brooks and Alison Knowles
An authoritative, easy-to-use guidebook that provides information on progressive programs in a wide variety of fields. Paperback, $8.95.

Kerala: Radical Reform as Development in an Indian State
Richard W. Franke and Barbara H. Chasin. Revised edition

In the last eighty years, the Indian state of Kerala has experimented in the use of radical reform that has brought it some of the Third World's highest levels of health, education, and social justice. Paperback, $10.95

Needless Hunger: Voices from a Bangladesh Village
James Boyce and Besty Hartmann

The global analysis of Food First is vividly captured here in a single village. Paperback, $6.95

A Quiet Violence: View from a Bangladesh Village
Betsy Hartmann and James Boyce

The root causes of hunger emerge through the stories of both village landowners and peasants who live at the margin of survival. Paperback, $17.95

Taking Population Seriously,
Frances Moore Lappé and Rachel Schurman

High fertility is a response to anti-democratic power structures that leave people with little choice but to have many children. Instead of repressive population control, the authors argue for education and improved standard of living. Paperback, $7.95

Video: *The Greening of Cuba,* Directed by Jaime Kibben

Cuba has combined time-tested traditional methods with cutting edge bio-technology, reminding us that developed and developing nations can choose a healthier environment and still feed their people. VHS videotape, $29.95.

Hunger Myths and Facts Sheet, $1 each, 5/$4, 20/$13, 100/$40.

Write or call to place book orders. All orders must be pre-paid. Please add $4.50 for the first book and $1.50 for each additional book for shipping and handling.

Food First Books

398 – 60th Street
Oakland, CA 94618
(510) 654-4400

ABOUT FOOD FIRST

(The Institute for Food and Development Policy)

Food First, also known as the Institute for Food and Development Policy, is a nonprofit research and education-for-action center dedicated to investigating and exposing the root causes of hunger in a world of plenty. It was founded in 1975 by Frances Moore Lappè, author of the bestseller *Diet for a Small Planet*, and food policy analyst Dr. Joseph Collins. Food First research has revealed that hunger is created by concentrated economic and political power, not by scarcity. Resources and decision-making are in the hands of a wealthy few, depriving the majority of land, jobs, and therefore food.

Hailed by *The New York Times* as "one of the most established food think tanks in the country," Food First has grown to profoundly shape the debate about hunger and development.

But Food First is more than a think tank. Through books, reports, videos, media appearances, and speaking engagements, Food First experts not only reveal the often hidden roots of hunger, they show how individuals can get involved in bringing an end to the problem. Food First inspires action by bringing to light the courageous efforts of people around the world who are creating farming and food systems that truly meet people's needs.

HOW TO BECOME A MEMBER OR INTERN OF FOOD FIRST

BECOME A MEMBER OF FOOD FIRST

Private contributions and membership gifts form the financial base of the Institute for Food and Development Policy. The success of the Institute's programs depends not only on its dedicated volunteers and staff, but on financial activists as well. Each member strengthens Food First's efforts to change a hungry world. We invite you to join Food First. As a member you will receive a twenty percent discount an all Food First books. You will also receive our quarterly publication, *Food First News and Views*, and timely *Backgrounders* that provide information and suggestions for action on current food and hunger crises in the United States and around the world. If you want so subscribe to our internet newsletter, *Food Rights Watch*, send us an e-mail at foodfirst@foodfirst.org. All the contributions are tax-deductible.

BECOME AN INTERN FOR FOOD FIRST

There are opportunities for interns in research, advocacy, campaigning, publishing, computers, media, and publicity at Food First. Our interns come from around the world. They are a vital part of our organization and make our work possible.

To become a member or apply to become an intern, just call, visit our website, or clip and return the attached coupon to

Food First/Institute for Food and Development Policy
398 60th Street, Oakland, CA 94618, USA
Phone: (510) 654–4400 Fax: (510) 654–4551
E-mail: foodfirst@foodfirst.org
Website: www.foodfirst.org

You are also invited to give a gift membership to others interested in the fight to end hunger.

JOINING FOOD FIRST

❒ I want to join Food First and receive a 20% discount on this and all subsequent orders. Enclosed is my tax-deductible contribution of:

❒ $100 ❒ $50 ❒ $30

NAME ______________________

ADDRESS ______________________

CITY/STATE/ZIP ______________________

DAYTIME PHONE (______) ______________________

E-MAIL ______________________

ORDERING FOOD FIRST MATERIALS

ITEM DESCRIPTION	QTY	UNIT COST	TOTAL

PAYMENT METHOD:

☐ CHECK

☐ MONEY ORDER

☐ MASTERCARD

☐ VISA

MEMBER DISCOUNT, 20% $ ________

CA RESIDENTS SALES TAX 8.25% $ ________

SUBTOTAL $ ________

POSTAGE: 15% UPS: 20% ($2 MIN.) $ ________

MEMBERSHIP(S) $ ________

ADDITIONAL CONTRIBUTION $ ________

TOTAL ENCLOSED $ ________

____
NAME ON CARD

____
CARD NUMBER EXP. DATE

____
SIGNATURE

MAKE CHECK OR MONEY ORDER PAYABLE TO:

Food First, 398 – 60th Street, Oakland, CA 94618

FOR GIFT MEMBERSHIPS & MAILINGS, PLEASE SEE COUPON ON REVERSE SIDE

FOOD FIRST GIFT BOOKS

Please send a Gift Book to (order form on reverse side):

NAME ____

ADDRESS ____

CITY/STATE/ZIP ____

FROM: ____

FOOD FIRST RESOURCE CATALOGS

Please send a Resource Catalog to:

NAME ____

ADDRESS ____

CITY/STATE/ZIP ____

NAME ____

ADDRESS ____

CITY/STATE/ZIP ____

NAME ____

ADDRESS ____

CITY/STATE/ZIP ____

FOOD FIRST GIFT MEMBERSHIPS

❐ Enclosed is my tax-deductible contribution of:

❐ $100 ❐ $50 ❐ $30

Please send a Food First membership to:

NAME ____

ADDRESS ____

CITY/STATE/ZIP ____

FROM: ____